WAVELET ANALYSIS

with Applications to

IMAGE PROCESSING

L. Prasad & S. S. Iyengar

CRC Press

Boca Raton   New York

Acquiring Editor:	Bob Stern
Project Editor:	Carol Whitehead
Marketing Manager:	Tim Pletscher
Cover design:	Dawn Boyd
PrePress:	Gary Bennett
Manufacturing:	Barbara Brownlee

Library of Congress Cataloging-in-Publication Data

Prasad, L. (Lakshman)
    Wavelet analysis with applications to image processing/ L. Prasad, S. S. Iyengar
        p.   cm.
    Includes bibliographical references and index.
    ISBN 0-8493-3169-2 (alk. paper)
    1. Image processing--Mathematics. 2. Wavelets (Mathematics) I. Iyengar, S. S.
(Sundararaja S.) II. Title.
    TA1637.P7 1997
    621.36'7--dc21

                                                                97-11042
                                                                CIP

Preface

Wavelet transforms have come to be an important tool of mathematical analysis, with a wide and ever increasing range of applications, in recent years. Among the areas where wavelet analysis finds application are differential equations, numerical analysis, and signal processing, to mention only a few. Wavelet analysis is among the newest and most powerful additions to the arsenals of mathematicians, scientists and engineers, and has succeeded in bringing together a vast pool of researchers by providing a novel and elegant "world-view". The versatility of wavelet analytic techniques has forged new interdisciplinary bonds by offering common solutions to apparently diverse problems and providing a new unifying perspective on several existing problems in varied fields. To be sure, wavelet analysis is a culmination of the attempts of workers in several fields to design new tools to solve problems in their areas. It has its origins in Calderon-Zygmund operator theory of harmonic analysis, in the theory of coherent states and renormalization group theory of quantum physics, and in subband filtering of signal processing. To some extent, this hybrid origin explains and endorses the flexibility and universal applicability of wavelet analytic methods.

Fourier analysis has been the dominant principal analytical tool in the mathematical sciences and engineering, offering a rich framework with powerful techniques to study properties of functions and operators, and signals via their spectra. In some problems, however, Fourier analytic techniques are inadequate or lead to extremely onerous computations. A case in point is the time-frequency analysis of signals. The wavelet transform of a function (signal) captures the localized time-frequency information of the function, unlike the Fourier transform, which sacrifices localization in one domain in order to secure it in the complementary domain. The property of time-frequency localization greatly enhances the ability to study the behaviors (e.g., smoothness, singularities) of signals as well as to change these features locally, without significantly affecting the state of the signals' characteristics in other regions of frequency or time. Also, unlike the Fourier transform, the continuous wavelet transform is closely linked to the discrete wavelet transform. This relation allows one to speak of a "wavelet series" of any finite energy signal, obtained by "discretizing" the continuous wavelet transform. The coefficients of such a wavelet series of a signal completely capture the time-frequency characteristics of the signal, with each coefficient corresponding to a discrete time and frequency window. This feature gives the discrete wavelet transform a multiresolutional property, which makes possible the study of a signal at varying resolutions. However, the algebraic properties of the wavelet transform are similar to those of the Fourier transform, and the discrete wavelet transform is defined and studied with the help of the Fourier transform. Thus, while wavelet transforms do not supplant the Fourier transform, they are relatively superior and more natural tools of analysis in certain settings.

Thus, the utility and versatility of wavelet analysis cannot be ignored by the student in the engineering and mathematical sciences. There is already a significant amount of material on the subject of wavelet analysis in the form of pioneering research papers, expository articles, books, etc., and the literature is growing rapidly. The books by C. K. Chui [3], I. Daubechies [9], and Y. Meyer [29] offer excellent expositions of various aspects and perspectives of wavelet analysis to the scientific community. However, there is still a need for books at the level of the undergraduate and graduate levels for a basic exposition of this fascinating subject, so that the student of engineering gets an early start in equipping himself/herself with the techniques of wavelet analysis. More often than not, the student of engineering is not exposed to a rigorous or full-blown course on Fourier or functional analysis, and this limits and impedes his/her access to the nuances of wavelet analysis. This book is an attempt to introduce the basic aspects of wavelet analysis to the students of electrical engineering, computer science and related disciplines as early as possible in their curriculum. This book evolved out of a series of informal lectures held by the first author to a small group of graduate students interested in the theory and applications of wavelet analysis.

The first four chapters introduce the basic topics of analysis that are vital to understanding the mathematics of wavelet transforms. These chapters are by no means a course in analysis, but serve to explain concisely the concepts of analysis needed to develop wavelet analysis in a legitimate fashion. Indeed, the authors feel that very often the student of engineering is asked to make many simplifying assumptions to compensate for the lack of a strong mathematical foundation, which may lead to a shallow or less than useful idea of the subject. The first chapter introduces notation and set-theoretic concepts used throughout the book. The second chapter briefly discusses the structure and properties of infinite dimensional linear manifolds, Banach and Hilbert spaces. Chapter three quickly develops the theory of Lebesgue integral, and introduces the spaces of absolutely integrable and square integrable functions. Chapter four deals with Fourier series and Fourier transforms.

Chapters five, six and seven are devoted to wavelet analysis and its applications to image processing. These chapters do not constitute a comprehensive treatment of all aspects of wavelet analysis, but trace an expository path through the major themes of wavelet transform theory and applications. Chapter five introduces the continuous wavelet transform in the context of time-frequency analysis, and the discrete wavelet transform as a means of expressing a function in terms of the coefficients of a vector in a Hilbert space with respect to a Riesz basis. Orthonormal wavelet bases associated with multiresolution analyses are discussed, and Mallat's fast wavelet algorithm for decomposing and reconstructing a signal is obtained.

Chapter six is dedicated to methods of construction of wavelet bases with various desirable properties. The Battle-Lemarié wavelets are constructed as an example of orthonormal wavelet bases associated with a multiresolution analysis. Next, a signal processing perspective is adapted to motivate the construction of compactly supported orthonormal and biorthogonal wavelet bases. This describes the work of Daubechies [8] and Cohen [6]. Chapter seven illustrates two applications of wavelet transform theory in image processing attributable to Mallat et al. [27, 28, 15]. The choice of image processing for illustrating the applications of wavelets was dictated by the authors' research interest in the area.

Acknowledgments

This book was made possible thanks to generous support from the Office of Naval Research Division of Electronics.

Our approach to wavelet analysis and its applications have been deeply influenced by the contributions of C. K. Chui, I. Daubechies, S. Mallat, and others in the field. Detailed comments from Dr. R. N. Madan and Professor R. L. Kashyap helped us to understand many of the implications of our work.

Invaluable comments from Professor K. R. Sreenivasan (Yale University) and Professor P. Vaidyanathan (California Institute of Technology) guided us in improving the content and scope of the book.

We are grateful to the following individuals in the LSU Robotics Research Laboratory who have commented and contributed to earlier drafts of the book: Ramana Rao, Daryl Thomas, Amit Nanavati, and John M. Zachary.

Also, a special thanks goes to John M. Zachary, a research associate in the LSU Robotics Research Laboratory, for his contribution to parts of Chapter 7 and his technical assistance in preparing the book in LATEX.

Over the years, we have received support in the form of grants from the Office of Naval Research Division of Electronics, the Naval Research Laboratory Remote Sensing Office, and the LEQSF-Board of Regents. We are deeply grateful for their support.

L. Prasad

S. S. Iyengar

Contents

List of Figures

Chapter 1

Introduction and Mathematical Preliminaries

1.1   Notation and Abbreviations

The following is a glossary of some of the symbols and abbreviations used freely in this book.

$\forall$	for all
$\exists$	there exists/ there exist
$\ni$	such that
$\therefore$	therefore
$\because$	because
$\in$	belongs to / belonging to, is an element of
$\not\ni$	does not belong to, or, is not an element of
$\cup$	union
$\cap$	intersection
$\subset$	is a subset of
$\supset$	is a superset of
$\not\subset$	is not a subset of

Φ	Phi, null set, empty set
$\leq$	less than or equal to
$\geq$	greater than or equal to
$\neq$	is not equal to / not equal to
$\equiv$	is identically equal to
$\pm$	plus or minus
$\mp$	minus or plus
$\times$	cross product / Cartesian product, cross
$\circ$	composition
$\Rightarrow$	implies
$\Leftarrow$	is implied by
$\Leftrightarrow$	implies and is implied by
$\mapsto$	goes to / is mapped onto
$\perp$	is perpendicular to
$\int$	integral of
$\sum$	summation of
$\prod$	product of
$\mathbf{F}$	any field of numbers; usually $\mathbf{R}$ or $\mathbf{C}$
$\mathbf{R}$	the field of real numbers
$\mathbf{R}^+$	the field of positive real numbers
$\mathbf{C}$	the field of complex numbers
$\mathbf{N}, \mathbf{N}^+$	the set of natural numbers
$\mathbf{Z}, \mathbf{Z}^+$	the ring of integers
$\mathbf{Q}, \mathbf{Q}^+$	the field of rational numbers
lim	limit of
max	maximum of
min	minimum of
sup	supremum of
inf	infimum of
supp	support of
$|n|$	absolute value of
iff	if and only if
w.r.t.	with respect to
w.l.g.	without loss of generality
dim	dimension of
codim	codimension of
f^{-1}	f inverse

1.2   Basic Set Operations

In this section, some elementary aspects of set theory used in this book will be introduced. The notion of a set has always eluded formal definition and, therefore, is taken to be a primitive notion in axiomatic set theory. However, in practice, one can easily work with sets without a formal definition because of their frequent occurrence in various branches of the mathematical sciences. It suffices, therefore, to say that the term "set" refers to a collection of objects. In actual mathematical usage this loose description becomes tighter once the context of the set and the category of objects in it is clear. For instance, all real numbers form a set. Likewise all integers, rationals between 1 and 3.7, etc., form sets. The term "set" is used synonymously with the equivalent terms "class," "family," "collection," and "system."

The objects contained in a set are called its "elements" or "members." If x, y, z, u, v, w belong to a set S, then this is written: $S = \{x, y, z, u, v, w\}$. *i.e.*, a set is represented by writing its elements within curly brackets.

To indicate that an object a is an element of a set A, the membership relation "$\in$" is used as follows: $a \in A$ (read "a belongs to A" or "a is a member of A" or "a is contained in A").

To indicate that an object a is not an element of a set A, the symbol "$\notin$" is used as follows: $a \notin A$ (read "a does not belong to A" or "a is not a member of A" or "a is not contained in A").

If A and B are two sets such that every element of A is also an element of B, then A is called a subset of B, written $A \subset B$ or $B \supset A$ (read " B contains A"). $A \subset B$ allows for the possibility that A and B coincide (i.e., they have identical set of elements in them). This is written $A=B$ (read "A is equal to B"). Clearly $A \subset B$ and $B \subset A$ implies $A=B$.

If $A \subset B$ but $A \neq B$, then A is said to be a proper subset of B. Consider the set of all odd integers divisible by two. Of course since there are no such odd integers, this "set" contains no elements. Such sets are called *empty* sets or *null* sets, denoted by 'Φ'. Thus (read "A is the null set") implies that A has no elements. For any set A, $\Phi \subset A$.

Given two sets A and B, one can construct a set C which contains all elements belonging to at least one of the sets A and B. The set C is called the *union* of the sets A and B, written $A \cup B$ (read "A union B").

Given two sets A and B, One can construct a set C which contains all elements belonging to both the sets A and B. The set C is called the *intersection* of the sets A and B, written $A \cap B$ (read "A intersection B"). If $A \cap B = \Phi$ then A and B are said to be mutually disjoint.

If A_α are sets, where the index α runs through an arbitrary set I called the indexing set (I could be the set of natural numbers, a finite set, or the set of real numbers. It is not necessary for α to take integer values), then one can define the union of the sets $A_\alpha, \alpha \in I$ as the set consisting of all elements which belong to at least one of the sets A_α. This set is written:

$$\bigcup_{\alpha \in I} A_\alpha \quad \text{or just} \quad \bigcup_\alpha A_\alpha$$

if the indexing set is clear from the context, or is left unspecified. Similarly, one can define the intersection of the sets $A_\alpha, \alpha \in I$ as the set consisting of all elements which belong to every one of the sets A_α This set is written:

$$\bigcap_{\alpha \in I} A_\alpha \quad \text{or just} \quad \bigcap_\alpha A_\alpha$$

if the indexing set is clear from the context, or is left unspecified.

The difference $A \backslash B$ of the set A from the set B is the subset of A consisting of all those elements in A not belonging to B. It is also called the *relative complement* of B in A. The following identities are easy to verify:

1. $(A \backslash B) \cap (B \backslash A) = \Phi$

2. $(A \backslash B) \cup (A \cap B) = A$

3. $(A \backslash B) \cup (A \cap B) \cup (B \backslash A) = A \cup B$

4. $(A \cup B) \backslash A = B$

5. $A \backslash B = A \backslash (A \cap B)$

The set $(A \cup B) \backslash (A \cap B) = (A \backslash B) \cup (B \backslash A)$ is called the *symmetric difference* of the sets A and B, written $A \triangle B$. The operations $\cup$, $\cap$ and $\triangle$ are symmetric, *i.e.*, $A \cup B = B \cup A$, $A \cap B = B \cap A$, and $A \triangle B = B \triangle A$, while '$\backslash$' is not; *i.e.*, $A \backslash B \neq B \backslash A$; in fact, as already noted, $(A \backslash B) \cap (B \backslash A) = \Phi$.

If $\{A_\alpha\}_{\alpha \in I}$ is a family of sets, and B is any other set, then the following distributive laws hold :

$$\left(\bigcup_\alpha A_\alpha\right) \cap B = \bigcup_\alpha (A_\alpha \cap B)$$

(*i.e.*, intersection distributes over union). Indeed,

$$x \in \left(\bigcup_{\alpha} A_\alpha\right) \cap B \Rightarrow x \in \bigcup_{\alpha} A_\alpha \text{ and } x \in B$$

i.e., $x \in A_\alpha$ for some α and $x \in B$, $\therefore x \in A_\alpha \cap B$ for some α, and $\therefore x \in \bigcup_\alpha (A_\alpha \cap B)$. Conversely, one can see that

$$x \in \bigcup_{\alpha}(A_\alpha \cap B) \Rightarrow x \in A_\alpha \cap B$$

for some α. $\therefore x \in \bigcup A_\alpha$ and $x \in B$. $\therefore x \in \left(\bigcup_\alpha A_\alpha\right) \cap B$

Similarly, it is easy to verify:

1.

$$\left(\bigcap_{\alpha} A_\alpha\right) \cup B = \bigcap_{\alpha}(A_\alpha \cup B)$$

(*i.e.*, union distributes over intersection),

2.

$$\left(\bigcup_{\alpha} A_\alpha\right) \backslash B = \bigcup_{\alpha}(A_\alpha \backslash B)$$

(*i.e.*, difference distributes over union).

If in a particular context every set referred to is a subset of a bigger set U, then U is called the *universal set* for all sets in that context. Then, for any set $A \subset U$, the set $U \backslash A$ is called the "complement of A" or "A complement," and represented by the symbol A^c. The following identities are easily verified:

1. $(A^c)^c = A$,

2. $U^c = \Phi$,

3. $\Phi^c = U$,

4. $(A \cup B)^c = A^c \cap B^c$,

5. $(A \cap B)^c = A^c \cup B^c$,

6. $(A \backslash B)^c = A^c \cup B$,

7. $(A^c \cup A) = U$.

A set may also be specified by means of a predicate :

$$A = \{x|P[x]\} \quad \text{or} \quad A = \{x : P[x]\}$$

(read "A is the set of all x such that $P[x]$ (is true)"). For example:

$$E = \{x|x \text{ is an integer divisible by 2}\}$$

is the set of all even integers. The predicate $P[x]$ here is "x is an integer divisible by 2". The predicate defining a set takes as input candidate elements for the set, and gives as output the values true or false.

1.3   Cardinality of Sets – Finite, Countable, and Uncountable Sets

The *cardinality* of a set is a measure of the size of that set. In the case of sets with finitely many elements in them, the cardinality of a set is the same as the number of elements in that set. If the number of elements in a set is infinite, then obviously it is futile to try and count them. Yet, not all infinite sets have the same "size." However, one can measure the relative sizes of sets:

Definition 1.1 *Two sets A and B are said to be* **equivalent** *or have the same* **cardinality** *(written $\#(A) = \#(B)$) if it is possible to show a one-to-one correspondence between their elements.*

Example: **N** and **Z** have the same cardinality: Consider the mapping $p : \mathbf{N} \to \mathbf{Z}$ given by $p(n) = (-1)^n[n/2]$, where $[x]$ is equal to the largest integer less than x. p is thus a one-to-one correspondence between **N** and **Z**.

Two finite sets have the same cardinality if they have an equal number of elements. It is clear that:

(a) $\#(A) = \#(B) and \#(B) = \#(C) implies \#(A) = \#(C)$.

(b) If $A = \bigcup_\alpha A_\alpha = B = \bigcup_\alpha B_{\alpha'}$ where the indexing set is the same in both the cases, and $\#(A_\alpha) = \#(B_\alpha)\forall\alpha$, then $\#(A) = \#(B)$.

Definition 1.2 *A set A is said to be countable if it has the same cardinality as the set of natural numbers N, i.e., if there is a one to one correspondence between the elements of A and those of N.*

In other words, countable sets are those infinite sets whose elements can be labeled uniquely and exhaustively using all natural numbers. The cardinality of countable sets is denoted by the symbol $\aleph_0$ (read "aleph-nought").

Taken together, the following two theorems show that countable sets are infinite sets of "smallest" cardinality:

Theorem 1.3 *Any infinite set A has a countable subset.*

Proof: Pick an element from A and label it a_1; pick another and label it a_2. Continuing this process indefinitely one has a countably infinite subset of A .

Theorem 1.4 *Every infinite subset of a countable set is countable.*

Proof: If $A = \{a_i\}_{i=1}^{\infty}$ is a countable set, and B is any infinite subset of A, arrange all the elements of B as a sequence with increasing indices:

$$a_{n_1}, a_{n_2}, ..., a_{n_k}, ...(n_1 < n_2 < .. < n_k < ..)$$

(This is possible since the elements of B are also in A, and hence are indexed already). Renumber the elements of B using the scheme $a_{n_k} \mapsto b_k$. Thus B is countable.

Theorem 1.5 *The union of a finite number of countable sets is countable.*

Proof: Let

$$A_1 = \{a_{1,1}, a_{1,2}, ..., a_{1,n}, ...\}$$
$$A_2 = \{a_{2,1}, a_{2,2}, ..., a_{2,n}, ...\}$$

$$.........................$$

$$A_m = \{a_{m,1}, a_{m,2}, ..., a_{m,n}, ...\}$$

be m countable sets. Consider the set $B = \{b_1, b_2, ..., b_n, ...\}$, obtained by labeling in sequence the elements of the above matrix column-wise, *i.e.*, $b_k = a_{k(mod\,m),[\frac{k}{m}]} \forall k \in \mathbf{N}$. Hence, $B = \bigcup_{i=1}^{m} A_i$, and B is countable.

In fact one can prove more:

Theorem 1.6 *The union of a countable number of countable sets is countable.*

Proof: If $A_i = \{a_{i,j}\}_{j=1}^{\infty} \forall i \in \mathbf{N}$, then the equation

$$a_{m,n} = b_{\frac{(m+n-1)(m_n-2)}{2}+n}$$

yields a set $B = \{b_n\}_{n=1}^\infty$, with $B = \bigcup_{n=1}^\infty$ and clearly, B is countable.

Theorem 1.7 Q: *the set of all rationals, is countable.*

Proof: If we agree to write each positive rational in its reduced form, then $\mathbf{Q}^+ \subset \{\frac{a}{b} \mid a, b \in \mathbf{N}\} = M$. Now $(a, b) \leftrightarrow \frac{a}{b}$ is a one-to-one correspondence between $\mathbf{N} \times \mathbf{N}$ and M , and $(a, b) \leftrightarrow \frac{(a+b-1)(a+b-2)}{2} + b$ is a one-to-one correspondence between $\mathbf{N} \times \mathbf{N}$ and $\mathbf{N}$. Therefore M is countable, and since $\mathbf{Q}^+ \subset M$, $\mathbf{Q}^+$; $\therefore$ $\mathbf{Q}$, is countable.

An uncountable set is an infinite set that cannot be exhaustively enumerated using natural numbers. In other words an uncountable set has a 'higher' cardinality than natural numbers. There are infinite sets that are uncountable:

Theorem 1.8 R: *the set of all real numbers, is uncountable.*

Proof: (by Cantor diagonalization) It is sufficient to demonstrate that a subset of $\mathbf{R}$ is uncountable.

Consider the closed interval [0,1]. Use base 2 (dyadic) representation to express each number in [0,1]. Then each number $\alpha \in [0, 1]$ is represented by a sequence $\alpha_1 \alpha_2 \alpha_3 ... \alpha_n...$, where each α_i is either a zero or a one. Thus

$$\alpha = \sum_{n=1}^\infty \frac{\alpha_n}{2^n}$$

The set of all numbers in [0,1] that have more than one dyadic representation are the dyadically rational numbers (i.e., the set of all numbers of the form:

$$\frac{a_0 + a_1 2 + ... + a_n 2^n}{b_0 + b_1 2 + ... + b_m 2^m}$$

with a_i's and b_i's $= 0$ or 1, and $n < m$, which have two equivalent dyadic expansions), which are countably many in number. We include both representations of each dyadic rational to the set D of dyadic representations of all real numbers between 0 and 1, and denote this set by D' . This inclusion does not change the cardinality of this set. (Why not?) Clearly D is countable iff D' is countable.

Now let the following be an enumeration of D' :

$$1 \quad a_{1,1} \quad a_{1,2} \quad a_{1,3} \quad ... \quad a_{1,m} \quad ...$$

$$2 \quad a_{2,1} \quad a_{2,2} \quad a_{2,3} \quad ... \quad a_{2,m} \quad ...$$

$$3 \quad a_{3,1} \quad a_{3,2} \quad a_{3,3} \quad \ldots \quad a_{3,m} \quad \ldots$$

....

$$n \quad a_{n,1} \quad a_{n,2} \quad a_{n,3} \quad \ldots \quad a_{n,m} \quad \ldots$$

....

where each row represents the dyadic expansion of a real number in [0,1], each entry in the left-most column is the index of the dyadic expansion to its right, and each $a_{i,j}$ is either 0 or 1. Now consider the dyadic expansion $b_1 b_2 b_3 ... b_n ...$ where $b_i = 1 - a_{i,i} \forall i \in \mathbf{N}$. This expansion represents a number in [0,1]. However, it is not present in the above enumeration, since it differs from the nth expansion at the nth place. This is contrary to the assumption that the above enumeration is exhaustive. Hence, the real numbers between 0 and 1 form an uncountable set. Therefore $\mathbf{R}$ is uncountable.

The cardinality of R is called the *power of the continuum*, denoted $\aleph_1$. The continuum hypothesis of Cantor states that there are no infinite sets whose cardinality is less than the power of the continuum and greater than that of a countable set.

1.4   Rings and Algebras of Sets

The following definitions introduce some algebraic systems of sets of interest to us later in this book:

Definition 1.9 *If S is any set, a nonempty system R of some subsets of S is called a* ring *of sets if $\forall A, B \in R$, the following conditions hold:*

(a) $A \cup B \in A$ *and*

(b) $A \backslash B \in R$

Remark: $\Phi \in R \because A \backslash A \in R$ and $A \cap B \in R \because A \cap B = A \backslash (A \backslash B)$.

Definition 1.10 *If S is any set, a nonempty system A of some subsets of S is called an* algebra *of sets over S if $\forall A, B \in A$, the following conditions hold:*

(a) $A \cup B \in A$ *and*

(b) $A^c = S \backslash A \in A$.

The following theorem is immediate:

Theorem 1.11 *A system A of subsets of a set S is an algebra over S, iff it is a ring over S, and $S \in A$.*

Definition 1.12 *If S is any set, a nonempty system R of some subsets of S is called a σ-ring over S, if it is closed w.r.t. set differences and countable unions; i.e.,*

(a) *If $\{A_n\}_{n=1}^{\infty}$ is a set of subsets of S, then $\bigcup_{n=1}^{\infty} A_n \in R$, and*

(b) *$\forall A, B \in R, A \backslash B \in R$.*

Definition 1.13 *If S is any set, a nonempty system A of some subsets of S is called a σ-algebra over S, if it is closed w.r.t. complementation and countable unions; i.e.,*

(a) *If $\{A_n\}_{n=1}^{\infty}$ is a set of subsets of S, then $\bigcup_{n=1}^{\infty} A_n \in R$, and*

(b) *$\forall A \in A, A^c = S \backslash A \in A$.*

The following theorem is immediate:

Theorem 1.14 *A is a σ-algebra over a set S iff A is a σ-ring over S, and contains S.*

Chapter 2

Linear Spaces, Metric Spaces, and Hilbert Spaces

2.1 Linear Spaces

Linearity is one of the most sought after properties in mathematics. It is also a profoundly simplifying conceptual tool in the study of mathematical entities from the geometric, algebraic, and analytic points of view. In this book the main structures of interest are spaces of functions, which are studied using certain illuminating analytic tools. Both the spaces studied and the tools used in their study are linear in nature in that the spaces of functions are linear spaces and the tools are linear functionals and operators on these linear spaces. This section introduces and discusses some elementary properties of linear spaces.

Definition 2.1. *A* **linear space** *is a quadruple* $(L,\mathbf{F},+,\cdot)$ *consisting of a nonempty set L of elements, a field $\mathbf{F}$ of scalars (e.g., $\mathbf{R}$ or $\mathbf{C}$), a binary operation '+' called* **vector addition**, *defined between every pair of elements in L, and a binary operation '$\cdot$' called* **scalar multiplication**, *defined between any element of $\mathbf{F}$ and any element of L. For brevity the quadruple $(L,\mathbf{F},+,\cdot)$ will be denoted by L itself.*

The linear space L satisfies the following properties: 1. Properties of

Vector Addition

   (a) $\forall x, y \in L, x + y \in L$ (closure under vector addition)

   (b) $\forall x, y \in L, x + y = y + x$ (commutativity)

   (c) $\forall x, y, z \in L, (x + y) + z = x + (y + z)$ (associativity)

   (d) $\exists 0 \in L$ called the *zero vector* such that $x + 0 = x \; \forall x \in L$

   (e) $\forall x \in L, \exists -x \in L$ called the inverse of x such that $x + (-x) = 0 \; \forall x \in L$

2. Properties of Scalar Multiplication

   (a) $\forall \alpha \in \mathbf{F}, \forall x \in L, \alpha \cdot x \in L$ (closure of L under scalar multiplication)

   (b) $\forall \alpha, \beta \in \mathbf{F}, \forall x \in L, \alpha(\beta \cdot x) = (\alpha\beta) \cdot x$

   (c) If '1' is the multiplicative identity of $\mathbf{F}$, then $1 \cdot x = x, \forall x \in L$

3. Distributivity

   (a) $\forall \alpha, \beta \in \mathbf{F}, \forall x \in L, (\alpha + \beta) \cdot x = \alpha \cdot x + \beta \cdot x$

   (b) $\forall \alpha \in \mathbf{F}, \forall x, y \in L, \alpha \cdot (x + y) = \alpha \cdot x + \alpha \cdot y$

The elements of L are often called vectors, in analogy with the vectors of two- and three-dimensional spaces, which satisfy the above mentioned properties. In general, however, the elements of L may look very different from two- or three-dimensional vectors. For instance, they could be functions, as they indeed are for the most part in this book. The linear spaces of interest in this book are the so-called infinite dimensional vector spaces.

The letters $x, y, z, u, v, w, \xi, \eta, \zeta$ or subscripted versions of them will be used to denoted elements of L, while the letters $\alpha, \beta, \gamma, \delta, \lambda, a, b, c, d, r, s$ or subscripted versions of them will be used to denote elements of $\mathbf{F}$, unless otherwise stated.

We will drop the notational distinction between 'O' and 'o', and '+' and '$+$' and write αx instead of $\alpha \cdot x$ henceforth, since the distinctions will be clear from the context of the symbols. There will be no scope for confusion, since addition of a scalar and a vector is undefined, as is the multiplication of two vectors using the scalar multiplication on $\mathbf{F}$. If $\mathbf{F} = \mathbf{R}$, then L is a *real* linear space, while if $\mathbf{F} = \mathbf{C}$, then L is a *complex* linear space.

Examples:

1. If $L = (L,F,+,\cdot)$ with $L=\mathbf{R}$, $F=\mathbf{R}$, and $+$ and $\cdot$ the usual addition and multiplication on $\mathbf{R}$, then L is a real linear space.

2. The set $L = \underbrace{\mathbf{R} \times \mathbf{R} \times \ldots \times \mathbf{R}} = \mathbf{R}^n = \{(x_1, x_2, \ldots, x_n) | x_i \in \mathbf{R} \; \forall 1 \leq i \leq n\}$, with the operations of vector addition and scalar multiplication defined by:

   $x = (x_1, x_2, \ldots, x_n), y = (y_1, y_2, \ldots, y_n) \Rightarrow x + y = (x_1 + y_1, x_2 + y_2, \ldots, x_n + y_n)$

   $\alpha x = \alpha(x_1, x_2, \ldots, x_n) = (\alpha x_1, \alpha x_2, \ldots, \alpha x_n) \; \forall \alpha \in \mathbf{R}$, with $O = (0, 0, \ldots, 0)$, and $-x = (-x_1, -x_2, \ldots, -x_n)$ is a real linear space.

3. $C[a, b]$, the set of all complex valued continuous functions of a real variable defined on the closed interval [a,b], is a complex linear space.

4. $L = l_2(\mathbf{R}) = \{(x_1, x_2, \ldots, x_n, \ldots) | x_i \in \mathbf{R} \; \forall 1 \leq i \leq n, \text{ and } \sum_{i=1}^{\infty} x_i^2 < \infty\}$ (the space of all square-summable, infinite sequences of real numbers) with operations of vector addition and scalar multiplication defined componentwise as in Example 2 is a real linear space.

It is easy to conceive of complex analogies of the examples 1, 2, and 4.

Lemma 2.2. If L is a linear space over a field $\mathbf{F}$, then $\forall x \in L, 0.x = O$, and $(-1).x = -x$, where 0 is the additive identity of $\mathbf{F}$, and -1 is the additive inverse of the multiplicative identity of $\mathbf{F}$.

Proof: $0 \cdot x = (0 + 0) \cdot x = 0 \cdot x + 0 \cdot x$. If $y = 0 \cdot x$, then $y = y + y$. Adding $-y$ to both sides of this equation, we have $O = y$. $\therefore 0 \cdot x = O$.

$\forall x, O = 0 \cdot x = (1-1) \cdot x = (1 + (-1)) \cdot x = x + (-1) \cdot x$, i.e., $(-1) \cdot x + x = O$. Adding $-x$ to both sides, we have $(-1) \cdot x = -x$.

Definition 2.3 *If L and L' are two linear spaces and h is a mapping of L onto L': $h{:}L \rightarrow L'$ given by $h(\alpha x + \beta y) = \alpha h(x) + \beta h(y)$. $\forall \alpha, \beta \in F$, $\forall x, y \in L$, then L is said to be a* homomorphism *of L onto L'.*

Remark: If $h{:}L \rightarrow L'$ is a homomorphism, then $h(0) = 0$ where $0 \in L$ and $0 \in L'$. Indeed, $h(0) = h(0 + 0) = 2h(0)$.

Definition 2.4 *If h: L→L' is a homomorphism such that $h(x) = 0 \Leftrightarrow x = 0$, then h is called an* isomorphism.

Remark: If h:L→L' is an isomorphism, then $\forall x, y \in L, h(x) = h(y) \Leftrightarrow x = y$.

If h:L→L' is an isomorphism, the correspondence $h(x) \leftrightarrow x$ between elements of L and L' is one-to-one and onto: $h(x) \leftrightarrow x$, $h(y) \leftrightarrow y \Rightarrow h(x + y) \leftrightarrow x + y$, $h(\alpha x) \leftrightarrow \alpha x$. Isomorphic linear spaces may be thought as different realizations or representations of the same linear space.

2.1.1   Subspaces

If L is a linear space over a field **F**, and L' is a nonempty subset of L such that $\forall x, y \in L'$, $\forall \alpha, \beta \in F, \alpha x + \beta y \in L'$, then L' is said to be a *subspace* of L, written $L' \subset_s L$.

A subspace L' of L is a linear space in its own right. The trivial space consisting of the zero element only and the whole space L are subspaces of L. A proper subspace of L is a subspace of L which is not the whole space L and contains at least one nonzero element of L.

Definition 2.5 *If $x_1, x_2, \ldots, x_n$ are elements of a linear space L such that there exist scalars $\alpha_1, \alpha_2, \ldots, \alpha_n \in$ F not all zero, satisfying $\alpha_1 x_1 + \alpha_2 x_2 + \ldots + \alpha_n x_n = 0$, then $x_1, x_2, \ldots, x_n$ are said to be* **linearly dependent** *. If there exists no such set of scalars, then $x_1, x_2, \ldots, x_n$ are said to be* **linearly independent** *.*

Definition 2.6 *A linear space L is said to be n-dimensional if n linearly independent elements can be found in L, but every set of (n+1) elements is linearly dependent. If, however, for every n there exist n linearly independent elements in L, then L is said to be infinite dimensional.*

Remark: Any set of n linearly independent elements of an n-dimensional linear space L is said to be a basis for L, and every element of L can be expressed uniquely as a linear combination of elements of the basis.

2.1.2 Factor spaces (quotient spaces)

If $L' \subset_s L$ then x,y $\in$ L belong to the same residue class generated by L' if
x-y $\in L'$. The set of all such residue classes is a linear space and is called
the *factor space* of L relative to L', written L/L'. If $\xi, \eta \in L/L'$ are two
residue classes , then choose $x \in \xi$, $y \in \eta$. Then the class $\xi + \eta$ is the
class to which x+y belongs, and L/L' is a linear space. If $L' \subset_s L$ then the
dimension of L/L' is called the *codimension* of L' in L.

Theorem 2.7 *If $L' \subset_s L$ then L' has finite codimension n iff $\exists$ linearly
independent elements $x_1, \ldots, x_n \in L$ and every $x \in L$ has a unique representation of the form $x = \alpha_1 x_1 + \ldots + \alpha_n x_n + y$ where $\alpha_1, \ldots, \alpha_n \in$ F and
y$\in L'$.*

    Proof: If $\xi \in L/L'$ and $x_k \in \xi_k$, then $\xi = \xi_1 \alpha_1 + \ldots + \xi_n \alpha_n$ $\forall \xi \in$
L/L'. Thus $\xi_1, \ldots, \xi_n$ is a basis for L/L' and dim $L/L' = n$. If codim L'
$= n$, $\exists$ basis of $L/L' = < \xi_1, \ldots, \xi_n >$. If x $\in$ L and x $\in \xi \in L/L'$, then
$\xi = \alpha_1 \xi_1 + \ldots + \alpha_n \xi_n$, i.e., $x = \alpha_1 x_1 + \ldots + \alpha_n x_n + y$ for fixed $x_1, \ldots, x_n \in$
$\xi_1, \ldots, \xi_n$ respectively, and y$\in L'$. Indeed, if $x = \alpha_1' x_1 + \ldots + \alpha_n' x_n + y'$,
then $(\alpha_i' - \alpha_i) x_i + y - y' = 0$, i.e., $(\alpha_i' - \alpha_i)\xi_i = 0$, a contradiction and hence
the theorem.

2.1.3 Linear functionals

Definition 2.8 *A mapping f :L $\to$ F is a **functional** on L. f is additive if
$f(x+y) = f(x)+f(y)$ and homogeneous if $f(\alpha x) = \alpha f(x)$. If f is defined on
a complex linear space then f is conjugate homogeneous if $f(\alpha x) = \bar{\alpha} f(x)$.
Accordingly a functional with the above two properties is a linear/conjugate-
linear functional.*

    Examples:

1. If $(a_1, \ldots, a_n) \in \mathbf{R^n}$ or $\mathbf{C^n}$ is a fixed n-tuple, then

    $f : \mathbf{R^n} \to \mathbf{R}$ given by $(x_1, \ldots, x_n) \longmapsto a_1 x_1 + a_2 x_2 \ldots + a_n x_n$

    or $f : \mathbf{C^n} \to \mathbf{C}$ given by $(x_1, \ldots, x_n) \longmapsto a_1 \bar{x}_1 + a_2 \bar{x}_2 \ldots + a_n \bar{x}_n$ are
    real or complex linear functionals

2. If $I(x) = \int_a^b \bar{x}(t)\varphi(t)dt \varphi(t) \in$ C[a,b] fixed and $x(t) \in$ C[a,b], then I is
    a linear functional on C[a,b] where C[a,b] is the space of all continuous
    functions on the closed interval [a,b].

2.1.4   Null space (kernel) of a functional – hyperplanes

If $f : L \to \mathbf{F}$ is a linear functional, consider the set $L_f = \{x \mid f(x) = 0\}$.

Since $f(0) = 0$ and $f(\alpha x + \beta y) = \alpha f(x) + \beta f(y)$, $f(x) = 0$, $f(y) = 0 \Rightarrow$ $f(\alpha x + \beta y) = 0$. Hence L_f is a subspace of L, written $L_f \subset L$.

Theorem 2.9 *Let x_0 be any fixed element of $L \setminus L_f$. Then every element $x \in L$ has a unique representation of the form $u = \alpha x_0 + y$, where $y \in L_f$.*

Proof: $f(x_0) \neq 0$ and $x_0 \neq 0$ (obvious). Without a loss of generality, assume $f(x_0) = 1$ (else replace x_0 by $x_0 / f(x_0)$). Let $x \in L$. Let $y = x - \alpha x_0$, where $\alpha = f(x)$. Now $y \in L_f$. $\therefore x = \alpha x_0 + y$. This representation is unique. Indeed, $x = \alpha' x_0 + y' \Rightarrow (\alpha - \alpha')x_0 = y - y'$ and $\alpha = \alpha' \Rightarrow y = y'$. $\alpha \neq \alpha' \Rightarrow x_0 \in L_f$, a contradiction and hence the theorem.

Corollary 2.9.1 x_1, x_2 are in the same residue class in L with respect to $L_f (i.e., x_1 - x_2 \in L_f)$ iff $f(x_1) = f(x_2)$.

Corollary 2.9.2 L_f has Codim=1.

Corollary 2.9.3 Two nontrivial linear functionals f and g with the same space are proportional.

Indeed, let $x_0 \in L \setminus L_f \ni f(x_0) = 1$, then $g(x_0) \neq 0$ since $x = f(x)x_0 + y$ $(y \in L_f)$ $g(x) = f(x)g(x_0)$. $\therefore g(x_0) = 0 \Rightarrow g$ is trivial. $\therefore g(x) = g(x_0)f(x)$

Definition 2.10. *If $L' \subset_s L \ni Codim=1$, then every class generated by L' is called a **hyperplane parallel** to L'.*

2.1.5   Geometric interpretation of linear functions

The following theorem gives a geometric interpretation of linear functionals by demonstrating a one-to-one correspondence between the set of all non-trivial linear functionals on a linear space L and the set of all hyperplanes not passing through the origin in L.

Theorem 2.11 *Given a linear space L and a nontrivial linear functional f on it, the set $M_f = \{x \mid f(x) = 1\}$ is a hyperplane parallel to L_f. Conversely, let $M' = L' + x_0 = \{y \mid y = x + x_0, x \in L\}$ and $(x_0 \notin L')$ be any hyperplane* *parallel to a subspace $L' \subset L$ of codim 1 and not passing through the origin. Then $\exists$ a unique linear functional f on $L \ni M' = \{x | f(x) = 1\}$.*

Proof: Given f, let $x_0 \in L \ni f(x_0) = 1$ (such an x_0 can always be found). Then $\forall x \in M_f \; x = x_0 + y, y \in L_f$.

Conversely, given $M' = L' + x_0 (x_0 \notin L')$, then $\forall x \in L, x = \alpha x_0 + y, y \in L'$. ($\because \; L'$ has codim $= 1$). Setting $f(x) = \alpha$ we get the desired linear functional f. f is unique and $g(x) \equiv 1$ for $x \in M' \Rightarrow g(y) = 0 \; \forall y \in L'$, so that $g(x) = g(\alpha x_0 + y) = \alpha = f(\alpha x_0 + y) = f(x)$. Thus, there is one-to-one correspondence between the class of all nontrivial linear functionals on L, and the set of all hyperplanes in L not passing through the origin.

A special kind of linear functional on a linear space is the *norm*. Geometrically, it represents the *"length"* of a vector in the space over which it is defined.

2.1.6 Normed linear spaces

Definition 2.12 *A functional p defined on a linear space L is a* **norm** *(in L) if it has the following properties:*

(i) $p(x) \geq 0 \; \forall x \in L, p(x) = 0 \Leftrightarrow x = 0$

(ii) $p(\alpha x) = |\alpha| p(x) \; \forall x \in L$ and $\forall \alpha \in \mathbf{F}$. ($\mathbf{F} = \mathbf{R} \, or \, \mathbf{C}$)

(iii) $p(x + y) \leq p(x) + p(y)$ (Convexity) $\forall x, y \in L$

The norm of an element of a linear space corresponds to the notion of the length of a vector in two- or three-dimensional vector spaces.

Definition 2.13 *A linear space L with a norm $p(x)$ denoted $||x||$ is called a* **normed linear space.**

Rewriting the properties of a norm, we have:

(i) $||x|| \geq 0 \; \forall x \in L, ||x|| = 0$ iff $x = 0$

(ii) $||\alpha x|| = |\alpha| \; ||x|| \; \forall x \in L$ and $\forall \alpha \in \mathbf{F}$.

(iii) $||x + y|| \leq ||x|| + ||y|| \; \forall x, y \in L$.

The norm may also be interpreted as the *distance* of any given element of a linear space from the zero element. Indeed, one may introduce the idea of distance between any two elements of a linear space through the norm. This is extremely important for the study of the analytic properties of functionals and operators on a linear space. To understand and exploit the idea of distance as an analytic tool, one will do well to study point spaces with distance defined between all pairs of points before focusing on linear spaces.

2.2   Metric Spaces

Definition 2.14 *A* **metric space** $\mathcal{M}$ *is an ordered pair* (χ, ρ)*, where* χ *is a set of points and* ρ *is a real, single-valued, and nonnegative function.* $\forall x, y, z \in \chi$

(a) $\rho(x, y) \geq 0, = 0 \; iff \; x = y$ *(positive definiteness)*

(b) $\rho(x, y) = \rho(y, x)$ *(symmetry)*

(c) $\rho(x, y) \leq \rho(x, y) + \rho(y, z)$ *(triangle inequality)*

The following are some examples of metric spaces:

1. $\mathcal{M} = (\chi, \rho)$, with $\chi = \mathbf{R}$, and $\rho(x, y) = |x - y|$, $\forall x, y \in \mathbf{R}$

2. $\mathcal{M} = (\chi, \rho)$, with $\chi = \mathbf{R}^n$ and $\forall x = (x_1, \ldots, x_n), y = (y_1, \ldots, y_n) \in \mathbf{R}^n$ $\rho(x, y) = [\sum_{i=1}^{n} (x_i - y_i)^2]^{1/2}$

3. $\mathcal{M} = (\chi, \rho)$, with $\chi = C[a, b]$: the space of all continuous functions on a closed interval [a,b] and $\rho(x, y) = max_{a \leq t \leq b} |f(t) - g(t)|$, $\forall f, g \in C[a, b]$.

4. $\mathcal{M} = (\chi, \rho)$, with $\chi = l_2(\mathbf{R})$, and $\rho(x, y) = [\sum_{i=1}^{n} (x_i - y_i)^2]^{1/2}$ $\forall x = (x_1, x_2, \ldots, x_n, \ldots), y = (y_1, y_2, \ldots, y_n, \ldots) \in l_2(\mathbf{R})$.

2.2.1   Continuous mappings

Definition 2.15 *If* $\mathcal{M} = (\chi, \rho)$ *and* $\mathcal{M}' = (\chi', \rho')$ *are two metric spaces, and* $f : \chi \to \chi'$ *is a mapping sending points of* χ *into points of* χ' *given by*

$x \mapsto f(x)$, *then f is said to be* continuous *at the point $x_0 \in \chi$ if, given any $\epsilon > 0, \exists \delta > 0$ such that $\rho'(f(x), f(x_0)) < \epsilon$, whenever $\rho(x, x_0) < \delta$.*

In other words if f is continuous at x_0, then points "close" to x_0 in χ are mapped by f to points "close" to $f(x_0)$ in χ'.

A mapping f is said to be continuous on χ if it is continuous at every point of χ.

The following examples illustrate the concept of continuous mappings:

1. If $\mathcal{M} = \mathcal{M}' = (\chi, \rho)$ where $\chi = \mathbf{R}$ and $\rho(x, y) = |x - y| \ \forall x, y \in \mathbf{R}$ then $f : \chi \to \chi$ is a real valued continuous function on $\mathbf{R}$.

2. If $\mathcal{M} = (\mathbf{R}, \rho)$ and $\mathcal{M}' = (\mathbf{R}^+, \rho)$, where ρ is the same as in the above example, and $\mathbf{R}^+$ is the set of all positive real numbers, then $f : \mathbf{R} \to \mathbf{R}^+$ given by $f(x) = e^x$ is a real valued continuous function.

Definition 2.16 *A one-to-one mapping f of $\mathcal{M} = (\chi, \rho)$ onto $\mathcal{M}' = (\chi', \rho')$ is said to be an* isometry *if $\rho(x, y) = \rho'(f(x), f(y)) \ \forall x, y \in \chi$.*

Two isometric spaces are thought of as different representations of the same space, and therefore no distinction is made between them.

2.2.2 Convergence

Definition 2.17 *The open sphere $S(x_0, r)$ in a metric space $\mathcal{M}$ is given by $S(x_0, r) = \{x \in \mathcal{M} | \rho(x, x_0) < r\}$ where x_0 is the center and r the radius of the open sphere. The closed sphere corresponding to it is $S[x_0, r] = \{x \in \mathcal{M} | \rho(x, x_0) \leq r\}$.*

Definition 2.18 *An open sphere of radius ϵ with center x_0 will be called an ϵ neighborhood of x_0, denoted by $N_\epsilon(x_0)$.*

Definition 2.19 *A point $x \in \mathcal{M}$ is called a* contact point *of a set $S \subset \mathcal{M}$, if every neighborhood of x contains at least one point of S. The set of all contact points of a set of S is denoted $\overline{S}$ and is called the* closure *of S. Clearly $S \subset \overline{S}$.*

Definition 2.20 *A sequence of points $\{x_n\}_{n=1}^\infty$ in a metric space $\mathcal{M}$ is said to* **converge** *to point $x \in \mathcal{M}$ if every neighborhood of x contains all except* *at most finitely many points of* $\{x_n\}_{n=1}^{\infty}$. *More precisely, if* $\forall \epsilon > 0 \exists N(\epsilon) \in$ $\mathbf{N} \ni x_n \in O_\epsilon(x)$ $\forall n > N(\epsilon)$, *then* x *is called the limit of* $\{x_n\}_{n=1}^{\infty}$ *written* $lim_{n \to \infty} x_n = x$.

Remarks:

(a) Clearly $\{x_n\}_{n=1}^{\infty}$ converges to $x \Leftrightarrow lim_{n \to \infty} \rho(x, x_n) = 0$.

(b) As an immediate consequence of the definition of a limit:

   1. No sequence can have two distinct limits.
   2. If a sequence $\{x_n\} \to x$, then so does every subsequence of $\{x_n\}$.

Theorem 2.21 *x is a contact point of the set* $\mathcal{S} \Leftrightarrow \exists \{x_n\}_{n=1}^{\infty} \subset \mathcal{S} \ni x_n \to x$ *as $n \to \infty$.*

Definition 2.22 *A point* $x \in \mathcal{M}$ *is called a* **limit point** *of a set* $\mathcal{S} \in \mathcal{M}$ *if every neighborhood of x contains infinitely many points of $\mathcal{S}$. (The limit point may or may not belong to $\mathcal{S}$.) A point $x \in \mathcal{S}$ is called an* isolated point *of $\mathcal{S}$ if $\exists$ a neighborhood of x such that no point of $\mathcal{S}$ is contained in it, other than x itself.*

Theorem 2.23 *x is a limit point of the set* $\mathcal{S} \Leftrightarrow \exists$ *a sequence* $\{x_n\}$ *of distinct points of $\mathcal{S}$ such that $x_n \to x$ as $n \to \infty$.*

2.2.3   Dense subsets

Definition 2.24 *If* $\mathcal{A}, \mathcal{B} \subset \mathcal{M}$, *then* $\mathcal{A}$ *is* **dense** *in* $\mathcal{B}$ *if* $\overline{\mathcal{A}} \supset \mathcal{B}$. $\mathcal{A}$ *is* everywhere dense *in* $\mathcal{M}$ *if* $\overline{\mathcal{A}} = \mathcal{M}$. *A set $\mathcal{A}$ is said to be* nowhere dense *if it is dense in no open sphere at all.*

The following are examples of dense subsets:

1. The set of rational numbers. $\mathbf{Q}$ is dense in $\mathbf{R}$.

2. $\mathbf{Q}^n$ is dense in $\mathbf{R}^n$.

3. If $q_n(\mathbf{Q}) = \{x \in l_2(\mathbf{R}) | x_i \in \mathbf{Q} \ \forall 1 \leq i \leq \infty$ and $x_i = 0 \ \forall i > n\}$, then $\cup_{n=0}^{\infty} q_n(\mathbf{Q})$ is dense in $l_2(\mathbf{R})$.

4. The set $\mathbf{Q}[x]$ of all polynomials in one variable with rational coefficients is dense in the set $C[a, b]$ of all continuous real valued functions on $[a, b]$.

Definition 2.25 *A metric space $\mathcal{M}$ is said to be* **separable** *if it has a countable and everywhere dense subset.*

Since $\mathbf{Q}, \mathbf{Q}^n, \cup_{n=0}^{\infty} q_n(\mathbf{Q})$, and $\mathbf{Q}[x]$ are countable, going by the above examples of dense subsets $\mathbf{R}, \mathbf{R}^n, l_2(\mathbf{R})$, and $C[a,b]$ are separable.

2.2.4   Closed sets

Definition 2.26 *A set $S \subset \mathcal{M}$ is closed if $S = \overline{S}$ (i.e., if it contains all its limit points).*

Examples: The empty set $\boldsymbol{\Phi}$ and the whole space $\mathcal{M}$ are closed; $[a,b] \subset \mathbf{R}$ is closed; any finite set of points is closed; intersection of closed sets are closed; finite union of closed sets are closed.

2.2.5   Open sets

Definition 2.27 *A point $x \in S$ is an* interior point *of S if x has a neighborhood $N_\epsilon(x) \subset S$ (i.e., a neighborhood consisting of points of S alone). A set $S \subset \mathcal{M}$ is* open *if all its points are interior points.*

Some examples of open sets are:

1. Every open interval $(a,b) \subset \mathbf{R}$ is open.

2. The set $\{f \in C[a,b] | f(x) < g(x) \quad \forall x \in [a,b]$ and for some fixed $g \in C[a,b]\}$ is open.

3. ϕ and M are open. Arbitrary unions and finite intersections of open sets are open.

Theorem 2.28 *$S \subset \mathcal{M}$ is open $\Leftrightarrow S^c$ is closed.*

Proof: S is open $\Leftrightarrow \forall x \in S$, x is interior to S $\quad \therefore x$ is not a contact point of S^c. $\therefore S^c$ is closed. If S^c is closed then every $x \in S$ interior to S for it would be a contact point of S^c otherwise. $\therefore S$ is open.

In general the structure of open and closed sets is complex. However, the structure of open and closed sets on $\mathbf{R}$ is quite simple.

Theorem 2.29 *Every open set on* **R** *is the union of at most countable mutually disjoint open intervals, and every closed set is the complement of an open set.*

2.2.6   Complex metric spaces

Definition 2.30 *A sequence* $\{x_n\}$ *in a metric space* $\mathcal{M} = (\chi, \rho)$ *is said to satisfy the* Cauchy *criterion if, given any* $\epsilon > 0 \exists N(\epsilon) \in \mathbf{N}$, *then* $\rho(x_n, x_{n'}) < \epsilon \ \forall n, n' > N(\epsilon)$.

Definition 2.31 *A subsequence* $\{x_n\}$ *of points in* $\mathcal{M}$ *is called a* Cauchy sequence *if it satisfies the Cauchy criterion.*

Theorem 2.32 *Every convergent sequence is a Cauchy sequence.*

Definition 2.33 $\mathcal{M}$ *is a* **complete metric space** *if every Cauchy sequence in* $\mathcal{M}$ *converges to a point in* $\mathcal{M}$. *Otherwise* $\mathcal{M}$ *is said to be* **incomplete.**

Example: **R**,...,**R**^{n} are complete metric spaces.

2.2.7   Completion of metric spaces

Every metric space can be completed.

Definition 2.34 *If* $\{x_n\}$ *and* $\{y_n\}$ *are two Cauchy sequences in* $\mathcal{M}$ *such that* $|x_n - y_m| \to 0$ *as* $n, m \to \infty$ *then they are said to be equivalent (denoted* $\{x_n\} \sim \{y_n\}$).

Consider the class of all possible Cauchy sequences in $\mathcal{M}$. Under the above defined equivalence relation the elements of this class fall into equivalence classes. Each equivalence class is called the limit of each of the sequences belonging to it. $\mathcal{M}$ along with all the equivalence classes of Cauchy sequences is called the completion of $\mathcal{M}$. In other words, each equivalence class is thought of as a point to be included in the metric space $\mathcal{M}$. in order to complete it.

2.2.8   Norm-induced metric and Banach spaces

Every normed linear space becomes a metric space if we set $\rho(x,y) = \| x - y \| \ \forall x, y \in L$.

Definition 2.35 *A complete normed linear space relative to metric $\rho(x,y) = \| x - y \|$ is called a* **Banach space**.

Definition 2.36 *Two norms $\| \cdot \|_1, \| \cdot \|_2$ on a linear space L are said to be equivalent if $\exists$ constants $a, b > 0 \ni a \| x \|_1 \leq \| x \|_2 \leq b \| x \|_1$. (If L is finite dimensional, then any two norms on it are equivalent).*

Other important functionals on linear spaces are bilinear functionals called *inner products* or *scalar products* . On a real linear space, the scalar product of two vectors divided by the product of their norms has a useful geometric significance: It gives the "angle" between the two vectors. We now turn our attention to linear spaces equipped with such bilinear functionals.

2.3   Euclidean Spaces

2.3.1   Scalar products, orthogonality and bases

Definition 2.37 *A* **scalar product** *in a real (complex) linear space L is a real (complex) function defined for every pair of elements $x, y \in L$, denoted by $\langle x, y \rangle$. A scalar product has the following properties:*

1. $\langle x, x \rangle \geq 0, \ \langle x, x \rangle = 0 \Leftrightarrow x = 0$

2. $\langle x, y \rangle = \langle y, x \rangle (\langle x, y \rangle = \overline{\langle y, x \rangle})$

3. $\langle \lambda x, y \rangle = \lambda \langle x, y \rangle$

4. $\langle x, y + z \rangle = \langle x, y \rangle + \langle x, z \rangle \ \forall x, y, z \in L \ and \ \forall \lambda \in \mathbf{R(C)}.$

Definition 2.38 *A linear space L with a scalar product is called a* **Euclidean space** *, denoted $\mathcal{E}$, i.e., $(L, \langle, \rangle) = \mathcal{E}$.*

The scalar product of a Euclidean space induces a norm on it. Indeed, define $\|x\| = \sqrt{\langle x, x \rangle}$. It is easy to verify that $\|\|$ is a norm on a Euclidean space with scalar product $\langle, \rangle$.

$\forall x, y \in \mathcal{E} |\langle x, y \rangle| \leq \|x\|\|y\|$ (Schwartz inequality) .

Indeed, $\varphi(\lambda) = \langle \lambda x + y, \lambda x + y \rangle \geq 0, i.e., \lambda^2 \|x\|^2 + 2Re\lambda(\langle x, y \rangle) + \|y\|^2 \geq 0.$

This implies that the quadratic form $|\lambda|^2 \|x\|^2 + 2|\lambda||\langle x, y \rangle| + \|y\|^2$ in $|\lambda|$ has either no real root or one real root of multiplicity two. Hence, the discriminant is nonpositive, implying that $|\langle x, y \rangle|^2 \leq \|x\|^2 \|y\|^2$. Hence, the inequality.

Definition 2.39 *Given any two elements $x, y \in \mathcal{E}$ define the angle θ between x and y by $cos^{-1}[\langle x, y \rangle / \|x\|\|y\|]$. $\langle x, y \rangle = 0 \Rightarrow \theta = \pi/2$, then x and y are said to be **orthogonal** to each other.*

Definition 2.40 *A set of nonzero vectors $\{x_\alpha\} \subset \mathcal{E}$ is an **orthogonal system** if $\langle x_\alpha, x_\beta \rangle = 0 \, \forall \alpha \neq \beta$, and an **orthonormal system** if*

$$\langle x_\alpha, x_\beta \rangle = \begin{cases} 0 & \text{if } \alpha \neq \beta \\ 1 & \text{otherwise} \end{cases}$$

If $\{x_\alpha\}$ is an orthogonal system, then clearly

$$\left\{ \frac{x_\alpha}{\| x_\alpha \|} \right\}$$

is an orthonormal system.

Theorem 2.41 *The vectors in an orthogonal system $\{x_\alpha\}$ are linearly independent.*

Proof: Indeed if $\sum_j c_j x_{\alpha_j} = 0$ then $\langle \sum_j c_j x_{\alpha_j}, x_{\alpha_i} \rangle = 0 \Rightarrow c_i \langle x_{\alpha_i}, x_{\alpha_i} \rangle = 0 \Rightarrow c_i = 0.$

Definition 2.42 *An orthogonal system $\{x_\alpha\}$ is an **orthogonal basis** if it is complete, i.e., if the smallest closed subspace of $\mathcal{E}$ containing $\{x_\alpha\}$ is the whole space $\mathcal{E}$, i.e., $span(\{x_\alpha\}) = \mathcal{E}$.*

The following are some examples of Euclidean spaces:

1. $\mathbf{R}^n$ with $\langle x, y \rangle = \sum_{i=1}^n x_i y_i$, or $\mathbf{C}^n$ with $\langle x, y \rangle = \sum_{i=1}^n x_i \bar{y}_i$ are Euclidean spaces. The corresponding norms are given by $\|x\| = [\sum_{i=1}^n x_i^2]^{1/2}$ on $\mathbf{R}^n$, and $\|x\| = [\sum_{i=1}^n |x_i|^2]^{1/2}$ on $\mathbf{C}^n$. The set of vectors $\{e_i = (\delta_{i1}, \delta_{i2}, ..., \delta_{in}), 1 \leq i \leq n\}$ with

$$\delta_{ij} = \begin{cases} 0 & \text{if } i \neq j \\ 1 & \text{otherwise} \end{cases}$$

is an orthonormal basis for these spaces.

2. $l_2(\mathbf{R})$ with $\langle x, y \rangle = \sum_{i=1}^{\infty} x_i \overline{y}_i$, or $l_2(c)$ with $\langle x, y \rangle = \sum_{i=1}^{\infty} x_i \overline{y}_i$ are Euclidean spaces. The corresponding norms are given by $|x| = [\sum_{i=1}^{\infty} x_i^2]^{1/2}$ on $l_2(\mathbf{R})$, and $|x| = [\sum_{i=1}^{\infty} |x_i|^2]^{1/2}$ on $l_2(\mathbf{C})$. The set of vectors $\{e_i = (\delta_{i1}, \delta_{i2}, ..., \delta_{in}, ...), 1 \le i < \infty\}$, with

$$\delta_{i,j} = \begin{cases} 0 & \text{if } i \ne j \\ 1 & \text{otherwise} \end{cases}$$

is an orthonormal basis for these spaces.

3. The space $C[a,b]$ with $\langle f, g \rangle = \int_a^b f(x)\overline{g(x)}dx$, & $\|f\| = [\int_a^b |f(x)|^2 dx]^{1/2}$ is a Euclidean space. An important orthonormal basis for $C[a,b]$ is the set of functions

$$\left\{ 1, \cos(\frac{2\pi n x}{b-a}), \sin(\frac{2\pi n x}{b-a}) \right\}_{n=1}^{\infty}$$

2.3.2 Existence of an orthogonal basis

We restrict our attention to separable Euclidean spaces, i.e., Euclidean spaces containing a countable everywhere dense subset.

Theorem 2.43 *Every orthogonal system* $\{x_\alpha\}$ *in a separable Euclidean space $\mathcal{E}$ has no more than countably many elements x_α.*

Proof: Without loss of generality assume $\{x_\alpha\}$ is orthonormal, then $\|x_\alpha - x_\beta\| = \sqrt{\langle x_\alpha - x_\beta, x_\alpha - x_\beta \rangle} = \sqrt{2}$ if $\alpha \ne \beta$. Consider the set of open spheres $S(x_\alpha, 1/2)$

$$S(x_\alpha, 1/2) \cap S(x_\beta, 1/2) = \phi \ \forall \alpha \ne \beta$$

Since $\mathcal{E}$ is separable, each sphere contains at least one element from some countable subset $\{y_n\}$ which is everywhere dense in $\mathcal{E}$. Therefore, there are no more than countably many such spheres $S(x_\alpha, 1/2)$ and, consequently, no more than countably many elements x_α.

The existence of an orthogonal basis in any separable Euclidean space is guaranteed by the following theorem and its corollary:

Theorem 2.44 (Orthogonalization Theorem) *Let $f_1, ..., f_n, ...$ be any countable set of linearly independent elements of a Euclidean space $\mathcal{E}$. Then $\mathcal{E}$ contains a set of elements $\varphi_1, \varphi_2, ..., \varphi_n, ...$ such that:*

(a) $\{\varphi_n\}$ *is an orthonormal system*

(b) $\varphi_n = a_{n1}f_1 + ... + a_{nn}f_n (a_{nn} \neq 0) \forall n$

(c) $\forall f_n = b_{n1}\varphi_1 + ... + b_{nn}\varphi_n (b_{nn} \neq 0)$

Moreover, every element of the system $\{f_n\}$ *is uniquely determined by this to within a factor* ± 1.

Proof: Construct φ_1:set $\varphi_1 = a_{11}f_1 \cdot \langle\varphi_1, \varphi_1\rangle = 1 \Rightarrow a_{11}^2 \langle f_1, f_1 \rangle = 1$ $\therefore$

$$a_{11} = \frac{1}{b_{11}} = \frac{1}{\sqrt{\langle f_1, f_1 \rangle}} \cdot r$$

This determines φ_1 uniquely except for the sign. Assume $\varphi_1, ..., \varphi_{n-1}$ satisfying conditions of the theorem have been constructed. Then $f_n = b_{n1}\varphi_1 + ... + b_{nn-1}\varphi_{n-1} + h_n$ where $\langle h_n, \varphi_k \rangle = 0 \ \forall 1 \leq k \leq n$ and $b_{nk} = \langle f_n, \varphi_k \rangle, 1 \leq k \leq n-1, \langle h_n, h_n \rangle > 0$ for otherwise $\{f_n\}$ would not be linearly independent. Let $\varphi_n = \frac{h_n}{\sqrt{\langle h_n, h_n \rangle}}$. $\therefore f_n = b_{n1}\varphi_1 + ... + b_{nn-1}\varphi_{n-1} + b_{nn}\varphi_n$, where $b_{nn} = \sqrt{\langle h_h, h_n \rangle}$, and $\varphi_n = -b_{n1}\varphi_1 - ... - b_{nn-1}\varphi_{n-1} - a_{nn}f_n$, where $a_{nn} = \frac{1}{b_{nn}} = \frac{1}{\sqrt{\langle h_h, h_n \rangle}}$ since $\varphi_k = \sum_{m=1}^{k} a_{km}f_m$, $\forall 1 \leq k \leq n - 1, \varphi_n = a_{n1}f_1 + ... + a_{nn-1}f_{n-1} + a_{nn}f_n$. Moreover $\langle \varphi_n, \varphi_k \rangle = 0 \ \forall k < n$ and $\langle \varphi_n, \varphi_n \rangle = 1$.

We have thus constructed φ_n given $\varphi_1, .., \varphi_{n-1}$. The proof follows by mathematical induction.

Corollary 2.44.1 Every separable Euclidean space $\mathcal{E}$ has a countable orthogonal basis.

Proof: Let $\{\psi_n\}$ be a countable everywhere dense subset of $\mathcal{E}$. Then a completer set of linearly independent elements $f_1, f_2, ..., f_n, ...$ can be selected from it. Eliminate from the sequence $\{\psi_n\}$ all elements ψ_n which can be written as linear combinations of elements $\psi_i, i < n$. Applying orthogonalization to $\{f_n\}$ we get an orthogonal basis.

2.3.3 Bessel's inequality and closed orthogonal systems

If $e_1, e_2, ..., e_n$ is an orthonormal basis for $\mathbf{R}^n$ then every vector $x \in \mathbf{R}^n$ can be written in the form $x = \sum_{i=1}^{n} c_i e_i$ with $c_i = \langle x, e_i \rangle$. This generalizes to the case of an infinite-dimensional Euclidean space $\mathcal{E}$. Let $\{\phi_n\}$ be an orthogonal system in $\mathcal{E}$ and $f \in \mathcal{E}$. Consider the sequence of numbers $c_k = \langle f, \phi_k \rangle k = 1, 2, \dots$ associated with f, called the *components* or *Fourier coefficients* of f with respect to the system $\{\phi_n\}$.

Consider the series $\sum_{k=1}^{\infty} c_k \phi_k$ (for the time being, purely formal) associated with f, called the *Fourier Series* of f with respect to the system $\{\phi_n\}$. Two questions arise in this context: Does this series converge? If so, does it converge to f? In the following, the conditions under which the Fourier series of an element of a Euclidean space converges are set forth.

Theorem 2.45 *Given an orthogonal system $\{\phi_n\}$ in a Euclidean space $\mathcal{E}$ let $f \in \mathcal{E}$ be an arbitrary element. Then the expression $\|f - \sum_{k=1}^{n} a_k \phi_k\|$ achieves its minimum for $a_k = c_k = \langle f, \phi_k \rangle \; \forall 1 \le k \le n \; \forall n$ and this minimum is $\|f\|^2 - \sum_{k=1}^{n} c_k^2$. Hence, $\sum_{k=1}^{n} C_k^2 \le \|f\|^2$. (Bessel's inequality).*

Proof: Let $S_n = \sum_{k=1}^{n} a_k \phi_k$. Then

$$
\begin{aligned}
\|f - S_n\|^2 &= \|f\|^2 - 2\langle f, \sum_{k=1}^{n} a_k \phi_k \rangle + \sum_{k=1}^{n} a_k^2 \\
&= \|f\|^2 - 2\sum_{k=1}^{n} a_k c_k + \sum_{k=1}^{n} a_k^2 \\
&= \|f\|^2 - \sum_{k=1}^{n} c_k^2 + \sum_{k=1}^{n} c_k^2 + -2\sum_{k=1}^{n} a_k c_k + \sum_{k=1}^{n} a_k^2 \\
&= \|f\|^2 - \sum_{k=1}^{n} c_k^2 + \sum_{k=1}^{n} (c_k - a_k)^2
\end{aligned}
$$

$\therefore$

$\|f - S_n\|^2$ is minimum when $a_k = c_k \; \forall 1 \le k \le n$ and this minimum is equal to $\|f\|^2 - \sum_{k=1}^{n} c_k^2$. Since $\|f - S_n\|^2 \ge 0$, we have $\|f\|^2 \ge \sum_{k=1}^{n} c_k^2 \; \forall n$.

Hence the series $\sum_{k=1}^{n} c_k^2$ is convergent and $\sum_{k=1}^{n} c_k^2 \le \|f\|^2$.

Geometrically, Bessel's inequality implies that the sum of the squares of the projection of a vector f onto a set of mutually perpendicular direction cannot exceed the square of the vector intself. The case where Bessel's inequality becomes an equality is particularly important.

Definition 2.46 *Suppose $\{\phi_n\}$ is an orthonormal system in $\mathcal{E}$ such that for every $f \in \mathcal{E}, \sum_{k=1}^{\infty} c_k^2 = \|f\|^2$, where $c_k = \langle f, \phi_k \rangle, \; \forall k$. Then the orthogonal* *system $\{\phi_n\}$ is said to be closed.*

Theorem 2.47 *An orthonormal system $\{\varphi_n\}$ in a Euclidean space $\mathcal{E}$ is closed iff $\forall f \in \mathcal{E}$ $f = \sum_{n=1}^{\infty} \langle f, \varphi_n \rangle \varphi_n$, i.e., f is the sum of its Fourier series.*

Proof: By definition above, $\mathcal{E}$ is closed iff $\|f\|^2 = \sum_{k=1}^{\infty} \langle f, \varphi_k \rangle^2$, $\forall f \in \mathcal{E}$. If $S_n = \sum_{k=1}^{n} c_k \varphi_k, c_k = \langle f, \varphi_k \rangle$, then $\|f - S_n\|^2 = \|f\|^2 - \sum_{k=1}^{n} c_k^2$ $\forall f \in E$. Letting $n \to \infty$ we have $\|f\|^2 = \sum_{k=1}^{\infty} c_k^2$ $\forall f$ iff $f = \sum_{k=1}^{\infty} c_k \varphi_k$ $\forall f \in \mathcal{E}$. This is called *Parseval's identity* .

If an orthonormal system $\{\varphi_n\}$ is closed, then by the above theorem every $f \in \mathcal{E}$ is the limit of partial sums of its Fourier series, i.e., linear combinations of elements of $\{\varphi_n\}$ are complete then every $f \in \mathcal{E}$ can be approximated closely by a linear combination: $\sum_{k=1}^{\infty} a_k \varphi_k$, but the partial Fourier series $\sum_{k=1}^{\infty} c_k \varphi_k$ is at least as good as its approximation. Hence, f is the sum of its own Fourier series and therefore, by the theorem above, $\{\varphi_n\}$ is closed.

Corollary 2.47.1 Every separable Euclidean space $\mathcal{E}$ contains a closed orthonormal system $\{\varphi_n\}$.

2.3.4   Complete Euclidean spaces and the Riesz-Fischer theorem

In a Euclidean space $\mathcal{E}$ if $\{\varphi_n\}$ is an orthogonal system (not necessarily complete), then it follows from Bessel's inequality that a necessary condition for the numbers, $c_1, \ldots, c_k, \ldots$ to be Fourier coefficients of $f \in \mathcal{E}$ is $\sum_{k=1}^{\infty} c_k^2 < \infty$. It turns out that this condition is also sufficient:

Theorem 2.48 (Riesz-Fischer Theorem) *Given an orthogonal system $\{\varphi_n\}$ in a complete Euclidean space $\mathcal{E}$ let the number $c_1, c_2, \ldots$ be such that $\sum_{k=1}^{\infty} c_k^2 < \infty$. Then $\exists f \in \mathcal{E} \ni c_i = \langle f, \varphi_i \rangle$ and $\|f\|^2 = \sum_{k=1}^{\infty} c_k^2$.*

Proof: Let $f_n = \sum_{k=1}^{n} c_k \varphi_k$. $\therefore \|f_{n+p} - f_n\|^2 = \|c_{n+1}\varphi_{n+1} + \ldots + c_{n+p}\varphi_{n+p}\|^2 = \sum_{k=n+1}^{n+p} c_k^2 \to 0$ as $n \to \infty$. i.e., $\{f_n\}$ is a Cauchy sequence. $\therefore f_n \to f \in \mathcal{E}$ since $\mathcal{E}$ is complete.

Moreover, $\langle f, \varphi_k \rangle = \langle f_n, \varphi_k \rangle + \langle f - f_n, \varphi_k \rangle$ where $\langle f_n, \varphi_k \rangle = c_k$ if $n \geq k$ and $\langle f - f_n, \varphi_k \rangle \to 0$ as $n \to \infty$ ($\because \|\langle f - f_n, \varphi_k \rangle| \leq \|f - f_n\|\|\varphi_k\|)$. $\therefore$ as $n \to \infty, \langle f_n, \varphi_k \rangle = c_k$, since the left hand side is independent of n.

Also $\|f - f_n\| \to 0$ as $n \to \infty$. $\therefore \|f - \sum_{k=1}^{n} c_k \varphi_k\|^2 = \|f\|^2 - \sum_{k=1}^{n} c_k^2 \to 0$ as $n \to \infty$.

Theorem 2.49 *If $\{\varphi_k\}$ is an orthogonal system in a complete Euclidean space $\mathcal{E}$ then $\{\varphi_k\}$ is complete iff $\mathcal{E}$ contains no nonzero element orthogonal to all the elements of $\{\varphi_k\}$.*

Proof: $\{\varphi_k\}$ is complete $\Rightarrow \{\varphi_k\}$ is closed. Let $\langle f, \varphi_k \rangle = 0 \ \forall \varphi_k$. Then all the Fourier coefficients vanish. $\therefore \|f\|^2 = \sum_{k=1}^{\infty} c_k^2 = 0$ (by the Riesz-Fischer theorem). $\therefore f = 0$. $\{\varphi_k\}$ is not complete $\Rightarrow \mathcal{E}$ contains $g \neq 0 \ni$ $\|g\|^2 > \sum_{k=1}^{\infty} c_k^2$ where $c_k = \langle g, \varphi_k \rangle \ \forall k \geq 1$. By the Riesz-Fischer theorem, $\exists f \in \mathcal{E}$ such that $\langle f, \varphi_k \rangle = c_k$ and $\|f\|^2 = \sum_{k=1}^{\infty} c_k^2$. But $\langle f - g, \varphi_k \rangle = 0 \ \forall k$, and since $\|f\|^2 = \sum_{k=1}^{\infty} c_k^2 < \|g\|^2$, we have $f - g \neq 0$.

2.4 Hilbert Spaces

Definition 2.50 *A **Hilbert space** $\mathcal{H}$ is a Euclidean space which is complete, separable and infinite dimensional. That is, a Hilbert space $\mathcal{H}$ is a set of elements such that:*

(a) $\mathcal{H}$ is a (real/complex) Euclidean space, i.e., a (real/complex) linear space equipped with a scalar product (bilinear/sesquilinear functional).

(b) $\mathcal{H}$ is complete with respect to the metric $\rho(f, g) = \|f - g\|$

(c) $\mathcal{H}$ is separable, i.e., $\mathcal{H}$ contains a countable everywhere dense subset.

(d) $\mathcal{H}$ is infinite dimensional.

Example: $l_2(\mathbf{R})$ is a Hilbert space, and so is $l_2(\mathbf{C})$

Definition 2.51 *Two Euclidean spaces $\mathcal{E}$ and $\mathcal{E}^*$ are isomorphic to each other if there exists a one-to-one correspondence $x \leftrightarrow x^*$, $y \leftrightarrow y^*$ between the elements of $\mathcal{E}$ and $\mathcal{E}^*$ preserving linear operations and scalar products*

$$x + y \leftrightarrow x^* + y^*, \alpha x \leftrightarrow \alpha x^*, \langle x, y \rangle = \langle x^*, y^* \rangle$$

Any two n-dimensional real(complex) Euclidean spaces are isomorphic, and they are all isomorphic to $\mathbf{R}^n(\mathbf{C}^n)$. However, two infinite-dimensional Euclidean spaces need not be isomorphic.

Example: $l_2(\mathbf{R}) = \{\{a_i\}_{i=1}^{\infty} : a_i \in \mathbf{R} \ \forall i \geq 1, \& \sum_{i=1}^{\infty} a_i^2 < \infty\}$ is complete, while $C(R)_{[a,b]}$ the space of all real valued continuous functions on $[a, b]$ is not.

Theorem 2.52 (Isomorphism Theorem) *Any two Hilbert spaces are isomorphic.*

Proof: Claim: Every Hilbert space $\mathcal{H}$ is isomorphic to l_2.

Let $\{\phi_k\}$ be a complete orthonormal system in $\mathcal{H}$ (such a system always exists). With every $f \in \mathcal{H}$ associate its Fourier coefficients $\{c_k\}$ with respect to $\{\phi_k\}$.

$$f \mapsto (c_1, \ldots, c_k, \ldots) = (\langle f, \phi_1 \rangle, \ldots, \langle f, \phi_k \rangle, \ldots)$$

Since $\sum_{k=1}^{\infty} c_k^2 \leq \infty, (\ldots, c_k, \ldots) \in l_2$.

Conversely, by Riesz-Fischer, to every element $(c_1, c_2, \ldots) \in l_2$, there corresponds $f \in \mathcal{H}$ with the c_i as its Fourier coefficients. This corresspondence between elements of $\mathcal{H}$ and those of l_2 is one-to-one. Moreover, if $f \leftrightarrow (c_1+, c_2, \ldots)$ and $\tilde{f} \leftrightarrow (\tilde{c}_1, \tilde{c}_2, \ldots)$, then clearly $f + \tilde{f} \leftrightarrow (c_1 + \tilde{c}_1, c_2 + \tilde{c}_2, \ldots)$ and $\alpha f \leftrightarrow (\alpha c_1, \alpha c_2, \ldots)$.

Finally, by Parseval's identity, $\langle f, f \rangle = \sum_{k=1}^{\infty} c_k^2, \langle \tilde{f}, \tilde{f} \rangle = \sum_{k=1}^{\infty} \tilde{c}_k^2$

$$\langle f + \tilde{f}, f + \tilde{f} \rangle = \langle f, f \rangle + \langle f, \tilde{f} \rangle + \langle \tilde{f}, f \rangle + \langle \tilde{f}, \tilde{f} \rangle = \sum_{k=1}^{\infty} |c_k| + \tilde{c}_k^{\,2}$$

Therefore, $\langle f, \tilde{f} \rangle = \sum_{k=1}^{\infty} c_k \tilde{c}_k$. Hence scalar products are preserved.

To within an isomorphism, there is only one Hilbert space and this space has l_2 as its coordinate realization, just as the space of all ordered n-tuples of real numbers with the scalar product $\sum_{k=1}^{n} x_k y_k$ is the coordinate realization of axiomatically defined Euclidean n-space.

2.4.1   Subspaces, orthogonal complements, and direct sums

Definition 2.53 *A linear manifold in a Hilbert space $\mathcal{H}$ is a set L of elements of $\mathcal{H}$ such that*

$$f, g \in L \Rightarrow \alpha f + \beta g \in L$$

for arbitrary numbers α and β.

Definition 2.54 *A subspace of $\mathcal{H}$ is a closed linear manifold of $\mathcal{H}$.*

Lemma 2.55 *If a metric space $\mathcal{M}$ has a countable everwhere dense subset, then so does every subset $\mathcal{M}' \subset \mathcal{M}$.*

Proof: Let $\{\xi_i\}$ be a countable everywhere dense subset of $\mathcal{M}$. Let $a_n = inf_{\eta \in \mathcal{M}'} \rho(\xi_n, \eta)$ be the shortest distance between ξ_n and $\mathcal{M}'$. Then, given any positive integers n and p, $\exists \eta_{n,p} \in \mathcal{M}' \ni \rho(\xi_n, \eta_{n,p}) < a_n + 1/p$. Given any $\epsilon > 0$ and $\eta \in \mathcal{M}'$ let $1/p < \epsilon/3$ and choose n such that $\rho(\xi_n, \eta) < \epsilon/3$ (this is possible since $\{\xi_i\}$ is dense in M'). Then $\rho(\xi_n, \eta_{n,p}) < a_n + 1/p < \epsilon/3 + \epsilon/3 = 2\epsilon/3$ $\therefore \rho(\eta, \eta_{n,p}) \leq \rho(\eta, \xi_n) + \rho(\xi_n, \eta_{n,p}) < \epsilon$, i.e., $\mathcal{M}'$ has an everywhere dense subset $\{\eta_{n,p}\}$ containing no more than countably many elements.

Theorem 2.56 *Every subspace $\mathcal{M}$ of a Hilbert space $\mathcal{H}$ is either a complete separable Euclidean space or itself a Hilbert space. Moreover $\mathcal{M}$ has an orthonormal basis like $\mathcal{H}$ itself.*

Proof: $\mathcal{M}$ is Euclidean space and is complete. Since it is a subspace of $\mathcal{H}$, $\mathcal{M}$ is separable by the above lemma. Since $\mathcal{M}$ is separable, it has an orthogonal basis.

Theorem 2.57 *If $\mathcal{M}$ is a subspace of a Hilbert space $\mathcal{H}$ and $\mathcal{M}^\perp$ is the orthogonal complement of $\mathcal{M}$, i.e., $\mathcal{M}^\perp = \{h' \in \mathcal{H} | \langle h', h \rangle = 0 \forall h \in \mathcal{M}\}$, then $\mathcal{M}'$ is also subspace of $\mathcal{H}$.*

Proof: $\mathcal{M}^\perp$ is a linear space: $\langle h_1', h \rangle = \langle h_2', h \rangle = 0 \Rightarrow \langle \alpha_1 h_1', \alpha_2 h_2', h \rangle = 0 \; \forall \alpha_1 \alpha_2$.

If $\{h_n'\}$ is a sequence in $\mathcal{M}^\perp$ such that $h_n' \to h$ as $n \to \infty$, then given any $h \in \mathcal{M} \langle h', h \rangle = lim_{n \to \infty} \langle h_n', h \rangle = 0$. $\therefore h' \in \mathcal{M}^\perp$. By definition, $h' \in \mathcal{M}^\perp$ iff $\langle h', h \rangle = 0 \; \forall h \in \mathcal{M}$. But then, $h \in \mathcal{M}$ iff $\langle h, h' \rangle = 0 \; \forall h' \in \mathcal{M}^\perp$. Therefore, $(\mathcal{M}^\perp)^\perp = \mathcal{M}$. Thus $\mathcal{M}$ and $\mathcal{M}^\perp$ are mutually orthogonal subspaces of $\mathcal{H}$, written $\mathcal{M} \perp \mathcal{M}^\perp$.

Theorem 2.58 *If $\mathcal{M}$ is a subspace of $\mathcal{H}$, then every $f \in \mathcal{H}$ has a unique representation of the form $f = h + h'$ where $h \in \mathcal{M}$ and $h' \in \mathcal{M}^\perp$.*

Proof: Let $\{\phi_k\}$ be an orthogonal basis in $\mathcal{M}$, and let $f \in \mathcal{H}$. Let $h = \sum_{k=1}^{\infty} c_k \phi_k, c_k = \langle f, \phi_k \rangle$. By Bessel's inequality $\sum_{k=1}^{\infty} c_k^2 < \infty$. Therefore, by the Riesz-Fischer theorem h exists and belongs to $\mathcal{M}$. Let $h' = f - h$, then obviously $\langle h', \phi_k \rangle = 0 \; \forall k$. Since for any $g \in \mathcal{M}$,

$$g = \sum_{k=1}^{\infty} a_k \phi_k a_k = \langle g, \phi_k \rangle$$

Then,

$$\langle h', g \rangle = \sum_{k=1}^{\infty} a_k \langle h', \phi_k \rangle = 0$$

$\therefore h' \in \mathcal{M}^{\perp}$. This proves the existence of the representation. Suppose $f = h_1 + h_1'$ is another representation of f, $h_1 \in \mathcal{M}, h_1' \in \mathcal{M}^{\perp}$. Then $\langle h_1, \phi_k \rangle = \langle f, \phi_k \rangle = c_k \; \forall k$. $\therefore h_1 = h, h_1' = h'$.

Corollory 2.58.1 Every orthonormal system $\{\phi_k\}$ in $\mathcal{H}$ can be enlarged to give a complete orthonormal system in $\mathcal{H}$.

Proof: Let $\mathcal{M}$ be the linear closure of $\{\phi_k\}$ so that $\{\phi_k\}$ is complete in $\mathcal{M}$. Let $\{\phi_k\}$ be a complete orthonormal system in $\mathcal{M}^{\perp}$ (such a system exists since $\mathcal{M}^{\perp}$ is a subspace, then $\{\phi_k\} \cup \{\phi_k'\}$ is a complete orthonormal system in $\mathcal{H}$).

Corollory 2.58.2 If $\mathcal{M}$ is a subspace in $\mathcal{H}$, then $codim \mathcal{M}^{\perp} = n$ if $dim \mathcal{M} = n$, and $dim \mathcal{M}^{\perp} = n$ if $codim \mathcal{M} = n$.

If $\mathcal{M} \subset_s \mathcal{H}$, since for every $f \in \mathcal{H}$, $f = h + h'$, for $h \in \mathcal{M}$ and $h' \in \mathcal{M}^{\perp}$, uniquely. $\mathcal{H}$ is said to be the direct sum of mutually orthogonal spaces $\mathcal{M}$ and $\mathcal{M}^{\perp}$ written: $\mathcal{H} = \mathcal{M} \oplus \mathcal{M}^{\perp}$. The concept of direct sum generalizes to the case of any finite or countable number of subspace. $\mathcal{H}$ is said to be a direct sum of subspaces $\mathcal{M}_1, \mathcal{M}_2, ..., \mathcal{M}_n, ...$, written : $\mathcal{M}_1 \oplus \mathcal{M}_2 \oplus .... \oplus \mathcal{M}_n \oplus ... = \oplus_{k=1}^{\infty} \mathcal{M}_k$, if

(a) $\mathcal{M}_i \perp \mathcal{M}_j \; \forall i \neq j$

(b) $\forall f \in \mathcal{H} \; f = h_1 + h_2 + ... + h_n$ where $h_i \in \mathcal{M}_i$ (this representation is unique and if it exists) and further $\|f\|^2 = \sum_{k=1}^{\infty} \|h_k\|^2$.

Besides direct sums of subspaces of $\mathcal{H}$, one can define the direct sum of finitely many or countably many Hilbert spaces: If $\mathcal{H}_1$ and $\mathcal{H}_2$ are two Hilbert spaces, their direct sum $\mathcal{H}$ is defined as follows: $\mathcal{H} = \mathcal{H}_1 \} \oplus \mathcal{H}_2 = \{(h_1, h_2) | h_1 \in \mathcal{H}_1 \text{ and } h_2 \in \mathcal{H}_\in \}$ such that

(a) $\alpha(h_1, h_2) + \beta(h_1', h_2') = (\alpha h_1 + \beta h_1', \alpha h_2 + \beta h_2')$

(b) $\langle (h_1, h_2), (h_1', h_2') \rangle = \langle h_1, h_1' \rangle + \langle h_2, h_2' \rangle$

The subspaces $\{(h_1, 0) | h_1 \in \mathcal{H}_1\}$ and $\{(0, h_2) | h_2 \in \mathcal{H}_2\}$ are orthogonal and can be identified in a natural way with $\mathcal{H}_\infty$ and $\mathcal{H}_\in$ respectively. Given Hilbert spaces

$$\mathcal{H}_1, \mathcal{H}_2, \ldots, \mathcal{H} = \mathcal{H}_1 \oplus \mathcal{H}_2 \oplus \ldots \oplus \mathcal{H}$$
$$= \{(h_1, h_2, \ldots, h_n, \ldots) | h_n \in \mathcal{H}_n\} \ni \sum_{n=1}^{\infty} \|h_n\|^2 < \infty$$

and if $h = (h_1, h_2, \ldots, h_n, \ldots)$, $g = (g_1, g_2, \ldots, g_n, \ldots)$, $\alpha h + \beta g \in \mathcal{H} = (\alpha h_1 + \beta g_1, \ldots, \alpha h_n + \beta g_n)$ and $\langle h, g \rangle = \sum_{n=1}^{\infty} \langle h_n, g_n \rangle$, is a Hilbert space.

2.5 Characterization of Euclidean spaces

Given a normed linear space $\mathcal{N} = (\mathcal{L}, \| \ \|)$, what are the conditions under which $\mathcal{N}$ is Euclidean, i.e., what conditions on the norm $\| \ \|$ of $\mathcal{N}$ guarantee that it is derivable from some suitably defined inner product on $\mathcal{N}$?

Theorem 2.59 *A necessary and sufficient condition for a normed linear space $\mathcal{N} = (\mathcal{L}, \| \ \|)$ to be Euclidean is that $\|f + g\|^2 + \|f - g\|^2 = 2(\|f\|^2 + \|g\|^2)$ $\forall f, g \in \mathcal{N}$.*

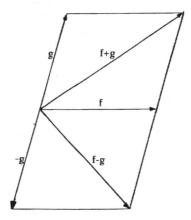

Figure 2.1: Geometric illustration of the characterization theorem

Proof: If $\mathcal{N}$ is Euclidean then there exists a scalar product:$\langle, \rangle$ such that $\langle f, f \rangle = \|f\|^2$. And $\|f + g\|^2 + \|f - g\|^2 = \langle f + g, f + g \rangle \langle f - g, f - g \rangle$

$= \|f\|^2 + 2\langle f,g \rangle + \|g\|^2 + \|f\|^2 - 2\langle f,g \rangle - \|g\|^2 = 2(\|f\|^2 + \|g\|^2)$. Therefore, the condition is necessary.

To prove sufficiency, set $\langle f,g \rangle = \frac{1}{4}(\|f+g\|^2 - \|f-g\|^2)$ and verify the properties of an inner product for $\langle f,g \rangle$. Clearly, $\langle , \rangle$ generates the norm.
$\therefore \langle f,g \rangle = \frac{1}{4}(\|f+g\|^2 - \|f-g\|^2)$,

1. $\langle f,f \rangle \geq 0$ and $\langle f,f \rangle = 0 \Leftrightarrow f = 0$
2. $\langle f,g \rangle = \langle g,f \rangle$.

The proofs of $\langle f+g,h \rangle = \langle f,h \rangle + \langle g,h \rangle$ (*) and $\langle \alpha f,g \rangle = \alpha \langle f,g \rangle$ (**) require a little work.

To prove (*), set $\phi(f,g,h) = 4[\langle f+g,h \rangle - \langle f,h \rangle - \langle g,h \rangle]$ and using the definition of $\langle , \rangle$, show that $\phi(f,g,h) \equiv 0$.

To prove (**), set $\varphi(c) = \langle cf,g \rangle - c\langle f,g \rangle$ $\forall c \in \mathbf{F}$, where f and g are fixed but arbitrary elements of $\mathcal{N}$.

It follows from this that $\varphi(0) = 0$. And $\varphi(-1) = 0$ $\because \langle -f,g \rangle = -\langle f,g \rangle$
$\therefore \forall n \in Z$,

$$\langle nf,g \rangle = \langle sgn(n)(\underbrace{f+f+...+f}_{|n| \text{ times}}),g \rangle$$

$$= sgn(n)\langle \underbrace{f+f+...+f}_{|n| \text{ times}},g \rangle$$

$$= |n|sgn(n)\langle f,g \rangle = n\langle f,g \rangle$$

therefore, $\varphi(n) = 0$.

$$\forall p,q \in Z, \langle \frac{p}{q}f,g \rangle = p\langle \frac{1}{q}f,g \rangle = \frac{p}{q} \cdot q\langle \frac{1}{q}f,g \rangle = \frac{p}{q}\langle f,g \rangle$$

i.e., $\varphi(c) = 0$, $\forall c \in \mathbf{Q}$, and $\varphi(c)$ is a continuous function of c (why?).
$\therefore \varphi(c) \equiv 0 \ \forall c \in R$.

Examples:

1. The n-dimension space $\mathbf{R}^n$ equipped with the norm $\|x\|_p = [\sum_{k=1}^{n} x_k^p]^{1/p}$ is a normed linear space if $p \geq 1$, and a Euclidean space if $p = 2$, but fails to be euclidean if $p \neq 2$. Let

$$\left\{ \begin{array}{l} f = (1,1,0,0,...,0) \\ g = (1,-1,0,0,...,0) \end{array} \right\} \in \mathbf{R}^n$$

$\therefore$, $f+g = (2,0,0,...,0)$ and $f-g = (0,2,0,0,...0)$, and $\|f\|_p = \|g\|_p = 2^{1/p}$.

Also, $\|f + g\|_p = \|f - g\|_p = 2$. $\therefore \|f + g\|_p^2 + \|f - g\|_p^2 = 8$ and $2(\|f\|_p^2 + \|g\|_p^2) = 4 \times 2^{2/p}$. Thus,

$$4 \times 2^{2/p} = 8 \Leftrightarrow p = 2$$

2. The space $C_{[0,\pi/2]}$ (all functions continuous on $[0,\pi/2]$), with $\|f\| = max_{0 \le t \le \pi/2}. |f(t)|$ is a normed linear space, but not a Euclidean space. Let

$$\left\{ \begin{array}{l} f(t) = \cos t \\ g(t) = \sin t \end{array} \right\} \in C_{[0,\pi/2]}. \|f\| = \|g\| = 1.$$

Then, $\|f+g\| = max_{[0 \le t \le \pi/2]} |\cos t + \sin t| = \sqrt{2}$, and $\|f-g\| = max_{[0 \le t \le \pi/2]} |\cos t - \sin t| = 1$, and $(\sqrt{2})^2 + 1^2 \ne 2(1^2 + 1^2)$.

Chapter 3

Integration

Integration is one of the most important and basic concepts of analysis. In this book it is a very important tool in the analysis and synthesis of functions. We give a brief description of measure theory and the Lebesgue integral on $\mathbf{R}$. But first we discuss very briefly the Riemann integral and its drawbacks.

3.1 The Riemann Integral

Definition 3.1 *If $[a, b]$ is a closed interval then its* measure *or* length *is given $s([a, b]) = b - a$.*

Definition 3.2 *By an **n-partition** of $[a, b]$ is meant any sequence of $n + 1$ points $x_0, x_1, ..., x_n$ such that $a = x_0 < x_1 < ... < x_n = b$, dividing $[a, b]$ into n subintervals $I_1, I_2, ..., I_n, I_i = [x_{i-1}, x_i] \ \forall 1 \leq i \leq n$. The n-partition is written $\Pi_n = \{a = x_0 < x_1 < x_2 < ... < x_n = b\}$.*

Let f be a bounded, real valued function defined on the closed interval $[a, b]$.

Let $\Pi_n = \{a = x_0 < x_1 < x_2 < ... < x_n = b\}$ be be an n-partition of $[a, b]$.

In each subinterval $I_k = [x_{k-1}, x_k]$ of $[a, b]$ choose an arbitrary point

$\xi_k, 1 \le k \le n$.

Then the Riemann sum of f w.r.t. Π_n is given by

$$R_{\Pi_n}(f) = \sum_{i=1}^{n} f(\xi_i)s(I_i)$$

Let $\delta(\Pi_n) = max_{1 \le i \le n}s(I_i)$: the length of the largest subinterval of $[a, b]$ w.r.t. Π_n.

Let $\Pi_1, \Pi_2, ..., \Pi_n, ...$ be a sequence of partitions of $[a, b]$ such that $\delta(\Pi_n) \to 0$ as $n \to \infty$. If the sequence $\{R_{\Pi_n}\}_{n=1}^{\infty}$ has a limit as $n \to \infty$ that is independent of the sequence of partitions (provided $\delta(\Pi_n) \to 0$ as $n \to \infty$), or of the choice of the points $\xi_i \in [x_{i-1}, x_i]$ $\forall 1 \le i \le n$, then the limit is called the *Riemann integral* of f and is written:

$$\int_a^b f(x)dx = \lim_{\delta(\Pi_n) \to 0} R_{\Pi_n}(f)$$

3.1.1   Upper and lower Riemann integrals

Definition 3.3 *If Π is a partition of $I = [a, b]$ into subintervals $I_1, ..., I_n$, and*

$$m_i = \inf_{x \in I_i} f(x), \quad M_i = \sup_{x \in I_i} f(x), \quad I_i = [x_{i-1}, x_i] \, \forall 1 \le i \le n,$$

then the expression

$$L_\Pi(f) = \sum_{i=1}^{n} m_i s(I_i)$$

*is called the **lower Darboux sum** of f corresponding to the partition Π, and*

$$U_\Pi(f) = \sum_{i=1}^{n} M_i s(I_i)$$

*is called the **upper Darboux sum** of f.*

Definition 3.4 *For any $\xi_i \in I_i, 1 \le i \le n$, and*

$$ms(I) \le L_\Pi(f) \le R_\Pi(f) \le U_\Pi(f) \le Ms(I),$$

where

$$m = \inf_{x \in I} f(x) \quad and \quad M = \sup_{x \in I} f(x),$$

if Π' is any other partition of I obtained by further partitioning each subin-terval I_i of I into n_i subintervals I_{i_j}, $1 \le j \le n_i$, $1 \le i \le n$, then Π' is called a **refinement** *of Π* .

If Π' is such a refinement of Π, then:

$$m_i s(I_i) \le \sum_{j=1}^{n_i} m_{i_j} s(I_{i_j}) \le \sum_{j=1}^{n_i} \xi_{i_j} s(I_{i_j})$$

$$\le \sum_{j=1}^{n_i} M_{i_j} s(I_{i_j}) \le M_i s(I_i) \quad \forall 1 \le i \le n$$

$$\sum_{j=1}^{n} m_i s(I_i) \le \sum_{i=1}^{n} \sum_{j=1}^{n_i} m_{i_j} s(I_{i_j}) \le \sum_{i=1}^{n} \sum_{j=1}^{n_i} \xi_{i_j} s(I_{i_j})$$

$$\le \sum_{i=1}^{n} \sum_{j=1}^{n_i} M_{i_j} s(I_{i_j}) \le \sum_{j=1}^{n} M_i s(I_i)$$

i.e., $L_\Pi(f) \le L_{\Pi'}(f) \le U_{\Pi'}(f) \le U_\Pi(f)$.

If Π and Π' are arbitrary partitions, and Π'' is a partition of I obtained by including all the points of the partitions Π and Π', then clearly Π'' is a refinement of both Π and Π''.

$$\therefore L_\Pi(f) \le L_{\Pi''}(f) \le U_\Pi \le U_{\Pi'}(f)$$

and

$$\therefore \quad L_\Pi(f) \le U_\Pi(f)$$

Hence, a lower Darboux sum is never greater than an upper Darboux sum. Writing:

$$\sup_\Pi L_\Pi(f) = \underline{\int_a^b f(x)dx} \text{ and } \inf_\Pi U_\Pi(f) = \overline{\int_a^b f(x)dx}$$

where the supremum and infimum are taken with respect to all partitions Π of $I = [a, b]$, clearly

$$\underline{\int_a^b f(x)dx} \le \overline{\int_a^b f(x)dx}$$

$\underline{\int_a^b f(x)dx}$ is called the lower Riemann integral and $\overline{\int_a^b f(x)dx}$ the upper Riemann integral of f.

If $\{\Pi_n\}_{n=1}^{\infty}$ is a sequence of partitions of $I \ni \delta(\Pi_n) \to 0$ as $n \to \infty$, then it can be shown that

$$L_{\Pi_n}(f) \to \overline{\int_a^b f(x)dx} \text{ and } U_{\Pi_n}(f) \to \overline{\int_a^b f(x)dx} \text{ as } n \to \infty$$

and that f is Riemann integrable iff

$$\underline{\int_a^b f(x)dx} = \overline{\int_a^b f(x)dx} \text{ and } \int_a^b f(x)dx = \underline{\int_a^b f(x)dx} = \overline{\int_a^b f(x)dx}$$

While the theory of Riemann integration is satisfactory for solving many problems of pure and applied mathematics, there are some serious short-comings to this theory that make it inadequate in many areas of mathematics and the mathematical sciences.

To illustrate one such drawback, we consider the family of bounded real functions $\{f_n\}_{n=1}^{\infty}$ defined on the closed interval $[0,1]$, where each f_n is defined by

$$f(x) = \begin{cases} 1 & \text{if } x = 0, \text{ or } x = \frac{p}{q} \text{ for some } p, q \in N \ni 0 < p \le q \le n \\ 0 & \text{otherwise} \end{cases}$$

Thus, for each $f_n(x)$

$$\int_0^1 f_n(x)dx$$

exists in the Riemann sense and is equal to zero. Further, we note that

$$\int_0^1 |f_n(x) - f_m(x)| \, dx = 0 \quad \forall m, n \in N$$

Now, $f_n \to f$ as $n \to \infty$, where

$$f(x) = \begin{cases} 1 & \text{if } x \text{ is rational} \\ 0 & \text{if } x \text{ is irrational} \end{cases}$$

but

$$\underline{\int_0^1 f(x)dx} = 0 \text{ while } \overline{\int_0^1 f(x)dx} = 1$$

Therefore, f is not Riemann integrable.

Now consider all Riemann integrable functions on $[0,1]$ with the metric

$$\rho(f,g) = \int_0^1 |f(x) - g(x)| \, dx$$

This is a metric space, and the functions f_n defined above form a Cauchy sequence in it. However, the limit f of this sequence does not belong to it. This lack of completeness is a serious shortcoming since completeness is an indispensable property in analysis.

Another drawback of the Riemann integral is that it requires the space over which the integral is evaluated to be uniform or homogeneous, in that the value of the integral of a function must be independent of any translation of the function in the space over which it is defined.

These defects can be remedied by introducing a different scheme of integration. This entails the study of abstract spaces using the concept of measure, which gives rise to a general theory of integration. We shall mainly be interested in the Lebesgue integral on **R**. In order to understand the Lebesgue integral, it is necessary to have a grasp of the Lebesgue measure on **R** and Lebesgue measurable functions on **R**. Before proceeding to these concepts, it will be illuminating to have an informal look at the essential difference between the Riemann integral and the Lebesgue integral, as this will help keep track of where we're headed as we go through a series of theorems and definitions leading to the Lebesgue integral on **R**.

3.1.2 Riemann integration vs Lebesgue integration

Let f be a nonnegative continuous function f on $[a, b]$. Let Π be a partition $\Pi = \{a = x_0 \leq x_1 \leq x_2 \leq < \ldots \leq x_n = b\}$ of the domain of f into n subintervals $[x_{i-1}, x_i] \ \forall 1 \leq i \leq n$. The Riemann sum of f w.r.t. Π is given by

$$R_\Pi(f) = \sum_{i=1}^{n} f(\xi_i)(x_i - x_{i-1})$$

where

$$\xi_i \in [x_{i-1}, x_i] \ \forall 1 \leq i \leq n$$

Now as $\delta(\Pi) \to 0$

$$R_\Pi(f) \to \int_a^b f(x)dx$$

In the Lebesgue scheme, instead of partitioning the domain of f, the range of f is subdivided: $min \ f \geq y_0 < y_1 < \ldots < y_n \geq max \ f$, to form the sum:

$$\sum_{i=1}^{n} y_{i-1} \cdot measure(\{x \mid f(x) \in [y_{i-1}, y_i]\})$$

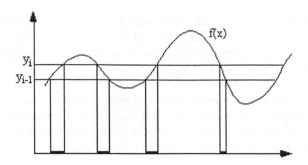

Figure 3.1: The bold intervals on the x-axis correspond to $\langle x \mid f(x) \in [y_{i-1}, y_i] \rangle$.

where

$$measure(\{x \mid f(x) \in [y_{i-1}, y_i]\})$$

is the sum of the lengths of those subintervals of $[a, b]$ on which $y_{i-1} \leq f(x) \leq y_i$. It is then verified that the above sum tends to

$$\int_a^b f(x)dx \text{ as } \to \infty$$

Simple as this idea may seem, it has a great impact on the concept of the integral. By extending the idea of measure from lengths of unions of disjoint subintervals of $[a, b]$ (Borel sets) to the wider class of (Lebesgue) measurable sets of $[a, b]$, a much larger class of functions can be integrated by the Lebesgue scheme than by the Riemann scheme (see Figure 3.1).

3.2   The Lebesgue Measure on R

In this section the concept of the length (measure) of an interval is generalized to obtain measures of more complicated sets of the real line, the so called measurable sets. This is crucial for evaluating the sum leading to the Lebesgue integral. In other words, one is interested in measuring the lengths of sets of the form $\{x \mid a \leq f(x) \leq b\}$, for suitable functions f on the real line.

Let $P = \{[a, b] \mid -\infty < a \leq b < \infty\}$ be the class of all bounded, left closed, right open intervals or semi closed intervals. The empty set is given by the semi closed intervals $[a, a) \; \forall a \in \mathbf{R}$.

Let R denote the class of all finite (disjoint) unions of sets in P, *i.e.*,

$$R = \{\cup_{i=1}^{n} [a_i, b_i) \mid [a_i, b_i) \in P \; \forall 1 \leq i \leq n\}$$

The choice of semiclosed intervals is a technical device to ensure closure of R under differences, unions, and intersections, since the set difference, intersection, and union of semiclosed intervals is a union of disjoint semiclosed intervals. R is a ring on $\mathbf{R}$.

On the class P, define a set function $\mu : P \to \mathbf{R}$ by $\mu([a, b)) = b - a$. If $a = b$ then $[a, b) = \Phi$, and $\mu(\Phi) = 0$. The set function μ satisfies the following properties on P :

1. If $\{E_1, ..., E_n\} \subset P \ni E_i \cap E_j = \Phi$ and $E_i \subset E_0 \in P$, then

$$\sum_{i=1}^{n} \mu(E_i) \leq \mu(E_0)$$

Let $E_i = [a_i, b_i) \; \forall \; 0 \leq 1 \leq n$ w. l. g., let

$$a_0 \leq a_1 \leq b_1 \leq a_2 \leq b_2 \leq ... \leq a_n \leq b_n \leq b_0,$$

then

$$
\begin{aligned}
\sum_{i=1}^{n} \mu(E_i) &= \sum_{i=1}^{n} (b_i - a_i) \leq \sum_{i=1}^{n-1} (a_{i+1} - b_i) + \sum_{i=1}^{n} (b_i - a_i) \\
&= b_n - a_1 \leq b_0 - a_0 = \mu(E_0)
\end{aligned}
$$

2. If a closed interval $F_0 = [a_0, b_0]$ is contained in the union of a finite number of bounded open intervals $U_1, ..., U_n$, $U_i = (a_i, b_i)$ $1 \leq i \leq n$ then

$$b_0 - a_0 < \sum_{i=1}^{n} (b_i - a_i)$$

3. If $\{E_n\}_{n=0}^{\infty}$ is a sequence of sets in P such that $E_0 \subset \bigcup_{i=1}^{\infty} E_i$, then

$$\mu(E_0) \leq \sum_{i=1}^{\infty} \mu(E_i)$$

where

$$E_i = [a_i, b_i) \; \forall \; 0 \leq i \leq \infty$$

If $a_0 = b_0$ then it is trivially true. If $b_0 - a_0 > \epsilon$ let $F_0 = [a_0, b_0 - \epsilon]$. For any $\delta > 0$ write

$$U_i = (a_i - \delta/2^i, b_i) \quad \therefore \, F_0 \subset \bigcup_{i=1}^{\infty} U_i$$

Now, the Heine-Borel theorem [34] states that if a closed interval is contained in a union of infinitely many open intervals, then it is also contained in the union of finitely many of those open intervals. Therefore,

$$\exists n \in \mathbf{N} \ni F_0 \subset \bigcup_{i=1}^{n} U_i$$

$$\therefore \, \mu(E_0) - \epsilon = (b_0 - a_0) - \epsilon \le \sum_{i=1}^{n}(b_i - a_i + \delta/2^i) \le \sum_{i=1}^{\infty} \mu(E_i) + \delta$$

Since ϵ and δ are arbitrary, statement 3 follows.

Definition 3.5 *A set function μ is* **countably additive** *if for any collection $\{E_i\}_{i=1}^{\infty}$ of disjoint sets of P,*

$$\mu(\bigcup_{i=1}^{\infty} E_i) = \sum_{i=1}^{\infty} \mu(E_i)$$

4. μ is countably additive on P. If $\{E_i\}_{i=1}^{\infty} \subset P$ is a sequence of disjoint sets in P such that

$$\bigcup_{i=1}^{\infty} E_i = E \in P$$

then since

$$\bigcup_{i=1}^{\infty} E_i \subset E$$

by 1,

$$\sum_{i=1}^{\infty} \mu(E_i) \le \mu(E) \quad \forall n \ge 1$$

$$\therefore \, \sum_{i=1}^{\infty}(E_i) \le \mu(E)$$

Since

$$E \subset \bigcup_{i=1}^{\infty} E_i, \text{ by 3. } \mu(E) \le \sum_{i=1}^{\infty} \mu(E_i)$$

$$\therefore \mu(\bigcup_{i=1}^{\infty} E_i) = \sum_{i=1}^{\infty} \mu(E_i)$$

Definition 3.6 *An* **extended real valued set function** *is a set function which can take the value* ∞ *on certain sets.*

Definition 3.7 *A* **measure** *is an extended real valued, nonnegative, and countably additive set function* μ, *defined on a ring R, such that* $\mu(\Phi) = 0$.

The set function μ on P can be extended to a measure $\bar{\mu}$ on R:

1. There exists a unique finite measure $\bar{\mu}$ on R such that whenever $E \in P$, $\bar{\mu}(E) = \mu(E)$. Every set $E \in R$ may be represented as a finite, disjoint union of sets in P.

   If

   $$E = \bigcup_{i=1}^{n} E_i \text{ and } E = \bigcup_{j=1}^{m} F_i$$

   are two such representations, then $\forall 1 \leq i \leq n$

   $$E_i = \bigcup_{j=1}^{m} (E_i \cap F_j)$$

Since μ is finitely additive,

$$\sum_{i=1}^{n} \mu(E_i) = \sum_{i=1}^{n} \sum_{j=1}^{m} \mu(E_i \cap F_j)$$

Similarly,

$$\sum_{j=1}^{m} \mu(F_i) = \sum_{j=1}^{m} \sum_{i=1}^{n} \mu(E_i \cap F_j)$$

Therefore $\forall E \in R$, $\bar{\mu}$ is unambiguously defined by

$$\bar{\mu}(E) = \sum_{i=1}^{n} \mu(E_i)$$

where

$$\{E_1, ...E_n\} \subset P \ni E_i \cap E_j = \phi \ \forall i \neq j \text{ and } \bigcup_{i=1}^{n} E_i = E$$

By definition $\bar{\mu}|_P = \mu$ $\therefore \bar{\mu}$ is unique.

2. $\bar{\mu}$ is countably additive on R .

If $\{E_i\}_{i=1}^{\infty} \ni E_i \cap E_j = \Phi \;\; \forall i \neq j$ and $\bigcup_{i=1}^{n} E_i = E \in R$,

then $\forall i$, E_i is a finite disjoint union of sets E_{ij} in R.

$$E_i = \cup_j E_{ij} \text{ and } \bar{\mu}(E_i) = \sum_j \mu(E_{ij})$$

If $E \in P$, then since μ is countably additive in P, and since $E_{ij} \cap E_{i\,j} \; \forall i \neq i$ or $j \neq j$,

$$\bar{\mu}(E) = \mu(E) = \sum_i \sum_j \bar{\mu}(E_{ij}) = \sum_i \bar{\mu}(E_i)$$

In general, however, E is a finite union of disjoint sets in R, $E = \cup_k F_k$. Therefore,

$$
\begin{aligned}
\bar{\mu}(E) &= \mu(E) \\
&= \sum_i \sum_i \bar{\mu}(F_k) = \sum_k \sum_i \bar{\mu}(E_i \cap F_k) = \sum_i \sum_k \bar{\mu}(E_i \cap F_k) \\
&= \sum_i \bar{\mu}(E_i)
\end{aligned}
$$

Henceforth, we will refer to $\bar{\mu}$ as μ.

The closure of R under *countable* unions is denoted S, which is a σ-ring. $\therefore$ S is a collection of subsets of $\mathbf{R}$ such that:

(a) $A, B \in S \Rightarrow A\backslash B \in S$,

(b) $\{A_i\}_{i=1}^{\infty} \subset S \Rightarrow \bigcup_{i=1}^{\infty} A_i \in S. \; P \subset S$.

Remark: If $\{A_i\}_{i=1}^{\infty} \subset S$, then $\bigcap_{i=1}^{\infty} A_i \in S$. Indeed, let $A = \bigcup_{i=1}^{\infty} A_i$.

Then,

$$\bigcap_{i=1}^{\infty} A_i = A\backslash \bigcup_{i=1}^{\infty} (A\backslash A_i) \in S$$

μ can now be extended to a countably additive set function on S.

If $\{E_i\}_{i=1}^{\infty} \subset S$ is a collection of pairwise disjoint sets, then $\forall i \exists$ a collection of pairwise disjoint sets $\{F_{ij}\}_{j=0}^{\infty} \subset P \ni E_i = \cup_{j=1}^{infty} F_{ij}$ and $\mu(E_i) = \mu\left(\cup_{j=1}^{infty} F_{ij}\right)$ since μ is countably infinite on P.

Hence,

$$\bigcup_{i=1}^{\infty} E_i = \bigcup_{i=1}^{\infty} \bigcup_{j=1}^{\infty} F_{ij},$$

and

$$\mu(\bigcup_{i=1}^{\infty} E_i) = \mu(\bigcup_{i=1}^{\infty} \bigcup_{j=1}^{\infty} F_{ij}) = \sum_{i=1}^{\infty} \sum_{j=1}^{\infty} \mu(F_{ij}) = \sum_{i=1}^{\infty} \mu(E_i)$$

Thus, defining

$$\mu(\bigcup_{i=1}^{\infty} E_i) = \sum_{i=1}^{\infty} \mu(E_i) \text{ for } \{E_i\}_{i=1}^{\infty} \subset S$$

we see that μ extends to S naturally.

Definition 3.8 *If μ is a measure on $S = S(\mathbf{R})$, then:*

(a) *μ is a **finite measure** if $\forall E \in S$, $\mu(E) < \infty$ and is a **totally finite measure** if, further, $\mathbf{R} \in S$.*

(b) *μ is a **σ-finite measure** if*

$$\forall E \in S \exists \{E_n\}_1^{\infty} \subset S \ni E \subset \bigcup_{i=1}^{\infty} E_n \text{ and } \mu(E_n) < \infty \ \forall n$$

*and is a **totally σ-finite measure** if, further, $\mathbf{R} \in S$.*

(c) *μ is a **complete measure** if $E \in S$, $F \subset E$ and $\mu(E) = 0 \Rightarrow F \in S$.*

Theorem 3.9 *If μ is a measure on a σ-ring S, then the class $\overline{S}$ of all sets of the form $E \Delta N$, where $E \in S$ and N is a subset of a set of measure zero in S, is a σ-ring, and the set function $\overline{\mu}$ defined by $\overline{\mu}(E \Delta N) = \mu(E)$ is a complete measure on $\overline{S}$. (The mesure $\overline{\mu}$ is called the completion of μ).*

Proof: If $E \in S$, $N \subset A \in S$, and $\mu(A) = 0$, then since

$$E \cup N = (E \backslash A) \Delta [A \cap (E \cup N)] \qquad (A \cap (E \cup N) \subset A)$$

and

$$E \Delta N = (E \backslash A) \cup [A \cap (E \Delta N)] \qquad (A \cap (E \Delta N) \subset A)$$

The class of $\overline{S}$ may also be described as the class of all sets of the form $E \cup N$, where $E \in S$ and N is a subset of a set of measure zero contained in S. Since this implies that the class is $\overline{S}$, using the union (instead of the symmetric difference) representation of sets in $\overline{S}$, it is easy to verify that

μ is a measure. The completeness of $\overline{\mu}$ is an immediate consequence of the fact that $\overline{S}$ contains all subsets of measure zero in $\overline{S}$.

Definition 3.10 *If* $\chi = \mathbf{R}$, *the sets of the* σ-*ring* S *generated by semiclosed intervals are called the* **Borel sets** *of* S. *If* $\overline{\mu}$ *is the measure on* S, *the sets of* $\overline{S}$ *are called* **Lebesgue measurable sets** *and* μ *is the* **Lebesgue measure**.

Definition 3.11 *A* **measurable space** *is a set* χ *and a* σ-*ring* S *of subsets of* χ *such that the union of all the elements of* $S = \chi$. *A set* $E \subset \chi$ *is* **measurable** *iff* $E \in S$. *A* **measure space** *is a measurable space* (χ, S) *with a measure* μ *on* S.

In conclusion, we make a few summarizing remarks on the Lebesgue measure on $\mathbf{R}$:

χ: $\mathbf{R}$

P: Class of all bounded, semiclosed intervals $[a, b) \subset \mathbf{R}$.

S: σ-ring generated by P.

μ: Additive set function on $\mathbf{R}$ defined by

$$\mu([a, b)) = b - a, \mu(\Phi) = 0$$

1. The elements of S are called the *Borel sets* of $\mathbf{R}$. By the foregoing extension theorem, μ is a measure defined for all Borel sets of $\mathbf{R}$.

2. If $\overline{\mu}$ on $\overline{S}$ is the completion of μ on S, the sets of $\overline{S}$ are the Lebesgue measurable sets of $\mathbf{R}$ and $\overline{\mu}$ is the Lebesgue measure, generally denoted by μ itself.

3. Since $\mathbf{R}$ is a countable union of sets in P, $\mathbf{R} \in S$. $\therefore A \in S \Rightarrow A^c S$. $\therefore S$ is a σ-algebra.

4. Since $\mu(\mathbf{R}) = \infty$ μ is not finite. Since μ is finite on P, both μ on S and $\overline{\mu}$ on $\overline{S}$ are totally σ-finite.

5. Every countable set is a Borel set of measure zero. Indeed, $\forall a \in \mathbf{R}$, $a = \bigcap_{n=1}^{\infty}(a, a + 1/n)$, $\therefore \mu(x) = \lim_{n \to \infty}(a, a + 1/n) = 0$.

   Since Borel sets form a σ-ring and μ is countably additive, the statement follows.

6. S coincides with the σ-ring generated by all open sets :

$$\{a\} \in S, \ (a,b) \in S \ \therefore \ (a,b) = (a,b) - \{a\} \in S$$

$$\{a\} = \bigcap_{n=1}^{\infty} (a - 1/n, a + 1/n), \ [a,b) = (a,b) \cup \{a\}$$

We now turn to those functions f for which sets of the form $\{x \mid a \leq f(x) \leq b\}$ are Lebesgue measurable.

3.3 Measurable Functions

In this section, we characterize those functions that are possible candidates for integration under the Lebesgue scheme, by means of their properties.

Definition 3.12 *If f is a real valued function on a set χ and E is any subset of $\mathbf{R}$, then the set inverse of f denoted $\mathbf{R}$ is defined on subsets of $\mathbf{R}$ by $f^{-1}(E) = \{x \in \chi \mid f(x) \in E\}$, and yields subsets of χ.*

Remark: $\forall f : \chi \to \mathbf{R}$,

(a) $f^{-1}(\bigcup_{n=1}^{\infty} E_n) = \bigcup_{n=1}^{\infty} f^{-1}(E_n)$, where $E_n \subset \mathbf{R} \ \forall 1 \leq n \leq \infty$

(b) $f^{-1}(E \backslash f) = f^{-1}(E) \backslash f^{-1}(F)$. $E, F \subset \mathbf{R}$.

Properties (a) and (b) imply that if $\sum$ is a σ-ring of subsets of $\mathbf{R}$ then so is

$$f^{-1}\left(\sum\right) := \{S \subset \chi \mid S = f^{-1}(E) \text{ for } E \subset \mathbf{R}\}$$

Let S be a σ-ring of subsets of χ so that (χ, S) is a measurable space. For each real valued function $f : \chi \to \mathbf{R}$, consider the subset of χ defined by $N(f) = \{x \mid f(x) \neq 0\}$.

Definition 3.13 *If f is a real valued function such that, for every Borel set E of $\mathbf{R}$ the set $N(f) \cap f^{-1}(E)$ is measurable (i.e., is contained in S), then f is called a **measurable function**.*

If $E = \mathbf{R}$ and f is measurable, then $\mathbf{N}(f)$ is measurable.

Theorem 3.14 *If f is a measurable function on χ, then for every measurable subset A of χ and every Borel set E of $\mathbf{R}$ the set $A \cap f^{-1}(E)$ is a measurable subset of χ.*

Proof: $A \cap f^{-1}(E) = [A \cap N(f) \cap f^{-1}(E)] \cup [(A \backslash N(f)) \cap f^{-1}(E)]$

Now $0 \in E \Rightarrow f^{-1}(E) \cap (A \backslash N(f)) = A \backslash N(f)$.

$0 \notin E \Rightarrow f^{-1}(E) \cap (A \backslash N(f)) = \Phi$.

In either case, $A \cap f^{-1}(E)$ is measurable.

Thus, a real valued function f defined on a set A is measurable on A if for every Borel set E of $\mathbf{R}$, $A \cap f^{-1}(E)$ is measurable.

In particular if χ is measurable (*i.e.*, $\chi \in S$), then a real valued function f is measurable on χ if $f^{-1}(E)$ is measurable for every Borel set E of $\mathbf{R}$.

Thus, if (χ, S) is a measurable space and B is the σ-ring of Borel sets on $\mathbf{R}$, then a real valued function $f : \chi \to \mathbf{R}$ is measurable if its inverse as a set function maps sets of B into sets of S.

Clearly the concept of measurability of a function $f : \chi \to \mathbf{R}$ depends on the σ-ring S.

If $\chi = \mathbf{R}$ and S and $\overline{S}$ are the class of Borel sets and the class of Lebesgue sets of $\mathbf{R}$ respectively, then a real valued function f of a real variable, $f : \mathbf{R} \to \mathbf{R}$, if measurable with respect to the σ-ring S is called a Borel measurable function, and if measurable with respect to the σ-ring $\overline{S}$ is called a Lebesgue measurable function.

Definition 3.15 *If f is infinite for some values of its argument, then it is called an extended real valued function, represented: $f : \chi \to \mathbf{R} \cup \{\infty, -\infty\}$.*

The concept of measurability for extended real valued functions is defined by regarding the singleton sets ∞ and $-\infty$ as Borel sets also. Thus an extended real valued function f is measurable, if, for every Borel set E of $\mathbf{R}$, each of the sets $f^{-1}(\{-\infty\})$, $f^{-1}(\{-\infty\})$, and $N(f) \cap f^{-1}(E)$ is measurable.

Remark: For this extended class of Borel sets it is no longer true that it is the σ-ring generated by semiclosed intervals.

Theorem 3.16 *A real valued function f on a measurable space (χ, S) is* *measurable iff, $\forall c \in \mathbf{R}$, the set $N(f) \cap x \mid f(x) < c$ is measurable.*

Proof: If $E = (-\infty, c) = \{y \in \mathbf{R} \mid y < c\}$, then E is a Borel set, and $f^{-1}(E) = \{x \mid f(x) < c\}$.

$\therefore$ If f is measurable, then $N(f) \cap \{x \mid f(x) < c\}$ is measurable $\forall c \in \mathbf{R}$.

Conversely, if $N(f) \cap x \mid f(x) < c$ is measurable $\forall c \in \mathbf{R}$, then let c, c' $\in \mathbf{R}$ such that $c \leq c'$. Since

$$\{x \mid f(x) < c\} - \{x \mid f(x) < c\} = \{x \mid c \leq f(x)c'\} = [c, c'), N(f) \cap f^{-1}(E)$$

is measurable for every semiclosed interval E. If C is the class of all subsets E of $\mathbf{R} \cup \{\infty, -\infty\}$ such $N(f) \cap f^{-1}(E)$ that is measurable, then C is a σ-ring and, further, every semiclosed interval of $\mathbf{R} \cup \{\infty, -\infty\}$. Hence, f is measurable.

Remarks:

(a) The above theorem is true even if $<$ is replaced by $\leq, >$ or $\geq$ since

$$\{x \mid f(x) \leq c\} = \bigcap_{n=1}^{\infty} \{x \mid f(x) < c + 1/n\}$$

$$\{x \mid f(x) > c\} = \mathbf{R}\{x \mid f(x) \leq c\} \text{ , and}$$

$$\{x \mid f(x) \geq c\} = \bigcap_{n=1}^{\infty} \{x \mid f(x) > c + 1/n\}$$

(b) The above theorem remains true if c is restricted to belong to an everywhere-dense subset of $\mathbf{R}$.

(c) If f is measurable, then $\forall c \in \mathbf{R}$ is measurable.

Theorem 3.17 *If f and g are extended real valued measurable functions on a measurable space (χ, S) and $c \in \mathbf{R} \cup \{\infty, -\infty\}$, then each of the following sets*

$$A = \{x \mid f(x) < g(x) + c\}$$

$$B = \{x \mid f(x) \leq g(x) + c\}$$

$$C = \{x \mid f(x) = g(x) + c\}$$

has a measurable intersection with every measurable set.

Proof: $A = \bigcup_{r \in Q}[\{x \mid f(x) < r\} \cap \{x \mid r - c < g(x)\}]$ (**Q**: set of all rational numbers) $\therefore$ A has the desired property.

$B = \chi \backslash \{x \mid g(x) < f(x) - c\}$ $\therefore$ B has the desired property.

$C = B \backslash A$ $\therefore$ C has the desired property.

Theorem 3.18 *If φ is an extended real valued Borel measurable function on the extended real line such that $\varphi(0) = 0$, and if f is an extended real valued function on a measurable space (χ, S), then the function $\tilde{f}(x) := \varphi(f(x))$, is a measurable function on χ.*

Proof: If E is a Borel set on $\mathbf{R} \cup \{\infty, -\infty\}$, then

$$N(\tilde{f}) \cap \tilde{f}^{-1}(E) = \{x \mid \varphi(f(x)) \in E \backslash \{0\}\} = \{x \mid f(x) \in \varphi^{-1}(E \backslash \{0\})\}$$

Since $\varphi(0) = 0$,
$$\varphi^{-1}(E \backslash \{0\}) = \varphi^{-1}(E \backslash \{0\}) \backslash \{0\}$$
Since φ is Borel measurable, $\varphi^{-1}(E \backslash \{0\})$ is a Borel set.

$$\therefore \{x \mid f(x) \in \varphi^{-1}(E \backslash \{0\})\} = N(f) \cap \{x \mid f(x) \in \varphi^{-1}(E \backslash \{0\})\}$$
$$N(f) \cap f^{-1}(\varphi^{-1}(E \backslash \{0\}))$$

$\therefore$ $\tilde{f}$ is measurable.

Corollary 3.18.1 Since for every $0 < a \in \mathbf{R}$, the function $\varphi : \mathbf{R} \to \mathbf{R}$ defined by $\varphi(t) = \mid t \mid^a$ is Borel measurable, wherever f is measurable, so is $\mid f \mid^\sigma$ $\forall \sigma > 0$, $\sigma \in \mathbf{R}$. Similarly f^n is measurable $\forall n \in \mathbf{N}$ whenever f is measurable.

Theorem 3.19 *If f and g are extended real valued measurable functions on a measurable space (χ, S), then so are $f + g$ and fg.*

Proof: Since

$$(f + g)^{-1}(\{\infty\}) = [f^{-1}(\{\infty\}) \backslash g^{-1}(\{-\infty\})] \cup [g^{-1}(\{\infty\}) \backslash f^{-1}(\{-\infty\})]$$

and

$$(f + g)^{-1}(\{-\infty\}) = [f^{-1}(\{-\infty\}) \backslash g^{-1}(\{\infty\})] \cup [g^{-1}(\{-\infty\}) \backslash f^{-1}(\{\infty\})]$$

and

$$\begin{aligned}
(fg)^{-1}(\{\infty\}) &= [f^{-1}(\{\infty\}) \backslash g^{-1}(\{x \mid g(x) < 0\})] \\
&\cup [g^{-1}(\{\infty\}) \backslash f^{-1}(\{x \mid f(x) < 0\})] \\
&\cup [f^{-1}(\{-\infty\}) \cap g^{-1}(\{x \mid g(x) < 0\})] \\
&\cup [g^{-1}(\{-\infty\}) \cap f^{-1}(\{x \mid f(x) < 0\})]
\end{aligned}$$

and

$$\begin{aligned}(fg)^{-1}(\{-\infty\}) \;=\; & [f^{-1}(\{\infty\})\cap g^{-1}(\{x\mid g(x)<0\})]\\ & \cup[g^{-1}(\{\infty\})\cap f^{-1}(\{x\mid f(x)<0\})]\\ & \cup[f^{-1}(\{-\infty\})\backslash g^{-1}(\{x\mid g(x)>0\})]\\ & \cup[g^{-1}(\{-\infty\})\backslash f^{-1}(\{x\mid f(x)>0\})]\end{aligned}$$

The left hand sides of the above identities are all measurable. Thus, we may restrict attention to finite valued functions. If f and g are finite functions and $c \in \mathbf{R}$, then

$$\{x\mid f(x)+g(x)<c\}=\{x\mid f(x)<-g(x)+c\}$$

We know that this set has measurable intersection with every measurable set and in particular the set $\{x\mid f(x)+g(x)>0\}=N(f+g)$ and $\therefore$, so also the set $\{x\mid f(x)+g(x)<c\}\cap N(f+g)$. Since

$$fg=\frac{1}{4}[(f+g)^2-(f-g)^2]$$

f_g is measurable.

Definition 3.20 *If f and g are finite valued functions, define the functions* $f\cup g$ *and* $f\cap g$ *by*

$$(f\cup g)(x)=max\{f(x),g(x)\},\ (f\cap g)(x)=min\{f(x),g(x)\}$$

Clearly, $f\cup g=(f+g+\mid f-g\mid)/2$ *and* $f\cap g=(f+g-\mid f-g\mid)/2.$

Corollary 3.20.1 From the above theorem, we have $f\cup g$ and $f\cap g$ measurable whenever f and g are measurable functions.

Writing $f^+=f\cup 0$ and $f^-=-(f\cap 0)$ (when 0 is the everywhere-zero function), $f=f^+-f^-$ and $\mid f\mid=f^+-f^-$. f^+ and f^- are called the positive part and negative part of f, respectively, and are both measurable.

If $\{f_n\}_{n=1}^{\infty}$ is a sequence of real valued function, we can define functions

$$\bigcup_{n=1}^{\infty}f_n\ \text{and}\ \bigcap_{n=1}^{\infty}f_n$$

as

$$(\bigcup_{n=1}^{\infty}f_n)(x)=\sup_{1\le n\le\infty}\{f_n(x)\}\ \text{and}\ (\bigcap_{n=1}^{\infty}f_n)(x)=\inf_{1\le n\le\infty}\{f_n(x)\}$$

where the sup and inf are the least upper bound and greatest lower bound of a set of real numbers. These are called the *supremum* and *infimum* functions of the given sequence of function $\{f_n\}_{n=1}^{\infty}$.

The limit supremum of the given sequence is given by

$$\lim \sup_{1 \le n < \infty} f_n = \inf_{1 \le n \le \infty} \sup_{1 \le n < \infty} f_m = \bigcap_{n=1}^{\infty} \bigcup_{m=n}^{\infty} f_m$$

The limit infimum of the given sequence is given by

$$\lim_{1 \le n < \infty} \inf f_n = \sup_{1 \le n < \infty} \inf_{1 \le n < \infty} f_m = \bigcup n = 1^{\infty} \bigcap m = n^{\infty} f_m$$

Theorem 3.21 *If f_n is a sequence of extended real valued, measurable functions on a measurable space (χ, S), then each of the functions*

$$h(x) \quad = \quad \sup_{n} \{f_n(x); 1 \le n < \infty\}$$

$$g(x) \quad = \quad \inf_{n} \{f_n(x); 1 \le n < \infty\}$$

$$f^*(x) \quad = \quad \lim \sup_{n} f_n(x)$$

$$f_*(x) \quad = \quad \lim \inf_{n} f_n(x)$$

is measurable.

Proof: As in the previous theorem, after examining a few cases one can reduce the general case to the case of finite valued functions. Consider

$$\{x \mid g(x) < c\} = \bigcup_{n=1}^{\infty} \{x \mid f_n(x) < c\}$$

Indeed,

$$y \in \{x \mid f_{n_0}(x) < c\} \Rightarrow f_{n_0}(y) < c \Rightarrow \inf_{n} \{f_n(y)\} < c$$

$$\Rightarrow g(y) < c \Rightarrow y \in \{x \mid g(x) < c\}$$

$$\therefore \{x \mid g(x) < c\} \supset \bigcup_{n=1}^{\infty} \{x \mid f_n(x) < c\}$$

If

$$\{x \mid g(x) < c\}$$

then
$$\inf_n \{f_n(x)\} < c$$

Now
$$\exists n_0 \in \mathbf{N} \ni f_{n_0}(x) < c$$

For otherwise,
$$f_n(y) \geq c \ \forall n \in \mathbf{N} \text{ and } \therefore c > \inf_n \{f_n(y)\} \geq c$$

which is impossible. $\therefore y \in \{x \mid f_{n_0}(x) < c\}$ for some n_0. Thus,

$$\{x \mid g(x) < c\} \subset \bigcup_{n=1}^{\infty} \{x \mid f_n(x) < c\}$$

The right hand side is measurable $\forall c \in \mathbf{R} \ \therefore \ g$ is a measurable function. Since $h(x) = -\inf_n\{-f_n(x)\}$, h is also measurable. Since

$$f^*(x) = \inf_{n \geq 1} \ \sup_{m \geq n} \{f_m(x)\}$$

and

$$f_*(x) = \sup_{n \geq 1} \ \inf_{m \geq n} \{f_m(x)\}$$

f^* and f_* are also measurable.

A sequence of functions $\{f_n\}$ is said to converge at a point x if $\limsup_n f_n(x) = \liminf_n f_n(x)$. This common value is denoted $\lim_n f_n(x)$.

From the above theorem the set of points of convergence of a sequence $\{f_n\}$ of measurable functions given by the set

$$\{x \mid \limsup_n f_n(x) = \liminf_n f_n(x)\} = \{x \mid f^*(x) = f_*(x)\}$$

has measurable intersection with every measurable set. Therefore the function f defined by $f(x) = \lim_n f_n(x)$ at every n for which the limit exists, is a measurable function.

3.3.1 Simple functions

Definition 3.22 *A function f, defined on a measurable space (χ, S) is called **simple** if there is a finite, disjoint class $\{E_1, ..., E_n\}$ of measurable sets and a finite set $\{\sigma_1, ..., \sigma_n\}$ of real numbers such that*

$$f(x) = \begin{cases} \sigma_i & \text{if } x \in E_i \ 1 \leq i \leq n \quad (\mid \sigma_i \mid < \infty \ \forall 1 \leq i \leq n) \\ 0 & \text{otherwise} \end{cases}$$

If

$$\chi_{E_i} f(x) = \begin{cases} 1 & \text{if } x \in E_i \\ 0 & \text{if } x \notin E_i \end{cases}$$

is the characteristic function of E_i, then it is easy to verify that χ_{E_i} is a measurable function if E_i is a measurable set. It is itself a simple function. f can then be expressed as a finite linear combination of such characteristic functions:

$$f(x) = \sum_{i=1}^{n} \sigma_i \chi_{E_i}(x)$$

The product of two simple functions and a finite linear combination of simple functions is again a simple function.

Theorem 3.23 *If f is an extended real valued measurable function, then there exists a sequence of simple functions $\{f_n\}$ such that f is the limit of this sequence.*

Proof: Let $f \geq 0$ $\forall n \in \mathbf{N}$ and $\forall x \in \chi$ define

$$f_n(x) = \begin{cases} \frac{i-1}{2^n} & \text{if } f(x) \in [\frac{i-1}{2^n}, \frac{i}{2^n}) \ \ 1 \leq i \leq 2^n n \\ n & \text{if } f(x) \geq n \end{cases}$$

f_n is a nonnegative simple function, and the sequence $\{f_n\}$ is increasing i.e., $f_{n+1}(x) \geq f_n(x)$ $\forall_x \in \chi$.

If $f(x) < \infty$, then $\exists n \ni 0 \leq f(x) - f_n(x) \leq \frac{1}{2^n}$.

If $f(x) = \infty$, then $f_n(x) = n$ $\forall n \in \mathbf{N}$

Since the difference of two simple functions is also a simple function, writing $f = f^+ - f^-$ and applying the result for nonnegative functions obtained above to f^+ and f^- separately, we see that the theorem follows for all measurable functions f.

3.4 Convergence of Measurable Functions

In this section we look at sequences of measurable functions and their convergence properties. These properties along with the above theorem help us define the Lebesgue integral for certain measurable functions.

Let (χ, S) be a measurable space and let $\mu : S \to \mathbf{R} \cup \{\infty\}$ be a measure on χ. Thus (χ, S, μ) is a measure space.

Definition 3.24 *If P is a proposition about the points of a measure space, which is true for all points of χ except at most for a set of points whose measure is zero, then the proposition P is said to be true almost everywhere, and is written "$P.a.e.$"*

Example: Consider the function

$$f(x) = \begin{cases} 1 & \text{if } x \text{ is irrational} \\ 0 & \text{if } x \text{ is rational} \end{cases}$$

then $f(x) = 1$ a.e.

Definition 3.25 *A function f is said to be* **essentially bounded** *if it is bounded a.e. More precisely, a function f is said to be essentially bounded if*

$$\exists 0 < c \in \mathbf{R} \ni \mu(\{x \mid \mid f(x) \mid > c\}) = 0$$

The quantity

$$\inf_{c \in \mathbf{R}} \ c \mid \mu(\{x \mid \mid f(x) \mid > c\}) = 0$$

is called the essential supremum of $\mid f \mid$, denoted ess. sup. $\mid f \mid$. Before proceeding further, we introduce a few notions about convergence of functions.

Definition 3.26 *A sequence $\{f_n\}$ of real valued functions is said to converge* **pointwise** *to a function f, if $\forall x$ given*

$$\epsilon > 0 \ \exists n_0 = n_0(x, \epsilon) \in \mathbf{N} \ni \mid f_n(x) - f(x) \mid < \epsilon \quad \forall n > n_0$$

Definition 3.27 *A sequence $\{f_n\}$ of real valued function is said to converge uniformly to a function f, if given $\epsilon > 0$, $\exists n_0 = n_0(\epsilon) \in \mathbf{N}$ (depending on ϵ only) $\ni \mid f_n(x) - f(x) \mid < \epsilon \ \forall n > n_0$ and $\forall x$.*

Definition 3.28 *A sequence $\{f_n\}$ of real valued functions is* **fundamental**, *if given $\epsilon > 0 \ \exists \ n_0(x, \epsilon) \in \mathbf{N} \ni \mid f_n(x) - f_m(x) \mid < \epsilon \ \forall n, m > n_0$.*

Definition 3.29 *A sequence $\{f_n\}$ of real valued function is* **uniformly fundamental**, *if given $\epsilon > 0 \exists \ \forall x, n_0 = n_0(\epsilon) \in \mathbf{N}$ such that $\mid f_n(x) - f_m(x) \mid < \epsilon \ \forall n, m > n_0$.*

The above notions will be reintroduced below in the context of measure theory:

Definition 3.30 *A sequence $\{f_n\}_{n=1}^{\infty}$ of extended real valued function is said to* **converge a.e.** *on the measure space (χ, S, m) to a limit function f if there exists a set E of measure zero (possibly empty) such that, if $x \notin E$*

and $\epsilon > 0,\ \exists n_0 = n_0(x, \epsilon) \in \mathbf{N} \ni$

$$\begin{cases} f_n(x) < \frac{-1}{\epsilon} & \text{if } f(x) = -\infty \\ |\, f_n(x) - f(x)\,| < \epsilon & \text{if } |\, f(x)\,| < \infty \\ f_n(x) > \frac{1}{\epsilon} & \text{if } f(x) = \infty \end{cases} \Bigg\} \forall n > n_0$$

Definition 3.31 *A sequence* $\{f_n\}_{n=1}^{\infty}$ *of real valued functions is* **fundamental a.e.,** *if a set* E *of measure zero such that if* $x \notin E$ *and* $\epsilon > 0, \exists n_0 = n_0(x, \epsilon) \in \mathbf{N} \ni$

$$|\, f_n(x) - f_m(x)\,| < \epsilon\ \forall n, m > n_0$$

Definition 3.32 *A sequence* $\{f_n\}_{n=1}^{\infty}$ *of real valued functions is said to* **uniformly converge** *to* f *a.e., if there exists a set* E *of measure zero* $\ni\ \forall\ \epsilon > 0, \exists n_0 = n_0(x, \epsilon) \in \mathbf{N} \ni |\, f_n(x) - f(x)\,| < \epsilon\ \forall n > n_0$ *and* $\forall x \notin E.$

Definition 3.33 *A sequence* $\{f_n\}$ *of a.e. finite valued measurable functions* **converges in measure** *to the measurable function* f, *if* $\forall \epsilon > 0,$

$$\lim_n \mu(\{x \mid |\, f_n(x) - f(x)\,| \geq \epsilon\}) = 0$$

Definition 3.34 *A sequence* $\{f_n\}$ *of a.e. finite valued measurable functions is* **fundamental in measure,** *if* $\forall \epsilon > 0$

$$\mu(\{x \mid |\, f_n(x) - f_m(x)\,| \geq \epsilon\}) \to 0 \text{ as } n, m \to \infty$$

We state the following theorem without proof:

Theorem 3.35 *If* $\{f_n\}$ *is a sequence of measurable functions which is fundamental in measure, then there exists a measurable function* f *such that* $\{f_n\}$ *converges in measure to* f.

3.5   Lebesgue Integration

In this section, the Lebesgue integral is defined for simple functions, and its properties are studied on convergent sequences of simple functions. These properties, coupled with the fact that it is possible to approximate measurable functions by sequences of simple functions, are used to define the Lebesgue integral for certain measurable functions.

Definition 3.36 *A simple function $f = \sum_{i=1}^{n} \sigma_i \chi_{E_i}$ on a measure space (χ, S, μ) is integrable if $\mu(E_i) < \infty \; \forall 1 \le i \le n \ni \sigma_i \ne 0$.*

Definition 3.37 *If $f = \sum_{i=1}^{n} \sigma_i \chi_{E_i}$ is an integrable simple function, then the integral of f, denoted $\int f d\mu$ (or $\int f(x) d\mu(x)$), is defined by*

$$\int f d\mu = \sum_{i=1}^{n} \sigma_i \mu(E_i)$$

Remark: If $f = \sum_{i=1}^{m} \beta_j \chi_{F_j}$ is another representation of f as a simple function, then consider the sets $G_{ij} = E_i \cap F_j \; \forall 1 \le i \le n, \, 1 \le j \le m$.

$$\mu(\bigcup_{j=1}^{m} G_{ij}) = \mu(E_i) \text{ and } \mu(\bigcup_{i=1}^{m} G_{ij}) = \mu(F_j)$$

Moreover, $G_{ij} \cap G_{i'j'} = \Phi$ if $i \ne i'$ or $j \ne j' \; \therefore \; \mu(\cup G_{ij}) = \sum \mu(G_{ij})$.

If $x \in E_i \cap F_j$, then $f(x) = \sigma_i = \beta_j = \gamma_{ij}$ say

$$\therefore f(x) = \sum_{i=1}^{n} \sum_{j=1}^{m} \gamma_{ij} \chi_{G_{ij}}$$

Now,

$$\int f d\mu = \sum_{i=1}^{n} \sigma_i \mu(E_i) = \sum_{i=1}^{n} \sigma_i \mu(\bigcup_{j=1}^{m} G_{ij}) = \sum_{i=1}^{n} \sigma_i (\sum_{j=1}^{m} \mu(G_{ij})) = \sum_{i=1}^{n} \sum_{j=1}^{m} \sigma_i \mu(G_{ij})$$

$$= \sum_{i=1}^{n} \sum_{j=1}^{m} \gamma_{ij} \mu(G_{ij}) = \sum_{j=1}^{m} \sum_{i=1}^{n} \gamma_{ij} \mu(G_{ij}) = \sum_{j=1}^{m} \sum_{i=1}^{n} \beta_{ij} \mu(G_{ij}) = \sum_{j=1}^{m} \beta_j \sum_{i=1}^{n} \mu(G_{ij})$$

$$= \sum_{j=1}^{m} \beta_j \mu(\bigcup_{i=1}^{n} G_{ij}) = \sum_{j=1}^{m} \beta_j \mu(F_j)$$

$\therefore \int f d\mu$ is unambiguously defined.

If $f = \chi_E$ then $\int f d\mu = \mu(E)$.

Theorem 3.38 *If f and g are integrable simple functions and if both a and $b \in \mathbf{R}$, then*

$$\int (af + bg) d\mu = a \int f d\mu + b \int g d\mu$$

Proof: Let

$$f = \sum_{i=1}^{n} \sigma_i \chi_{E_i} \text{ and } g = \sum_{j=1}^{m} \beta_j \chi_{F_{i'}}$$

and let

$$E = \bigcup_{i=1}^{n} E_i, F = \bigcup_{j=1}^{m}, \text{ and } G_{ij} = E_i \cap F_j \quad \forall 1 \le i \le n, \ 1 \le j \le m$$

Also let

$$E_i' = E_i \backslash f_j \ \forall 1 \le i \le n \text{ and } F_j' = F_j \backslash E \ \forall 1 \le j \le m$$

Then,

$$af + bg = \sum_{i=1}^{n} \sum_{j=1}^{m} (a\sigma_i + b\beta_j) \chi_{G_{ij}} + \sum_{i=1}^{n} a\sigma_i \chi_{E_{i'}} + \sum_{j=1}^{m'} b\beta_j \chi_{F_{j'}}$$

where

$$\{G_{ij}\}_{i=1,j=1}^{n,m} \bigcup \{E_i'\}_{i=1}^{n} \bigcup \{F_j'\}_{j=1}^{m}$$

is a class of mutually disjoint sets. Therefore,

$$
\begin{aligned}
\int (af + bg) d\mu &= \sum_{i=1}^{n} \sum_{j=1}^{m} (a\sigma_i + b\beta_j) \mu(G_{ij}) + \\
&\quad \sum_{i=1}^{n} a\sigma_i \mu(E_i') + \sum_{j=1}^{m} b\beta_j \mu(F_j') \\
&= \sum_{i=1}^{n} \sum_{j=1}^{m} a\sigma_i \mu(G_{ij}) + \sum_{i=1}^{n} a\sigma_i \mu(E_i') + \\
&\quad \sum_{j=1}^{m} \sum_{i=1}^{n} b\beta_j \mu(G_{ij}) + \sum_{j=1}^{m} b\beta_j \mu(F_i') \\
&= \sum_{i=1}^{n} a\sigma_i \mu(E_i \cap F) + \sum_{i=1}^{n} a\sigma_i \mu(E_i \backslash F) + \\
&\quad \sum_{j=1}^{m} b\beta_j \mu(F_j \cap E) + \sum_{j=1}^{m} b\beta_j \mu(F_j \backslash E) \\
&= \sum_{i=1}^{n} a\sigma_i \mu(E_i) + \sum_{j=1}^{m} b\beta_j \mu(F_j') \\
&= a \int f d\mu + b \int g d\mu
\end{aligned}
$$

Theorem 3.39 *If f is an integrable simple function that is nonnegative a.e., then $\int f d\mu \geq 0$.*

Proof: $f = \sum_{i=1}^{n} \sigma_i \chi_{\{E_i\}}$ and $f \geq 0$ a.e., implies that

$$\mu(E_i) > 0 \Rightarrow \sigma_i \geq 0 \; \forall 1 \leq i \leq n$$

now

$$\int f d\mu = \sum_{i=1}^{n} \sigma_i \mu(E_i)$$

If $\mu(E_i) > 0$ then $\sigma_i \geq 0$ and $\sigma_i \mu(E_i) \geq 0$.

If $\mu(E_i) = 0$ then $\sigma_i \mu(E_i) = 0$ $\therefore$ $\int f d\mu \geq 0$.

Corollary 3.39.1 *If f and g are integrable simple functions, such that $f \geq g$ a.e., then $\int f d\mu \geq \int g d\mu$.*

Proof: Apply the above theorem to $f - g$.

Corollary 3.39.2 *If f and g are integrable simple functions, then $\int |f+g| \, d\mu \leq \int |f| \, d\mu + \int |g| \, d\mu$.*

Proof: Observe that if $f = \sum_{i=1}^{n} \sigma_i \chi_{\{E_i\}}$ then $|f| = \sum_{i=1}^{n} |\sigma_i| \chi_{\{E_i\}}$, and apply the above theorem to $|f| + |g| - |f+g|$.

Definition 3.40 *If f is an integrable simple function, then the* **indefinite integral** *of f is a set function $I_f : \mathbf{S} \to \mathbf{R}$ defined by*

$$I_f(E) = \int_E f d\mu = \int \chi_E f d\mu$$

Remark: I_f is countably additive.

Definition 3.41 *If f and g are integrable simple functions, then the distance between f and g denoted $\rho(f,g)$ is defined by*

$$\rho(f,g) = \int |f - g| \, d\mu$$

Remark: $\rho(f,f) = 0, \rho(f,g) = \rho(g,f), \rho(f,g) \leq \rho(f,h) + \rho(h,g)$ but $\rho(f,g) = 0$ does not imply $f = g$. Thus ρ is not quite a metric.

Definition 3.42 *A sequence $\{f(n)\}_{n=1}^{\infty}$ of integrable simple functions is* **mean fundamental,** *if $\rho(f_n, f_m) \to 0$ as $n, m \to \infty$.*

Theorem 3.43 *A mean fundamental sequence $\{f_n\}$ of integrable simple functions is fundamental in measure.*

Proof: For any $\epsilon > 0$, define $E_{nm} = \{x :\mid f_n(x) - f_m(x) \mid \geq \epsilon\}$.

Then

$$\rho(f_n, f_m) = \int \mid f_n - f_m \mid d\mu \geq \int_{E_{nm}} \mid f_n - f_m \mid d\mu \geq \epsilon\mu(E_{nm})$$

$$\therefore \mu(E_{nm}) \to 0 \text{ as } n, m \to \infty$$

Theorem 3.44 *If $\{f_n\}$ is a mean fundamental sequence of integrable simple functions and I is a finite valued and countably additive set function on S, then $I(E) = lim_n I_{\{f_n\}}(E)$ exists for every measurable set E.*

Proof:

$$
\begin{aligned}
\mid I_{f_n}(E) - I_{f_m}(E) \mid &= \mid \int_E f_n d\mu - \int_E f_m d\mu \mid \\
&\leq \int_E \mid f_n - f_m \mid d\mu \\
&\leq \int \mid f_n - f_m \mid d\mu \to 0, n, m \to 0
\end{aligned}
$$

implies that the limit exists and is uniform. Finite additivity of limits implies the finite additivity of I. If $\{E_n\}_{n=1}^{\infty}$ is a sequence of measurable sets that show that $E_i \cap E_j = \Phi \;\; \forall i \neq j$ and $\bigcup_{i=1}^{\infty} E_i = E$, then for any pair of positive integers n, m,

$$\mid I(E) - \sum_{i=1}^{m} I(E_i) \mid$$
$$\leq \;\; \mid I(E) - I_{f_n}(E) \mid + \mid I_{f_n}(E) - \sum_{i=1}^{m} I_{f_n}(E_i) \mid + \mid I_{f_n}(\bigcup_{i=1}^{m} E_i) - I(\bigcup_{i=1}^{m} E_i) \mid$$

given

$$\epsilon > 0 \; \exists n_0 \in \mathbf{N} \ni \mid I(E) - I_{f_n}(E) \mid < \frac{\epsilon}{3}$$

and

$$\mid I_{f_n}(\bigcup_{i=1}^{m} E_i) - I(\bigcup_{i=1}^{m} E_i) \mid < \epsilon \; \forall n > n_0$$

and

$$\exists m_0 \in \mathbf{N} \ni \mid I_{f_n}(E) - \sum_{i=1}^{m} I_{f_n}(E_i) \mid + < \frac{\epsilon}{3} \;\;\;\; \forall m > m_0 \text{ (fixing n)}$$

$$\therefore \; \mid I(E) - \sum_{i=1}^{m} I(E_i) \mid < \epsilon \;\;\;\; \forall m > m_0$$

Hence,

$$I(E) = I(\bigcup_{i=1}^{\infty} E_i) = \sum_{i=1}^{\infty} I(E_i)$$

Definition 3.45 *A sequence of finite valued set functions* $\{\lambda_n\}$ *defined on a* σ-*ring* S *of measurable sets is said to be* **uniformly absolutely continuous,** *if for every* $\epsilon > 0$, $\exists \delta > 0 \ni |\lambda_n(E)| < \epsilon \ \forall n$ *whenever* $\mu(E) < \delta$.

Theorem 3.46 *If* $\{f_n\}$ *is a mean fundamental sequence of integrable simple functions, then the set functions* $\{I_n\}$ *are uniformly absolutely continuous where*

$$I_n(E) = \int_E f_n d\mu = \int \chi_E f_n d\mu$$

Proof: Given $\epsilon > 0$, let $n_0 \in N \ni \int |f_n - f_m| \, d\mu < \epsilon/2 \ \forall n, m \geq n_0$

and let

$$\delta > 0 \ni \int_E |f_n| \, d\mu < \epsilon/2 \quad \forall \mu(E) < \delta \ \forall 1 \leq n \leq n_0$$

If $E \in S \ni \mu(E) < \delta$ and $n \leq n_0$ then $|I_n(E)| \leq \int_E |f_n| \, d\mu < \epsilon$.

If $n > n_0$ then $|I_n(E) \leq \int |f_n - f_{n_0}| \, d\mu < \epsilon$.

Theorem 3.47 *If* $\{f_n\}$ *and* $\{g_n\}$ *are mean fundamental sequences of integrable simple functions which converge in measure to the same measurable function* f, *if the indefinite integrals of* f_n *and* g_n *are* I_{f_n} *and* I_{g_n} *respectively, and if for every measurable set* E $I_f(E) = \lim_n I_{f_n}(E)$ *and* $I_g(E) = \lim_n I_{g_n}$, *then the set functions* I_f *and* I_g *are identical.*

Proof: Given $\epsilon > 0$, since

$$E_n = \{x : |f_n(x) - g_n(x)| \geq \epsilon\} \subset \{x : |f_n(x) - f(x)| \geq \epsilon/2\} \cup \{x : |f_n(x) - g_n(x)| \geq \epsilon\}$$

We have $\mu(E_n) \to 0$ as $n \to \infty$.

Letting E be measurable with $\mu(E) \leq \infty$, we have

$$\int_E |f_n - g_n| \, d\mu \leq \int_{E \setminus E_n} |f_n - g_n| \, d\mu + \int_{E \cap E_n} |f_n| \, d\mu + \int_{E \cap E_n} |g_n| \, d\mu$$

Now,

$$\int_{E \setminus E_n} |f_n - g_n| \, d\mu \leq \epsilon \mu(E)$$

Choosing n sufficiently large, we can make

$$\int_{E \cap E_n} | \ f_n \ | \ d\mu \ \text{and} \ \int_{E \cap E_n} | \ g_n \ | \ d\mu$$

arbitrarily small since I_{f_n} and I_{g_n} are uniformly absolutely continuous by the previous theorem.

$$\therefore \int_E | \ f_n - g_n \ | \ d\mu \to 0 \ \text{as} \ n \to \infty$$

Since

$$| \ I_{f_n}(E) - I_{g_n}(E) \ | \leq \int_E | \ f_n - g_n \ | \ d\mu$$

we have $I_f(E) = I_g(E)$.

Since both I_f and I_g are countably additive, $I_f(E) = I_g(E)$ for every measurable set E of σ-finite measure.

Since f_n and g_n are simple functions, they are each defined over finitely many sets of finite measure. If F is the union of all these sets of all simple functions f_n, g_n, then F is a measurable set of σ-finite measure. Hence, for every measurable set E,

$$I_{f_n}(E \backslash F) = I_{g_n}(E \backslash F) = 0$$

$$\therefore I_f(E \backslash F) = I_f(E \backslash F) = 0$$

$$\therefore I_f(E) = I_f(E \cap F) \ \text{and} \ I_g(E) = I_g(E \cap F)$$

Definition 3.48 *A measurable function f which is finite valued a.e. on a measure space (χ, S, μ) is **integrable** if there exists a mean fundamental sequence $\{f_n\}$ of integrable simple functions which converges in measure to f. The integral of f denoted $\int f d\mu$ is defined by $\int f d\mu = \lim_n \int f_n d\mu$.*

Remark: From the theorem it is clear that the value of the integral of f is uniquely determined by any particular sequence satisfying the conditions in the above definition.

Remark: The properties of mean convergence and convergence in measure imply that the absolute values, constant multiples, and finite sums of integrable functions are integrable functions. The functions

$$f^+ = \frac{1}{2}(| \ f \ | + f) \ \text{and} \ f^- = \frac{1}{2}(| \ f \ | - f)$$

imply that f^+ and f^- are integrable.

If E is a measurable set and $\{f_n\}$ is a mean fundamental sequence of integrable simple functions converging in measure to the integrable function f, then clearly the sequence $\{\chi_E f_n\}$ is mean fundamental and converges in measure to $\chi_E f$.

The integral of f over E is defined by

$$\int_E f d\mu = \int \chi_E f d\mu$$

The theorems proved for integrable simple functions can easily be proved for general integrable functions also. These are left for the interested reader to complete as an exercise.

Definition 3.49 *A sequence $\{f_n\}$ of integrable functions* **converges in the mean** *to an integrable function f if $\rho(f_n, f) = \int | f_n - f | d\mu \to 0$ as $n \to 0$.*

Theorem 3.50 *If a sequence of integrable functions $\{f_n\}$ converges in the mean to f, then it also converges to f in measure.*

Proof: Given $\epsilon > 0$, define $E_n = \{x :| f_n(x) - f(x) |> \epsilon\}$. Then

$$\int | f_n - f | d\mu \geq \int_{E_n} | f_n - f | d\mu \geq \epsilon\mu(E_n)$$

$$\therefore \int | f_n - f | d\mu \to 0 \Rightarrow \mu(E_n) \to 0 \text{ as } n \to \infty$$

Theorem 3.51 *If f is integrable and nonnegative a.e., the $\int f d\mu = 0 \Leftrightarrow f = 0$ a.e.*

Proof: If $f = 0$ a.e., consider the sequence of functions $\{f_n\}$ where $f \equiv 0 \; \forall n$. This is a mean fundamental sequence of integrable simple functions which converges in measure to f, and it follows that

$$\int f d\mu = 0$$

Conversely, if $\{f_n\}$ is a mean fundamental sequence of integrable simple functions which converges in measure to f, then, without a loss of generality, we can assume that $f_n \geq 0$ since each f_n may be replaced by its absolute value.

$$\int f d\mu = 0 \Rightarrow \lim_n \int f_n d\mu = 0 \Rightarrow \{f_n\} \text{ converges in the mean to } 0$$

By the above theorem, this implies that $\{f_n\}$ converges in measure to 0.

Theorem 3.52 *If f is an integrable function and E is a set of measure zero, then*

$$\int_E f d\mu = 0$$

Proof: $\int_E f d\mu = \int \chi_E d\mu$ since $\mu(E) = 0$, $\chi_E = 0$ a.e. $\therefore \chi_E f = 0$ a.e.,

$\therefore$ by the above theorem $\int_E f d\mu = 0$.

Theorem 3.53 *If f is an integrable function which is ≥ 0 a.e. on a measureble set E, and if $\int_E f d\mu = 0$, then $\mu(E) = 0$.*

Proof: Let $E_0 = \{x \mid f(x) > 0$, $E_n = \{x \mid f(x) \geq \frac{1}{n}$, $\forall n \in \mathbf{N}$.

Since $f \geq 0$ a.e., $\mu(E \backslash E_0) = 0$,

$$\int_{E \cap E_n} f d\mu = 0 \; ; \text{ also } \int_{E \cap E_n} f d\mu \geq \frac{1}{n} \mu(E \cap E_n) \geq 0. \quad \therefore \mu(E \cap E_n) = 0$$

$$E_0 = \bigcup_{n=1}^{\infty} \Rightarrow \mu(E \cap E_0) = 0 \text{ since } \mu(E \cap E_0) \leq \sum_{n=1}^{\infty} \mu(E \cap E_n)$$

$$\therefore \mu(E) = 0$$

Theorem 3.54 *If f is an integrable function such that $\int_E f d\mu = 0$ for every measurable set E, then $f = 0$ a.e.*

Proof: If $E = \{x \mid f(x) > 0\}$, then $\int_E f d\mu = 0$ $\therefore$ by the previous theorem, $\mu(E) = 0$.

Similarly, if $F = \{x \mid -f(x) > 0\}$, $\mu(F) = 0$. $\therefore f = 0$ a.e.

Theorem 3.56 *If f is an integrable function, then $N(f) = \{x \mid f(x) \neq 0\}$ has σ-finite measure.*

Proof: If $\{f_n\}$ is a mean fundamental sequence of integrable simple functions converging in measure to f, then for each n, $N(f_n)$ is a set of finite measure.

If $E = N(f) \backslash \bigcup_{n=1}^{\infty}$ and F is any measurable subset of E, then, since

$\int_F f d\mu = \lim_n \int_F^{n=1} f_n d\mu = 0$, by the previous theorem, $f = 0$ a.e., on E.

$$\therefore \mu(E) = 0 \quad N(f) = \bigcup_{n=1}^{\infty} N(f_n) \cup E$$

Remark: If f is an extended real valued measurable function which is nonnegative a.e., and f is not integrable, then $\int f d\mu = \infty$. The most general class of functions f for which one can define $\int f d\mu$ is the class of all extended real valued measurable functions f for which at least one of the two functions f^+ and f^- is integrable.

Then, $\int f d\mu = \int f^+ d\mu - \int f^- d\mu$.

Thus, $\int f d\mu = +\infty, -\infty$ or finite.

Theorem 3.57 *If* $\{f_n\}$ *is a mean fundamental sequence of integrable simple functions which converges in measure to the integrable function* f, *then* $\rho(f, f_n) = \int | f - f_n | d\mu \to 0$ *as* $n \to \infty$. *Therefore, for every integrable function* f *and for every* $\epsilon > 0$, $\exists$ *an integrable simple function* $g \in \rho(f, g) < \epsilon$.

Proof: For any fixed $m \in \mathbf{N}$, $\{| f_n - f_m |\}$ is a mean fundamental sequence of integrable simple functions converging in measure to $| f - f_m |$. $\therefore \int | f - f_m | d\mu = \lim_n \int | f_n - f_m | d\mu$.

Since $\{f_m\}$ is mean fundamental, the theorem follows.

Theorem 3.58 *If* $\{f_m\}$ *is a mean fundamental sequence of integrable functions, then there exists an integrable function* f *such that* $\rho(f_n, f) \to 0$ *as* $n \to \infty$.

Proof: By the above theorem, for each $n \in \mathbf{N}$, $\exists$ an integrable simple function $g_n \ni \rho(f_n, g_n) < 1/n$ $\rho(f_n, g_n) < 1/n$. $\therefore \{g_n\}$ is a mean fundamental sequence of integrable simple functions. If f is a measurable function $\ni \{g_n\}$ converges in measure to f, then f is integrable. Since

$$0 \leq | \int f_n d\mu - \int f d\mu | \leq \int | f_n - f | d\mu = \rho(f_n, f) \leq \rho(f_n, g_n) + \rho(g_n, f)$$

and $\rho(f_n, g_n), \rho(g_n, g) \to 0$ as $n \to \infty$ the theorem follows.

Theorem 3.59 (Lebesgue's dominated convergence theorem) . *If* $\{f_m\}$ *is a sequence of integrable functions converging in measure to* f *(or converging to* f *a.e.) and* g *is an integrable function* $\ni | f_n(x) | \leq | g(x) |$ *a.e., then* f *is integrable and* $\{f_m\}$ *converges to* f *in the mean.*

3.5.1   Some properties of the Lebesgue integral

Theorem 3.60 *If f is measurable, g is integrable, and $\mid f \mid \leq \mid g \mid$ a.e., then f is integrable.*

Proof: Suffices to prove the theorem for nonnegative f ($\because f = f^+ - f^-$). If f is a simple function then there is nothing to prove. In general, $\exists$ an increasing sequence $\{f_n\}$ of nonnegative simple functions $\ni \lim_n f_n(x) = f(x) \; \forall x \in \chi \; 0 \leq f_n \leq \mid g \mid \Rightarrow f_n$ is integrable $\forall n \in \mathbf{N}$.

By Lebesgue's dominated convergence theorem, f is integrable.

Theorem 3.61 *If $\{f_m\}$ is an increasing sequence of extended real valued nonnegative measurable functions and if*

$$\lim_n f_n(x) = f(x) \text{ a.e., then } \lim_n \int f_n d\mu = \int f d\mu$$

Proof: If f is integrable, then the theorem follows by Lebesgue dominated convergence theorem.

Theorem 3.62 *A measurable function is integrable iff its absolute value is integrable.*

Proof: f is integrable $\Rightarrow \mid f \mid$ is integrable. If $\mid f \mid$ is integrable, since $\mid f \mid \leq \mid f \mid$, by the first theorem f is integrable.

Theorem 3.63 (Fatou's lemma). *If f_m is a sequence of nonnegative integrable function for which*

$$\liminf_n \int f_n d\mu < \infty \text{ , then } f(x) = \liminf_n f_n(x)$$

is integrable, and $\int f d\mu \leq \liminf_n \int f_n d\mu$

Proof: If $g_n(x) = \inf\{f_i(x) \mid n \leq i < \infty\}$ then $g_n \leq f_n$ and $\{g_n\}$ is an increasing sequence. Since

$$\int g_n d\mu \leq \int f_n d\mu, \quad \lim_n \int g_n d\mu \leq \liminf_n \int f_n d\mu < \infty$$

Since $\lim_n g_n = \liminf_n f_n = f$, then by the second theorem above, f is integrable and $\int d\mu = \lim_n \int g_n d\mu \leq \liminf_n \int f_n d\mu$.

Chapter 4

Fourier Analysis

The study of phenomena in physics and engineering is almost always done by representing them by means of functions of one or more variables. Often, the understanding or manipulation of these phenomena require a decomposition of their salient features into components at different scales.

For example, in signal processing one often comes across the problem of separating *noise* (unwanted components) from *information* (useful components) in a given signal (for instance, the noise may be due to certain *high frequency* components introduced by some interfering source). In order to reject the undesirable components, one must first be able to *resolve* the signal at hand into its components. Similarly, in image processing, one often encounters the problem of *filtering* an image in order to suppress some of its features while enhancing others.

Features manifest themselves at various resolutions, and hence it is important to be able to break down an image (signal/function) into component images (signals/functions) representing various resolutions (frequencies/scales). We have already seen that certain functions belonging to Hilbert spaces can be broken into and therefore written as superpositions of certain other *special* (basis) functions. In this chapter and in the rest of this book, we will see how these special functions can be chosen to represent various scales or resolutions, and how functions can be analysed and synthesized using these special functions.

This decomposition of a function into various components is sometimes called *spectral analysis*, analogous to the splitting of electromagnetic radiation (e.g., light) into individual electromagnetic waves of various frequencies (colors), collectively called the *spectrum* of the radiation. For this reason, sometimes the word "frequency" is also used to refer to any parameter representing scale or resolution level in such analyses.

Fourier analysis is a well-known technique of spectral analysis using trigonometric functions of various periods to represent functions at various scales. The techniques of Fourier analysis are powerful and find wide-ranging applications not only in the mathematical sciences but also within pure mathematics. In fact, later in this book, when we introduce wavelet analysis, the *algebra* of wavelet analysis closely resembles that of Fourier analysis. Wavelets, themselves being an alternative and more efficient tool for analyzing functions across scales, are constructed using Fourier analysis.

From a statistical point of view, obtaining the Fourier spectrum of a function is the same as obtaining the least square fit of sines and cosines of various frequencies in one or more dimensions. Multiple regression using trigonometric functions is very elegant and simple, since the trigonometric sines and cosines are mutually orthonormal, and the coefficients of regression (Fourier coefficients) are written as simple sums of products (in the discrete case) or as integrals of products of functions (in the continuous case). This process is called Fourier transformation.

4.1   The Spaces $\mathbf{L^1(\chi)}$ and $\mathbf{L^2(\chi)}$

Before we delve into Fourier analysis, we wish to isolate the functions that we are most interested in Fourier analyzing. These functions are mathematical equivalents of those found often associated with phenomena studied in physical and engineering sciences. These are most often finite energy processes and, hence, the associated functions have certain appropriate constraints on them. For instance, a signal with finite energy is represented by a square integrable function. We, therefore, rigorously formulate two important classes of functions below.

4.1.1 The space $L^1(\chi)$

If (χ, S, μ) is a measure space, consider the set Γ^1 of all complex valued functions f whose absolute values are integrable over χ, *i.e.*,

$$\int |f| \, d\mu < \infty.$$

Γ^1 is a linear space, since a finite linear combination of integrable functions is integrable. Indeed, $\int |\alpha f + \beta g| \, d\mu \le |\alpha| \int_\chi |f| \, d\mu + |\beta| \int_\chi |g| \, d\mu$.

Define a real valued function p on Γ^1 as follows:

$$p : \Gamma^1 \to \mathbf{R} \ni p(f) = \int_\chi |f| \, d\mu = \int |f| \, d\mu$$

p has the following properties:

(i) $p(f) \ge 0$

(ii) $p(\alpha f) = |\alpha| \, p(f)$

(iii) $p(f_1 + f_2) \le p(f_1) + p(f_2)$

These properties suggest that p is a norm on Γ^1. However, on closer inspection we see that p is not quite a norm. Indeed, even if $f \ne 0$, but $f = 0$ a.e., then $p(f) = 0$, contrary to the requirement $p(f) = 0 \Leftrightarrow f = 0$ for p to be a norm. If instead of the space Γ^1, we consider the set of equivalence classes of elements of Γ^1 defined by the equivalence relation "$\sim$", where $f \sim g$ if $f = g$ a.e., then on this set, p redefined as $p(\tilde{f}) = p(f)$, where $\tilde{f}$ is the equivalence class to which f belongs, is a norm as can be verified easily. More precisely, consider the subset Γ_0 of Γ^1, defined by $\Gamma_0 = \{f \in \Gamma^1 \mid f = 0\}$ a.e., Γ_0 is a subspace of Γ^1. The factor space Γ^1/Γ_0 is the set of all equivalence classes under the relation "$\sim$" mentioned above. Since two functions that are equal a.e., are indistinguishable in Γ^1/Γ_0, the equivalence classes will be treated as individual functions, and p is a norm on Γ^1/Γ_0. Γ^1/Γ_0 is denoted $L^1(\chi, \mu)$, or just $L^1(\chi)$, and $p(f)$ is replaced by the notation $\| f \|_1$. Thus, the zero element of $L^1(\chi)$ is the equivalence class of all those functions of Γ^1 that are zero a.e., and any element of this class is indistinguishable from the everywhere zero function in $L^1(\chi)$.

$L^1(\chi)$ is a metric space with the metric given by $\rho(f, g) = \| f - g \|_1 \; \forall f, g \in L^1(\chi)$.

Theorem 4.1 *The space* $\mathbf{L^1}(\chi)$ *is complete.*

Proof: Indeed, by a previous theorem, if $\{f_n\}_{n=1}^{\infty}$ is a mean fundamental sequence of absolutely integrable functions, then there exists an integrable function f such that $\rho(f_n, f) \to 0$ as $n \to \infty$. Hence, $f \in L^1(\chi)$, and $L^1(\chi)$ is complete.

Theorem 4.2 *The subspace of* $L^1(\chi)$ *consisting of all integrable simple functions is everywhere dense in* $L^1(\chi)$.

Proof: By Theorem 3.57, for every integrable function f and every $\epsilon > 0$, there exists an integrable simple function g such that

$$\rho(f, g) < \epsilon$$

$L^1(\chi)$ with the norm $\| \ \|_1$ is a Banach space. However, it is not a Euclidean space, since the norm $\| \ \|_1$ cannot be derived from a scalar product.

4.1.2   The space $\mathbf{L^2}(\chi)$

If (χ, S, μ) is a measure space, consider the set Γ^2 of all complex valued functions f whose squares are integrable, *i.e.*, $\int | f |^2 \, d\mu < \infty$.

Theorem 4.3 Γ^2 *is linear space.*

Proof: If $f \in \Gamma^2$, then clearly $\alpha f \in \Gamma^2$.

If $f, g \in \Gamma^2$, then since $| fg | \leq \frac{1}{2}[| f |^2 + | g |^2]$, fg is integrable. *i.e.*, $\int | fg | \, d\mu < \infty$. $\therefore | f + g |^2 \leq | f |^2 + 2 | fg | + | g |^2 \to f + g \in \Gamma^2$.

Remark: The set Γ_0 of all square-integrable functions that are zero a.e. is a subspace of Γ^2 .

The factor space Γ^2/Γ_0 is denoted $L^2(\chi)$ and is a linear space. As in the case of $L^1(\chi)$, the equivalence classes of functions in Γ^2 with respect to this factoring are elements of $L^2(\chi)$ and will be treated as individual functions in $L^2(\chi)$, each equivalence class represented by any one of its members in Γ^2.

The functional $\langle , \rangle : L^2(\chi) \times L^2(\chi) \to \mathbf{R}$ defined by $\langle f, g \rangle = \int f\bar{g} d\mu$ is a scalar product on $L^2(\chi)$. Indeed, it is easy to verify that:

(i) $\langle f, f \rangle \geq 0$, $\langle f, f \rangle = 0 \Leftrightarrow f = 0$.

(ii) $\langle f, g \rangle = \langle \overline{f, g} \rangle$.

(iii) $\langle \alpha f, g \rangle = \alpha \langle f, g \rangle$.

(iv) $\langle f, g_1 + g_2 \rangle = \langle f, g_1 \rangle + \langle f, g_2 \rangle$.

Hence, $L^2(\chi)$ is a Euclidean space, with the norm induced by the scalar product, given by $\| f \|_2 = \sqrt{\langle f, f \rangle}$.

$L^2(\chi)$ is a metric space, with the metric induced by the norm, given by $\rho(f, g) = \| f - g \|_2$.

Theorem 4.4 $L^2(\chi) \subset L^1(\chi)$ iff $\mu(\chi) < \infty$.

Proof: The Schwartz inequality in $L^2(\chi)$ is given by $| \langle f, g \rangle | \leq \| f \|_2 \| g \|_2$. Hence, replacing f by $| f |$ and g by 1, we have $\int | f | \, d\mu \leq \sqrt{\mu(\chi)} \| f \|_2$.

$\therefore$ If $\mu(\chi) < \infty$, then $f \in L^2(\chi) \Rightarrow f \in L^1(\chi)$.

However, if $\mu(\chi) = \infty$, then $f \in L^2(\chi)$ does not imply that $f \in L^1(\chi)$

Example: Let

$$f(x) = \frac{1}{\sqrt{1 + x^2}} \cdot f \in L^2(\mathbf{R}) \text{ since } \int_{-\infty}^{\infty} \frac{dx}{1 + x^2} = \pi < \infty$$

But

$$\int_{-\infty}^{\infty} \frac{dx}{\sqrt{1 + x^2}} = \infty$$

$$\therefore f \notin L^1(\mathbf{R})$$

Our aim is to show that $L^2(\chi)$ is a Hilbert space. In order to do this we have to show that:

(i) $L^2(\chi)$ is infinite dimensional

(ii) $L^2(\chi)$ is complete

(iii) $L^2(\chi)$ is separable.

In the following discussion, we will set out to prove (i) through (iii).

Theorem 4.5 $L^2(\chi)$ *is infinite dimensional.*

Proof: Let A, B be any two disjoint measurable subsets of χ such that $\mu(A)$, $\mu(B) < \infty$. Then, the functions χ_A and χ_B which are the characteristic functions of the sets A and B, are mutually orthogonal in $L^2(\chi)$.

Theorem 4.6 $L^2(\chi)$ *is complete.*

Lemma 4.7 If $\{f_n\}$ converges to f in the L^2-norm, then there is a subsequence f_{n_i} such that f_{n_k} converges to f pointwise a.e. as $k \to \infty$.

Proof: Since $\lim_{n \to \infty} \| f_n - f \|_2 = 0$, choose $n_1 < n_2 < L$ such that $\| f_{n_k} - f \|_2^2 \le 2^{-k} \forall k$. By the monotone convergence theorem,

$$\int_\chi \sum_{k=1}^{\infty} |f_{n_k} - f|^2 \, d\mu = \sum_{k=1}^{\infty} \int_\chi |f_{n_k} - f|^2 \, d\mu \le \sum_{k=1}^{\infty} 2^{-k} = 1 < \infty$$

$$\therefore \sum_{k=1}^{\infty} |f_{n_k} - f|^2 < \infty \text{ a.e.}$$

This is possible only if $\lim_{k \to \infty} f_{n_k} = f$.

Verify that if $f_n \to f$ in the L^2-norm and it also converges to g pointwise a.e., then $f = g$ a.e.

Verify using Lebesgue's dominated convergence theorem that if $f_n \to f$ pointwise a.e. and if $\sup_{n \ge 1} |f_n|^2$ is integrable, then f_n also converges to f in the L^2-norm.

Proof of Completeness: Let $\{f_n\}$ be a sequence in $L^2(\chi)$ which is fundamental (Cauchy) in the L^2-norm. Choose $n_1 < n_2 < \ldots$ such that $\| f_{n_i} - f_{n_j} \|_2^2 \le 2^{-i} \forall j \ge i$.

Define $A_j = \{ x \mid |f_{n_{j+1}}(x) - f_{n_j}(x)| \ge 2^{-j/3} \}$.

Set $B = \bigcap_{i \ge 1} \bigcup_{j \ge i} A_j$. B is the set of all those points which belong to infinitely many A_j's.

If $x \in \chi \backslash B$, then $f_{n_j}(x)$ converges as $j \to \infty$, since

$$x \in \chi \backslash B \Rightarrow |f_{n_{j+1}}(x) - f_{n_j}(x)| < 2^{-j/3} \forall j$$

is sufficiently large. Now,

$$\mu[\{x \mid |f(x)| \ge \epsilon\}] \le \epsilon^{-2} \int_\chi |f|^2 \, d\mu \text{ (Tchebyshev's inequality)}.$$

(Note that $\epsilon^{-2} \mid f \mid^2 \geq 1$ whenever $\mid f \mid \geq \epsilon$

$$\therefore \mu[\{x \mid \mid f(x) \mid \geq \epsilon\}] \geq \epsilon^{-2} \int_{\mu[\{x \mid \mid f(x) \mid \geq \epsilon\}]} \mid f \mid^2 d\mu \leq \epsilon^{-2} \int_{\chi} \mid f \mid^2 d\mu)$$

$$\mu(A_j) \leq 2^{2j/3} \parallel f_{n_{j+1}} - f_{n_j} \parallel_2^2 \leq 2^{-j/3} \forall j$$

We now require a lemma:

Lemma 4.8 (Borel-Cantelli lemma) . If $\{B_n\}_{n=1}^\infty$ is a sequence of sets in χ such that $\sum_{n=1}^\infty \mu(B_n) < \infty$, then almost no point $x \in \chi$ belongs to an infinite number of the sets B_n, i.e., .

$$\mu(\bigcap_{n \geq 1} \bigcup_{k \geq n} B_k) = 0$$

Proof:

$$\mu(\bigcap_{n \geq 1} \bigcup_{k \geq n} B_k) \leq \mu(\bigcup_{k \geq n} B_k) \leq \sum_{k \geq n} \mu(B_k) \ \forall n \geq 1$$

Since the right-most term in the above inequality is the tail of a convergent sum, it can be made as small as possible by letting $n \to \infty$. Hence, the lemma.

$\therefore$ By the above lemma, $\mu(B) = 0$.

$\therefore$ $C = \{x \mid f_{n_j}(x)$ fails to converge as $j \to \infty\} \subset B$ is of measure zero.

Now define a measurable function f as follows:

$$f(x) = \begin{cases} 0 & \text{if } x \in C \\ \lim_{j \to \infty} f_{n_j}(x) & \text{otherwise} \end{cases}$$

Claim: $f_n \to f$ as $n \to \infty$ in the L^2 - norm.

Since $\mu(C) = 0$, $f_{n_j} \to f$ pointwise a.e. By Fatou's lemma

$$\parallel f_n - f \parallel_2^2 = \int_\chi \lim_{j \to \infty} \mid f_n - f_{n_j} \mid^2 d\mu \leq \lim_{j \to \infty} \inf \int_\chi \mid f_n - f_{n_j} \mid^2 d\mu =$$

$$\lim_{j \to \infty} \inf \parallel f_n - f_{n_j} \parallel_2^2 \leq 2^{-i} \ \forall n \geq n$$

Since $\parallel f \parallel_2 \leq \parallel f_{n_1} \parallel_2 + \parallel f_{n_1} - f \parallel_2 \leq \parallel f_{n_1} \parallel_2 + 2^{-1/2} < \infty$, we have $f \in L^2(\chi)$.

Further, since $2^{-i} \to 0$ as $i \to \infty$, $f_n \to f$ as $n \to \infty$ in the L^2-norm. This proves that $L^2(\chi)$ is complete.

Theorem 4.9 $L^2(\chi)$ *is separable.*

We shall prove this theorem for the case when χ is a finite/infinite subinterval of **R**.

Proof: It is sufficient to show that every function $f \in L^2(\chi)$ can be approached as closely as desired by functions from a fixed countable family of functions $C \subset L^2(\chi)$ only. Further, it is sufficient to show this for real valued functions f only.

Let f be a real valued function in $L^2(\chi)$.

For any integers i, j, and $k \geq 1$ define

$$f_1 = \begin{cases} f & \text{if } | x | \leq i \\ 0 & if \ | x | > i \end{cases}$$

$$f_2 = \begin{cases} f_1 & \text{if } | f_1 | \leq j \\ 0 & if \ | f_1 | > j \end{cases}$$

and $f_3 = k^{-1}[kf_2]$ where $[f]$ is the largest integer $\leq f$.

f_1, f_2, and f_3 are measurable functions, and

$$\| f - f_1 \|_2^2 = \int_{\chi \cap \{x | |x| > i\}} | f |^2 \, d\mu$$

$$\| f - f_1 \|_2^2 = \int_{\chi \cap \{x | |f_1| > i\}} | f_1 |^2 \, d\mu \leq \int_{\chi \cap \{x | |f| > i\}} | f |^2 \, d\mu$$

$$\| f_2 - f_3 \|_2^2 = \int_{\chi \cap \{x | |f| \leq i\}} k^{-2} \leq 2ik^{-2}$$

and since

$$\| f - f_3 \|_2 \leq \| f - f_1 \|_2 + \| f_1 - f_2 \|_2 + \| f_2 - f_3 \|_2$$

$\| f - f_3 \|_2$ can be made as small as possible by choosing $i < j < k$ sufficiently large.

Consider the family C of piecewise constant functions f with compact support (i.e., the domain over which f is nonzero can be contained completely in a closed interval), taking rational values only, and having discontinuities only at a finite number of rational points.

C is a countable subfamily of $L^2(\chi)$ and is closed under finite rational linear combinations.

Now $f_3 = m/k$ on the set S given by

$$S = \chi \cap \{x \mid |x| \leq i\} \cap \{x \mid m/k \leq f_2(x) < (m+1)/k\}$$

$\therefore$ f_3 is the sum of characteristic functions of sets S with coefficients m/k, where $\mu(S) \leq 2i < \infty$.

Since S is Lebesgue measurable, given $\epsilon > 0$, one can find a countable family of intervals $\{I_n\}$ with rational endpoints such that

$$S \subset \bigcup_{n=1}^{\infty} I_n \text{ and } \sum_{n=1}^{\infty} \mu(I_n) - \mu(S) < \epsilon$$

Thus, the characteristic function of $\bigcup_{n=1}^{\infty} I_n$ is in C. Hence, the theorem.

The above theorems prove that $L^2(\chi)$ is a Hilbert space. Hence, $L^2(\chi)$ has an orthonormal basis.

If $f \in L^2(\chi)$ and $\{e_n\}_{n=1}^{\infty}$ is an orthonormal basis, then the Fourier coefficients of f with respect to it are given by $\hat{f}(n) = \langle f, e_n \rangle$ $n = 1, 2, \ldots$.

Thus, with each $f \in L^2(\chi)$ one can associate a sequence of complex number $(\hat{f}(1), (\hat{f}(2), \ldots)$. Conversely, by the Riesz-Fischer theorem, given an element $(c_1, c_2, \ldots)$ of $t^2(\mathbf{C})$, there exists $f \in L^2(\chi)$ such that $c_n = \{f, e_n\} \forall n \geq 1$. Moreover,

$$\sum_{n=1}^{\infty} |c_n|^2 = \|f\|_2^2$$

By Parseval's theorem, if $\{\hat{f}(n)\}_{n=1}^{\infty}$ are the Fourier coefficients of $f \in L^2(\chi)$ and $\{\hat{g}(n)\}_{n=1}^{\infty}$ are the Fourier coefficients of $g \in L^2(\chi)$, then

$$\langle f, g \rangle = \sum_{n=1}^{\infty} \hat{f}(n) \overline{\hat{g}(n)}$$

$$\therefore \|f\|_2^2 = \sum_{n=1}^{\infty} |\hat{f}(n)|^2 .$$

Denoting $(\hat{f}(1), \hat{f}(2), \ldots)$ by $\hat{f}$, it is clear that $\hat{f} \in l^2(\mathbf{C})$ is the linear space of all square-summable sequences of complex numbers. Indeed, as proved earlier, the map $\wedge : L^2(\chi) \to l^2(\mathbf{C})$ given by $f \mapsto \hat{f}$ is an isometric isomorphism of $L^2(\chi)$ onto $l^2(\mathbf{C})$. i.e.,

$$f, g \in L^2(\chi) \Rightarrow (\alpha f + \beta g)^\wedge = \alpha \hat{f} + \beta \hat{g}, \text{ and } \hat{f} = (0, 0, ...),$$

the zero sequence $\hat{f} \Rightarrow f \equiv 0$ and $\| f \|_2^2 = \| \hat{f} \|_2^2$ where

$$\| \hat{f} \|_2^2 = \sum_{n=1}^{\infty} | \hat{f}(n) |^2$$

is the norm on $l^2(\mathbf{C})$.

Thus, if $\{e_n\}_{n=1}^{\infty}$ is an orthonormal basis of $L^2(\chi)$ and $\hat{f}(n) = \langle f, e_n \rangle$, then the following hold:

(i) $\| f \|_2^2 = \int_\chi | f |^2 \, d\mu \geq \| \hat{f} \|^2 = \sum_{n=1}^{\infty} | \hat{f}(n) |^2$ (Bessel's inequality)

(ii) $\| f \|_2^2 = \| \hat{f} \|^2$ (Parseval's identity)

(iii) $\langle f, g \rangle = \int_\chi f \bar{g} d\mu = \langle \hat{f}, \hat{g} \rangle = \sum_{n=1}^{\infty} \hat{f}(n) \overline{\hat{g}(n)}$ (Parseval's formula)

(iv) $\wedge$ is an isometric isomorphism of $L^2(\chi)$ onto $l^2(\mathbf{C})$ (Riesz-Fischer theorem)

Example of an orthonormal basis of $L^2([0, 1])$.

Consider the functions:

$$e_0^0(x) = \begin{cases} 1 & \text{if } 0 \leq x \leq 1 \\ 0 & \text{otherwise} \end{cases}$$

$$e_n^k(x) = \begin{cases} 2^{n/2} & \text{if } \frac{k-1}{2^n} \leq x < \frac{k-\frac{1}{2}}{2^n} \\ -2^{n/2} & \text{if } \frac{k-\frac{1}{2}}{2^n} \leq x < \frac{k}{2^n} \\ 0 & \text{otherwise} \end{cases}$$

$\{e_n^k\}_{n=1,k=1}^{\infty,\infty}$ form an orthonormal basis of $L^2([0, 1])$, called the Haar basis.

Thus, if $\{e_n\}_{n=1}^{\infty}$ is an orthonormal basis of $L^2(\chi)$, then for any $f \in L^2(\chi)$,

$$f = \sum_{n=1}^{\infty} < f, e_n > e_n = \sum_{n=1}^{\infty} \hat{f}(n) e_n$$

This representation of f is called a Fourier series representation of f.

4.2   Fourier Series

We focus our attention now on periodic functions on $\mathbf{R}$. A function $f :$ $\mathbf{R} \to \mathbf{C}$ is periodic with period T if $f(x + T) = f(x) \ \forall x \in \mathbf{R}$.

The most familiar periodic functions are the circular trigonometric functions: $\sin x$, $\cos x$, $tanx$, etc.. One of the reasons for studying periodic functions on $\mathbf{R}$ is the amazing fact that any square integrable function of a real variable with compact support can be expressed as a superposition of sines and cosines of at most countably many frequencies, *i.e.*, if f is a square integrable function with support contained in the closed interval $[a, b]$, then it can be expressed as an infinite series obtained by linear combinations of the functions

$$1, \sin(\frac{2\pi n x}{b - a}), \cos(\frac{2\pi n x}{b - a}); \quad n = 1, 2, 3, ...,$$

over $[a, b]$. The infinite series however has an independent existence of its own as a function, and being made up of periodic functions whose periods are submultiples of the same period $(b - a)$, it is itself a periodic function of period $(b - a)$, defined over all of $\mathbf{R}$, and is, in fact, a periodic continuation of f onto the whole of $\mathbf{R}$. Thus, the study of square integrable functions on $\mathbf{R}$ with compact support is equivalent to the study of periodic functions that are square integrable on intervals of lengths equal to the corresponding periods.

If f has period T, then by writing $g(x) = f(Tx/2\pi)$ we have a function g of period 2π. The thesis of the following is that every periodic function can be expanded as a linear combination of countably many sines and cosines. These circular trigonometric functions have *natural* period 2π . In order to avoid scaling, we may, without loss of generality, study functions that are periodic with period 2π. In order to do this we study functions defined on the unit circle S^1.

Let S^1 be the unit circle centered at the origin. Let each point p of S^1 be parameterized by the angle θ subtended by the radius vector joining the point p to the origin 0, and the positive direction of the x- axis, the angle θ being measured in the counterclockwise direction.

We may, in fact, identify the points p with their parameter and put $p(\theta) = \theta$ where $0 \leq \theta < 2\pi$. Thus, each point on the circle is identified by the angle subtended by it and the positive x-axis at the origin in a unique fashion.

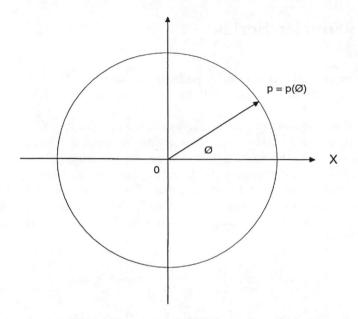

Figure 4.1: Parameterization of the unit circle.

One may, therefore, think of S^1 as the semi-closed interval $[0, 2\pi)$ and functions f on S^1 as periodic functions on $\mathbf{R}$ by putting $f(x + 2\pi) = f(x) \ \forall x \in \mathbf{R}$.

4.2.1   Fourier series of square integrable functions

The space $L^2(S^1)$ of all square-integrable functions f on S^1 is a Hilbert space of complex measurable functions on S^1 with the scalar product given by

$$\langle f, g \rangle = \frac{1}{2\pi} \int_0^{2\pi} f \bar{g} d\mu$$

where μ is the Lebesgue measure on S^1, and the norm induced by this inner product given by

$$\| f \|_2 = \frac{1}{2\pi} \int_0^{2\pi} | f |^2 \, d\mu]^{1/2} < \infty \ \ \forall f \in L^2(S^1)$$

The isomorphism between $L^2([0, 2\pi))$ and $L^2(S^1)$ is obvious.

The functions $\sin x$ and $\cos x$ belong to $L^2(S^1)$.

Consider the function e^{ix} defined by $e^{ix} = \cos x + i \sin x$. Clearly, e^{ix} is periodic of period 2π, $\frac{e^{ix} + e^{-ix}}{2} = \cos x$, and $\frac{e^{ix} - e^{-ix}}{2i} = \sin x$.

Also, $e^{i(x+y)} = e^{ix}e^{iy}$, as can be verified easily.

Consider the functions $\{e^{inx}\}_{x=-\infty}^{\infty}$. Let $e_n(x) = e^{inx}$. Then, $e_n \in L^2(S^1) \forall n \in \mathbf{Z}$.

It is easy to verify that

$$\langle e_n, e_m \rangle = \begin{cases} 1 & \text{if } n = m \\ 0 & \text{otherwise} \end{cases}$$

Indeed, we have the following theorem:

Theorem 4.10 *The family of functions $e_n(x) = e^{inx} \forall n \in \mathbf{Z}$ is an orthonormal basis of $L^2(S^1)$. Hence any function $f \in L^2(S^1)$ can be expanded into a Fourier series*

$$f = \sum_{n=-\infty}^{\infty} \hat{f}(n)e_n$$

with

$$\hat{f}(n) = \langle f, \bar{e}_n \rangle = \int_S f e_n d\mu = \frac{1}{2\pi} \int_0^{2\pi} f(x)e^{-inx} dx$$

The map $f \to \hat{f}$ is an isomorphism of $L^2(S^1)$ onto $l^2_{\leftrightarrow}(\mathbf{C})$, where $l^2_{\leftrightarrow}(\mathbf{C}) = \{(..., c_{-1}, c_0, c_1, ...) \mid c_i \in \mathbf{C} \forall i \in \mathbf{Z}\}$ is the space of all bi-infinite sequences of complex numbers. $l^2_{\leftrightarrow}(\mathbf{C})$ is isomorphic to $l^2(\mathbf{C})$ with the isomorphism given by $(..., c_{-1}, c_0, c_1, ...) \mapsto (c_0, c_{-1}, c_1, ...))$. Moreover,

$$\| f \|_2^2 = \frac{1}{2\pi} \int_0^{2\pi} | f |^2 \, dx = \| \hat{f} \|^2 = \sum_{n=-\infty}^{\infty} | \hat{f}(n) |^2$$

From now on, unless otherwise specified, $e_n(x) = e^{inx}, \forall n \in \mathbf{Z}$ and the term *Fourier series* will refer to Fourier series with respect to this family of basic functions.

Before proceeding further, we need to introduce some important spaces of functions.

The space $C^n(S^1)$ of $n(< \infty)$ times continuously differentiable functions on S^1 is a linear space consisting of functions on S^1, whose first n derivatives are continuous functions on S^1. The space of infinitely differentiable functions on S^1 is denoted $C^\infty(S^1)$. Following the proof of separability of $L^2(\chi)$, it is easy to show that $C^\infty(S^1)$ is dense in $L^2(S^1)$.

In order to prove the above theorem, it is sufficient to show that the Fourier series of a function $f \in C^p(S^1)$ actually converges to f for each $1 \leq p < \infty$. This is shown for $p = 1$ below:

Theorem 4.11 *For any $f \in C^1(S^1)$, the partial sums $S_n = S_n(f) = \sum_{|k|\leq n} \hat{f}(k)e_k$ converge to f uniformly, as $n \to \infty$.*

Proof: Let

$$
\begin{aligned}
D_n(x) &= \sum_{|k|\leq n} e_k(x) \\
&= \sum_{|k|\leq n} e^{ikx} \\
&= e^{-inx} \sum_{k=1}^{2n} e^{ikx} \\
&= e^{-inx} \left(\frac{e^{i(2n+1)x} - 1}{e^{ix} - 1} \right) \\
&= \frac{\sin(2n+1)x/2}{\sin x/2}
\end{aligned}
$$

then

$$
D_n(0) = 2n + 1
$$

D_n is an even function, and

$$
\frac{1}{2\pi} \sum_0^{2\pi} D_n(x)dx = \frac{1}{2\pi} \int_0^{2\pi} \sum_{|k|\leq n} e^{ikx} dx = 1
$$

D_n is called the Dirichlet kernel.

Now,

$$
S_n(x) = \sum_{|k|\leq n} \hat{f}(k)e_k(x) = \sum_{|k|\leq n} e_k(x) \int_0^{2\pi} f(y)\bar{e}_k(y)dy
$$

$$
= \frac{1}{2\pi} \int_0^{2\pi} \sum_{|k|\leq n} e_k(x-y)f(y)dy = \frac{1}{2\pi} \int_0^{2\pi} D_n(x-y)f(y)dy
$$

$$
= \frac{1}{2\pi} \int_{-x}^{2\pi-x} D_n(y)f(x+y)dy
$$

(by substituting $y - x$ by y)

$$= \frac{1}{2\pi} \int_{-\pi}^{+\pi} f(x+y) dy D_n(y) dy$$

(since $f(x+y)D_n(y)$ is periodic of period 2π).

$$\therefore S_n(x) = \frac{1}{2\pi} \int_{-\pi}^{+\pi} f(x+y) D_n(y) dy$$

Since $f \in C^1(S^1)$, the first derivative, f', of f exists and is continuous. Therefore,

$$(f')^\wedge(n) = \frac{1}{2\pi} \int_0^{2\pi} f' \bar{e}_n dx = -\frac{1}{2\pi} \int_0^{2\pi} f \bar{e}'_n dx = in \hat{f}(n)$$

(by integration by parts). $\therefore$ If $n \leq m < \infty$, then

$$| S_n - S_m | \leq \sum_{|k|>n} | \hat{f}(k) | = \sum_{|k|>n} |(f')^\wedge(k)| \, | \, k \, |^{-1}$$

$$\leq \left[\sum_{|k|>n} | (f')^\wedge(k) |^2 \right]^{1/2} [\sum_{|k|>n} k^{-2}]^{1/2}$$

(by Schwartz's inequality)

$$\leq \| f' \|_2 \left[\sum_{|k|>n} k^{-2} \right]^{1/2}$$

(Bessel's inequality)

$$\left[\sum_{|k|>n} k^{-2} \right]^{1/2} = \left[2 \left(\frac{1}{(n+1)^2} + \frac{1}{(n+2)^2} + \dots \right) \right]^{1/2} = \left[\frac{2}{n} \sum_{r=1}^{\infty} \frac{1/n}{(1+r/n)^2} \right]^{1/2}$$

$$\leq \left[\frac{2}{n} \int_1^\infty \frac{dx}{(1+x)^2} \right]^{1/2} = \left[\frac{1}{n} \right]^{1/2}$$

$$| s_n - s_m | \leq \| f' \|_2 \, n^{-1/2}$$

This implies that S_n converges uniformly. It is sufficient to show that $S_n \to f$ as $n \to \infty$. Since

$$\frac{1}{2\pi} \int_{-\pi}^{+\pi} D_n(x) dx = 1$$

$$S_n(x) - f(x) = \frac{1}{2\pi} \int_{-\pi}^{+\pi} [f(x+y) - f(x)] D_n(y) dy$$

$$= \frac{1}{2\pi} \int_{-\pi}^{+\pi} Q(x,y) \sin((2n+1)y/2) dy$$

where

$$Q(x,y) = \frac{[f(x+y) - f(x)]}{\sin(y/2)}$$

and is equal to $2f'(x)$ at $y = 0$. Fix $-\pi \le x < \pi$ as a function of $y \in (-\pi, \pi)$, $Q(x,y) \in L^2(-\pi, \pi)$, and

$$S_n(x) - f(x) = \frac{1}{2\pi} \int_{-\pi}^{+\pi} Q(x,y) \frac{e^{iny} e^{iy/2} - e^{-iny} e^{-iy/2}}{2i} dy$$

$$= (4\pi i)^{-1} \left[(Q^+)^\wedge(-n) - (Q^-)^\wedge(n) \right]$$

where $Q^+ = e^{iy/2}$, and $Q^- = e^{-iy/2}$.

By Bessel's inequality,

$$\sum_{n=-\infty}^{\infty} |(Q^{\pm})^\wedge(n)|^2 \le \|Q\|_2^2 < \infty$$

$$\therefore (Q^{\pm})^\wedge(n) \to 0 \text{ as } n \to \infty$$

Hence, $\lim_{n \to \infty} S_n = f$ for each fixed $x \in (-\pi, \pi)$. Letting $m \to \infty$ in the estimate

$$|S_n - S_m| \le \|f'\|_2 \, n^{-1/2}, \quad n \le m < \infty$$

we have

$$|S_n - f| \le Cn^{-1/2} \forall x \in [-\pi, \pi)$$

$\therefore S_n \to f$ uniformly as $n \to \infty$.

Corollary 4.11.1 Since the linear space $C^1(S^1)$ is dense in $L^2(S^1)$ ($\because C^\infty(S^1) \subset C^1(S^1)$) the orthogonal system $\{e_n \mid e_n(x) = e^{inx} \forall n \in Z\}$ is a basis for $L^2(S^1)$; hence the earlier theorem.

Remark: In order to generalize the above theorem's proof observe that

$$(f^{(p)})^\wedge(n) = (in)^p \hat{f}(n)$$

This results in the bound $\|S_n - F\|_\infty \le Cn^{-p+1/2}$ for some constant C, (where $\|f\|_\infty = ess. \sup_{x \in X} |f| \ge f$ a.e.).

Theorem 4.12 For $f \in C(S^1)$ (continuous functions on S'), the arithmetic means $n^{-1}(S_0 + \ldots + S_{n-1})$ of the partial sums $S_n = \sum_{|k| \le n} \hat{f}(k) e_k$ converge uniformly to f.

Proof: Observe that if $\{x_n\}_{n=0}^{\infty}$ is a sequence converging to y then the sequence $\{(x_0 + ... + x_{n-1})n^{-1}\}_{n=1}^{\infty}$ also converge to y.

Indeed, given $\epsilon > 0, \exists N_0 \in \mathbf{N} \ni |x_n - y| < \epsilon \forall n \geq n_0$.

Let $n > n_0$, then

$$|\frac{x_0 + ... + x_{n-1}}{n} - y| = |\frac{1}{n} \sum_{i=0}^{n-1}(x_i - y)|$$

$$\leq \frac{1}{n} \sum_{i=0}^{n-1}|x_i-y| = \frac{1}{n} \sum_{i=0}^{n_0-1}|x_i-y|+\frac{1}{n} \sum_{i=n_0}^{n}|x_i-y| < \frac{1}{n} \sum_{i=0}^{n_0-1}|x_i-y|+\frac{\epsilon(n - n_0 + 1)}{n}$$

$$= \frac{C(\epsilon)}{n} + \frac{\epsilon(n - n_0 + 1)}{n}$$

where $C(\epsilon)$ is a constant dependent on ϵ.

Now, letting $n \to \infty$ we see that

$$|\frac{x_0 + ... + x_{n-1}}{n} - y| < \epsilon$$

where ϵ is arbitrary.

$$\therefore \{(x_0 + ... + x_{n-1})n^1\}_{n=1}^{\infty}\} \mapsto y$$

Let $F_n(x) = \frac{1}{n}(D_0(x) + ... + D_{n-1}(x))$, where $D_i(x)$ is the i-th Dirichlet kernel. F_n is called the Fejer kernel .

$$\therefore F_n(x) = \frac{1}{n} \sum_{i=0}^{n-1} \frac{\sin(2k + 1)x/2}{\sin x/2} = \frac{1}{n} \left[\frac{\sin nx/2}{\sin x/2}\right]^2 \left(\text{using } D_n(x) \sum_{|k|\leq n} e_k(x)\right)$$

$$F_n \geq 0 \text{ and } \frac{1}{2\pi} \int_{-\pi}^{+\pi} F_n(x)dx = n^{-1} \sum_{k=0}^{n-1} \int_{-\pi}^{+\pi} D_n(x)dx = 1$$

To prove the theorem, it is sufficient to check that

$$|n^{-1} \sum_{k=0}^{n-1} S_k(x) - f(x)|$$

tends to zero uniformly as $n \to \infty$.

$$n^{-1} \sum_{k=0}^{n-1} S_k(x) - f(x) = n^{-1} \sum_{k=0}^{n-1} \frac{1}{2\pi} \int_{-\pi}^{+\pi} f(x + y)D_k(y)dy - f(x) =$$

$$\frac{1}{2\pi}\int_{-\pi}^{+\pi} f(x+y)F_n(y)dy - f(x) =$$

$$\frac{1}{2\pi}\int_{-\pi}^{+\pi} [f(x+y) - f(x)]F_n(y)dy$$

Let $0 < \epsilon < \pi$ be a small positive number, then

$$\frac{1}{2\pi}\int_{-\pi}^{+\pi} [f(x+y) - f(x)]F_n(y)dy$$

$$\frac{1}{2\pi}\int_{|y|<\epsilon} [f(x+y) - f(x)]F_n(y)dy + \frac{1}{2\pi}\int_{\pi\geq|y|\geq\epsilon} [f(x+y) - f(x)]F_n(y)dy$$

The first integral on the right hand side is bounded as below:

$$|\frac{1}{2\pi}\int_{|y|<\epsilon} [f(x+y) - f(x)]F_n(y)dy|$$

$$\leq \frac{1}{2\pi}\int_{|y|<\epsilon} |f(x+y) - f(x)|F_n(y)dy$$

$$\leq \max_{|\leq\epsilon} \max_{|x|<\pi} |[f(x+y) - f(x)]|\frac{1}{2\pi}\int_{|y|<\epsilon} F_n(y)dy$$

$$\leq \max_{|y|\leq\epsilon} \| f_y - f \|_\infty \quad \text{where} \quad \| f \|_\infty = ess.\sup_{x\in\chi} |f| \geq |f| \text{ a.e.}$$

is the essential supremum norm of f and $f_y(x) = f(x+y)$.

$\| f_y - f \|_\infty \to$ as $\epsilon \to 0$ since f is continuous (and, hence, uniformly continuous since S' is compact). The second integral on the right hand side is bounded as below:

$$|\frac{1}{2\pi}\int_{\pi\geq|y|\geq\epsilon} [f(x+y) - f(x)]F_n(y)dy|$$

$$\leq \frac{4}{n}\|f\|_\infty \frac{1}{2\pi}\int_\epsilon^\pi \left[\frac{\sin nx/2}{\sin x/2}\right]^2 dx \leq \frac{2}{n}(\sin(\epsilon/2))^{-2}\|f\|_\infty$$

for fixed $0 < \epsilon < \pi$. This can be made arbitrarily small by taking n sufficiently large.

4.2.2   Fourier series of absolutely integrable functions

Unlike the Fourier series of functions in $L^2(S^1)$, the Fourier series of functions in $L^1(S^1)$ don't always converge. In fact, there exists a function in

$L^1(S^1)$ whose Fourier series diverges everywhere on **R**. However, convergence is valid: The Cesaro means of partial sums of the Fourier series of a function in $L^1(S^1)$ converges to the function in the L^1-norm.

$$L^1(S^1) = \left\{ f \mid \frac{1}{2\pi} \int_0^{2\pi} |f| \, dx < \infty \right\}$$

is a normed linear space with the norm

$$\| f \|_1 = \frac{1}{2\pi} \int_0^{2\pi} |f(x)| \, dx$$

We have seen before that if $\mu(\chi) < \infty$, then $L^2(\chi) \subset L^1(\chi)$. Since $\mu(S^1) = 1$, $L^2(S^1) \subset L^1(S^1)$. This inclusion is proper. Indeed, if

$$f(x) = x^{-1/2} \text{ then } \frac{1}{2\pi} \int_0^{2\pi} |f| \, dx = \frac{1}{2\pi} \int_0^{2\pi} x^{-1/2} dx = \sqrt{\frac{2}{\pi}}$$

but

$$\frac{1}{2\pi} \int_0^{2\pi} |f|^2 \, dx = \frac{1}{2\pi} \int_0^{2\pi} \frac{1}{x} dx = \infty$$

$$\therefore f(x) = x^{-1/2} \in L^1(S^1) \text{ but } \notin L^2(S^1)$$

If $f \in L^1(S^1)$, then $f\bar{e}_n$ is absolutely integrable. One can, therefore, construct a formal Fourier series for f as follows:

$$S(f) = \sum \hat{f}(n) e_n$$

with

$$\hat{f}(n) = \frac{1}{2\pi} \int_0^{2\pi} f\hat{e}_n dx = \frac{1}{2\pi} \int_0^{2\pi} f(x) e^{-inx} dx$$

Theorem 4.13. *The arithmetic (Cesaro) means* $n^{-1}(S_0 + ... + S_{n-1})$ *of the partial sums*

$$S_n = \sum_{|k| \leq n} \hat{f}(k) e_k$$

of $S(f)$ for $f \in L^1(S^1)$ converges to f in the L^1-norm, i.e., $\lim_{n \to \infty} n^{-1}(S_0 + ... + S_{n-1} - f \|_1 = 0$. *In particular, the map $f \mapsto \hat{f}$ where $\hat{f} = (..., \hat{f}(-1), \hat{f}(0), \hat{f}(1), ...)$ is one-to-one.*

Proof:

$$n^{-1}(S_0 + ... + S_{n-1}) - f = \frac{1}{2\pi} \int_{-\pi}^{+\pi} [f(x+y) - f(x)] F_n(y) dy$$

$$\| n^{-1}(S_0+...+S_{n-1})-f \| \le \frac{1}{2\pi} \int_{-\pi}^{+\pi} \frac{1}{2\pi} \int_{-\pi}^{+\pi} | f(x+y)-f(x) | F_n(y)dydx =$$

$$\frac{1}{2\pi} \int_{-\pi}^{+\pi} \left(\frac{1}{2\pi} \int_{-\pi}^{+\pi} | f(x+y) - f(x) | dx \right) F_n(y)dy =$$

$$\frac{1}{2\pi} \int_{-\pi}^{+\pi} \| f_y - f \|_1 F_n(y)dy \text{ , where } f_y(x) = f(x+y)$$

Now,

$$\lim_{y \to 0} \| f_y - f \|_1 = 0$$

Indeed,

$$\frac{1}{2\pi} \int_0^{+\pi} | f_y(x) - f(x) | dx \le \max_{x \in [0,2\pi]} | f_y(x) - f(x) | \text{ for } f \in C(S^1)$$

(*i.e.*, f is uniformly continuous on S^1)

$$= \| f_y - f \|_\infty$$

and $\| f_y - f \|_\infty \to 0$ uniformly as $y \to 0$.

Since $C(S^1)$ is done in $L^1(S^1)$, this holds for $f \in L^1(S^1)$ also. The rest of the proof is the same as that of the previous theorem.

In general, it is not true that $\sum | \hat{f}(n) |^2 < \infty$ for $\in L^1(S^1)$. In fact, it is true only if $f \in L^2(S^1)$, but each Fourier coefficient is bounded:

$$| \hat{f}(n) | = | \frac{1}{2\pi} \int_0^{2\pi} f \hat{e}_n dx | \le \frac{1}{2\pi} \int_0^{2\pi} | f | dx = \| f \|_1$$

In fact, the Riemann-Lebesgue Lemma gives more: **Lemma 4.14** (Riemann-Lebesgue). The Fourier coefficient $\hat{f}(n)$ of any $f \in L^1(S^1)$ tends to 0 as $| n | \to \infty$.

Proof:

$$\hat{f}(n) = -\frac{1}{2\pi} \int_0^{2\pi} f(x)e^{-in(x-\pi/n)}dx = -\frac{1}{2\pi} \int_0^{2\pi} f(x+\pi/n)e^{-inx}dx$$

$$\therefore \hat{f}(n) = \frac{1}{2} \left\{ \frac{1}{2\pi} \int_0^{2\pi} f(x)e^{-inx}dx - \frac{1}{2\pi} \int_0^{2\pi} f(x+\pi/n)e^{-inx}dx \right\}$$

$$= \frac{1}{2} \cdot \frac{1}{2\pi} \int_0^{2\pi} [f(x) - f(x+\pi/n)]e^{-inx}dx$$

$$\therefore | \hat{f}(n) | \le \frac{1}{2} \cdot \frac{1}{2\pi} \int_0^{2\pi} | f(x) - f_{\pi/n}(x) | dx = \frac{1}{2} \| f - f_{\pi/n} \|_1$$

As shown in the proof of the previous theorem,

$$\| f - f_{\pi/n} \|_1 \to 0 \text{ as } | n | \to \infty$$

4.2.3 The convolution product on $L^1(S^1)$

If $f, g \in L^1(S^1)$, define

$$f * g = \frac{1}{2\pi} \int_0^{2\pi} f(x-y)g(y)dy$$

(The convolution of f with g) To check if this "product" makes sense, consider

$$I = \int_0^{2\pi} \int_0^{2\pi} | f(x-y)g(y) | \, dxdy$$

$| f(x-y)g(y) |$ is a nonnegative measurable function of two variables over a finite domain. $\therefore I \leq \infty$. Therefore, by Fubini's theorem :

$$I = \int_0^{2\pi} |g(y)|dy \int_0^{2\pi} |f(x-y)|dx = \|g\|_1 \|f\|_1 < \infty$$

$$\therefore \|f * g\|_1 \leq \frac{1}{2\pi} I < \infty$$

The convolution product is easily checked to be associative and commutative, *i.e.*,

$$(f * g) * h = f * (g * h) \text{ and } f * g = g * f$$

If $f \in L^1(S^1)$ and $g \in L^2(S^1)$, then $f * g \in L^2(S^1)$. In the light of the convolution product, it is easily seen that

$$S_n(f) = f * D_n, n^{-1} \sum_{n=0}^{n-1} S_k(f) = f * F_n$$

If $\hat{f} = (..., \hat{f}(-1), \hat{f}(0), \hat{f}(1), ...)$ and $f, g \in L^1(S^1)$ then $(f * g) = \hat{f}\hat{g}$.

Indeed,

$$
\begin{aligned}
(f * g)^{\wedge}(n) &= \frac{1}{2\pi} \int_0^{2\pi} \left[\frac{1}{2\pi} \int_0^{2\pi} f(x-y)g(y)dy \right] e^{-inx} dx \\
&= \frac{1}{2\pi} \int_0^{2\pi} \frac{1}{2\pi} \int_0^{2\pi} f(x-y)e^{-in(x-y)} g(y)e^{-iny} dxdy \\
&= \frac{1}{2\pi} \int_0^{2\pi} g(y)e^{-iny} dy \left[\frac{1}{2\pi} \int_0^{2\pi} f(x-y)e^{-in(x-y)} dx \right] \\
&= \frac{1}{2\pi} \int_0^{2\pi} g\bar{e}_n dx \frac{1}{2\pi} \int_0^{2\pi} f\bar{e}_n dx = \hat{f}(n)\hat{g}(n)
\end{aligned}
$$

For $f, g, h \in L^1(S^1)$ we have:

(i) $\alpha f + \beta g \in L^1(S^1) \forall \alpha, \beta \in C$

(ii) $f * g \in L^1(S^1)$

(iii) $f * g = g * f$

(iv) $(f * g) * h = f * (g * h)$

Thus $L^1(S^1)$ is an *Algebra* under $*$. However, $L^1(S^1)$ does not have a multiplication identity e such that $e * f = f * e = f$, for this would imply

$$(e * f)^\wedge(n) = \hat{e}(n)\hat{f}(n) = \hat{f}(n) \, i.e., \, \hat{e}(n) = 1 \forall n \in \mathbf{Z}$$

which is contrary to the Riemann-Lebesgue lemma which states

$$\hat{f}(n) \to 0 \text{ as } \mid n \mid \to \infty \forall f \in L^1(S^1)$$

So far the discussion of Fourier analysis has been restricted to the study of Fourier series of functions on S^1 or periodic functions of period 2π on **R**, which is the same. We now direct our attention to certain aperiodic functions of **R**. In particular, the Fourier analysis of functions of $L^1(\mathbf{R})$ and $L^2(\mathbf{R})$. The functions in $L^1(\mathbf{R})$ and $L^2(\mathbf{R})$ are very different from those in $L^1([a, b])$ and $L^2([a, b])$, hence the Fourier analysis of these function spaces is bound to be very different from what we have seen so far. However, we will begin with an informal discussion which illustrates the analogy between Fourier transforms and Fourier coefficients.

4.3   Fourier Transforms

Fourier series of functions on S^1 may be thought of as the expression of a periodic function of period 2π in terms of simple harmonics of the same period. The choice of the period is just a matter of convenience. Functions with an arbitrary finite period T can be treated similarly.

Indeed, the space $L^2([0, T])$ of square-integrable functions of period T on **R** has as the appropriate basis of harmonics given by $e_n(x) = e^{2\pi i n/T}$ with the norm

$$\|f\|_2 = \frac{1}{T} \int_0^T \mid f \mid^2 dx \text{ , and } \hat{f}(n) = \frac{1}{T} \int_0^T f(x)e^{-2\pi i n x/T} dx$$

with

$$f(x) = \sum_{-\infty}^{\infty} \left[\frac{1}{T} \int_0^T f(y)e^{-2\pi i n y/T} dy \right] e^{2\pi i n x/T}$$

The expression

$$f(x) = \sum_{-\infty}^{\infty} \left[T^{-1} \int_{-T/2}^{T/2} f(y)e^{-2\pi iny/T} dy \right] e^{2\pi inx/T} = \sum_{-\infty}^{\infty} \hat{f}(n)e_n(x)$$

for a continuous periodic function f of period T can be looked upon as a Riemann sum over $\mathbf{R}$ with subdivisions of width $2\pi/T$.

We can look upon each periodic function f of period T as a waveform with 'wavelength' equal to the period T. We can assume that all waveforms travel with the same 'velocity', equal to that of a waveform representing a circular function of period 2π. We further normalize this velocity by accepting the time taken by this waveform to travel distance 2π as unit time. Then, the quantity $2\pi/T$ gives the relative ('circular') frequency of a waveform of period (wavelength) T with respect to the waveform of period 2π. The quantities $\frac{n \cdot T}{2\pi}$ give the frequencies of higher harmonics of this waveform.

An aperiodic function f on $\mathbf{R}$ can be looked upon as a function with 'infinite' period (i.e., $T = \infty$ and, therefore, zero frequency)

In the expression of f as a Fourier series, if we let $\frac{2\pi}{T} = \Delta\omega$ and $n\Delta\omega = \omega_n$, then

$$f(x) = \sum_{-\infty}^{\infty} \left[\int_{-\pi/\Delta\omega}^{-\pi/\Delta\omega} f(y)e^{-i\omega_n y} dy \right] \Delta\omega . e^{-i\omega_n x}$$

where

$$\left[\int_{-\pi/\Delta\omega}^{-\pi/\Delta\omega} f(y)e^{-i\omega_n y} dy \right] \Delta\omega$$

is the approximate contribution of waveforms of frequencies ω with

$$\omega_n - \frac{\Delta\omega}{2} \leq \omega \leq \omega_n + \frac{\Delta\omega}{2}$$

to the amplitude of f over $\mathbf{R}$.

For an aperiodic function, let $\Delta\omega = 2\pi/T \to 0$ ($\because T \to \infty$).

We have under 'proper' conditions,

$$f(x) = \frac{1}{2\pi} \int_{-\infty}^{\infty} \left(\int_{-\infty}^{\infty} f(y)e^{-i\omega y} dy \right) e^{-i\omega x} d\omega$$

i.e., the Riemann sum in the limit is the above integral. The quantity in parentheses in the above 'formula' is a function of ω and is the component of f with frequency ω, contributing to f. Designating

$$\hat{f}(x) = \frac{1}{\sqrt{2\pi}} \int_{-\infty}^{\infty} f(y)e^{-i\omega y}dy$$

("Fourier transform" of f), we have

$$(\hat{f})^{\vee}(x) = f(x) = \frac{1}{\sqrt{2\pi}} \int_{-\infty}^{\infty} \hat{f}(\omega)e^{i\omega x}dx$$

("inverse Fourier transformation" of f). The quantity $\hat{f}(\omega)$ is the "Fourier coefficient" of f corresponding to the complex waveform $e^{i\omega x}$. $\hat{f}(\omega)$ is defined for each $\omega \in \mathbf{R}$. However, it does not make sense to talk about Fourier series with respect to trigonometric functions for functions in $L^1(\mathbf{R})$ or $L^2(\mathbf{R})$, since, the functions $e^{i\omega x}$ are not in $L^1(\mathbf{R})$ or $L^2(\mathbf{R})$, hence any nontrivial linear combination of such functions is also not in these spaces. Also, one requires uncountably many frequencies to describe the spectra of functions in $L^1(\mathbf{R})$ or $L^2(\mathbf{R})$. Nevertheless, the quantity $\hat{f}(\omega)$ as a function of ω describes the spectral characteristics of f completely, and is called the Fourier transform of f.

The above informal discussion will be dealt with rigorously in the following.

Definition 4.15 *The class of infinitely differentiable, rapidly decreasing function on $\mathbf{R}$ is denoted $C_{\downarrow}^{\infty}(\mathbf{R})$.*

More precisely, $C_{\downarrow}^{\infty}\mathbf{R} =$

$$\{f \mid \frac{d^n f}{dx^n}\} \text{ exists } \forall n \in \mathbf{N} \text{ and } \forall n \in \mathbf{R} \text{ and}$$

$$x^p \frac{d^q f}{dx^q} \to 0 \text{ as } \mid x \mid \to \infty \forall p, q \geq 0 \in \mathbf{N}\}$$

Theorem 4.16 $C_{\downarrow}^{\infty}(\mathbf{R})$ *is dense in both* $L^1(\mathbf{R})$ *and* $L^2(\mathbf{R})$.

Proof: It was shown earlier that the family C of piecewise constant functions f vanishing outside a compact set (closed interval), taking only rational values and having discontinuities at only a finite number of rational points, is dense in $L^2(\mathbf{R})$ (the same is true for $L^1(\mathbf{R})$). One only has to smooth out the discontinuous 'jumps' in an infinitely differentiable way. A prototype function for doing this can be constructed using the function

$$f(x) = \begin{cases} 0 & \text{if } x \leq 0 \\ e^{-\frac{1}{x^2}} & \text{if } x > 0 \end{cases}$$

Exercise: Construct an infinitely differentiable function g with compact support such that

$$g(x) = \begin{cases} 0 & \text{if } |x| \geq 1 \\ 1 & \text{if } |x| \leq \frac{1}{2} \end{cases}$$

(Hint: use the above defined function f). For a function $f \in L^2(\mathbf{R})$, the integral (Fourier transform of f)

$$\hat{f}(\omega) = \frac{1}{\sqrt{2\pi}} \int_{-\infty}^{\infty} f(x)e^{-i\omega x}\,dx$$

does not make sense outright, for the integral may not converge. This problem does not occur if $f \in L^1(\mathbf{R})$, but $\hat{f}$ may not be in $L^1(\mathbf{R})$, in which case one cannot invert $\hat{f}$ to obtain f. Since this problem does not arise for $f \in C^\infty_\downarrow(\mathbf{R})$, and $C^\infty_\downarrow(\mathbf{R})$ is dense in $L^1(\mathbf{R})$ and $L^2(\mathbf{R})$, we will define Fourier transforms of functions in $L^1(\mathbf{R})$ and $L^2(\mathbf{R})$ through Fourier transforms of functions in $C^\infty_\downarrow(\mathbf{R})$.

Theorem 4.17 *The Fourier transform* $\wedge$ *on* $C^\infty_\downarrow(\mathbf{R})$ *is defined by*

$$\hat{f}(\omega) = \frac{1}{\sqrt{2\pi}} \int_{-\infty}^{\infty} f(x)e^{-i\omega x}\,dx$$

which maps $C^\infty_\downarrow(\mathbf{R})$ *onto itself, and the inverse Fourier transform* $\vee$ *on* $C^\infty_\downarrow(\mathbf{R})$ *defined by*

$$\vee g(x) = \frac{1}{\sqrt{2\pi}} \int_{-\infty}^{\infty} g(\omega)e^{i\omega x}\,d\omega$$

satisfies

$$(\hat{f})^\vee = f$$

Furthermore, Parseval's identity $\|f\|_2 = \|\hat{f}\|_2$ *holds.*

Proof: Let $f \in C^\infty_\downarrow(\mathbf{R})$. Integration by parts yields:

$$\begin{aligned} (f')^\wedge(\omega) &= \frac{1}{\sqrt{2\pi}} \int_{-\infty}^{\infty} f'(x)e^{-i\omega x}\,dx \\ &= -\frac{1}{\sqrt{2\pi}} \int_{-\infty}^{\infty} f(x)(e^{-i\omega x})'\,dx = i\omega \hat{f}(\omega) \end{aligned}$$

Since f is rapidly decreasing,

$$\begin{aligned} (-ixf(x))^\wedge &= \frac{1}{\sqrt{2\pi}} \int_{-\infty}^{\infty} -ixf(x)e^{-i\omega x}\,dx \\ &= \frac{d}{d\omega} \frac{1}{\sqrt{2\pi}} \int_{-\infty}^{infty} f(x)e^{-i\omega x}\,dx = (\hat{f})' \end{aligned}$$

Therefore, by induction we have

$$(i\omega)^p \frac{d^q \hat{f}}{d\omega^q} = \left(\frac{d^p}{dx^p} (-ix)^q f \right)^\wedge \quad \forall \text{ integers } p, q \geq 0$$

$$\therefore |\omega|^p \left| \frac{d^q \hat{f}}{d\omega^q} \right| \leq \left\| \frac{d^p x^q f}{dx^p} \right\|_1 < \infty \quad \forall \text{ integers } p, q \geq 0$$

$$\therefore \hat{f} \in C_\downarrow^\infty(\mathbf{R})$$

If f has compact support (i.e., if the set of values for which f is nonzero can be enclosed in a closed and bounded subset of $\mathbf{R}$), then without loss of generality, assume that it is nonzero only within the closed interval $[-T/2, T/2]$. Then,

$$f(x) = \sum_{n=-\infty}^{\infty} \frac{1}{T} \int_{-T/2}^{T/2} f(y) e^{-2\pi i n y/T} dy \, e^{2\pi i n x/T}$$

Let

$$g(2\pi n/T) = \int_{-T/2}^{T/2} f(y) e^{-2\pi i n y/T} dy$$

$$\therefore f(x) = \sum_{n=-\infty}^{\infty} \frac{1}{T} g(2\pi n/T) e^{2\pi i n x/T} \quad (*)$$

letting $2\pi/T = \Delta\omega$ and $2\pi n/T = \omega_n \in \frac{2\pi}{T}\mathbf{Z}$ we have

$$f(x) \to \frac{1}{2\pi} \int_{-\infty}^{\infty} g(\omega) e^{i\omega y} \text{ as } T \to \infty, \quad (**)$$

where

$$g(\omega) = \int_{-\infty}^{\infty} f(y) e^{-i\omega y} dy = \sqrt{2\pi} \hat{f}(\omega)$$

i.e., $f(x) = (\hat{f})^\vee(x)$. Provided the above sum (*) converges to the integral (**), as $T \to \infty$.

Similarly,

$$\|f\|_2^2 = \frac{1}{T} \int_{-T/2}^{T/2} |f|^2 \, dx = \sum_{n=-\infty}^{\infty} T^{-1} |g(2\pi n/T)|^2$$

leads to Parseval's identity

$$\| f \|_2 = \| \hat{f} \|_2 = \left(\int_{-\infty}^{\infty} |\hat{f}(\omega)|^{1/2} \cdot d\omega \right) \text{ as } T \to \infty$$

if the sum

$$\sum_{n=-\infty}^{\infty} T^{-1} \mid g(2\pi n/T) \mid^2$$

converges to the integral

$$\int_{-\infty}^{\infty} \mid g(\omega) \mid^2 d\omega \text{ as } T \to \omega$$

The convergence of the sums to the respective integrals is easily shown in view of the fact $\hat{f} \in C_{\downarrow}^{\infty}(\mathbf{R}) \Rightarrow \mid \hat{f}(\omega) \mid \leq c/(1 + \omega^2)$.

In the case f is not a function with compact support, choose $u(x) \in C_{\downarrow}^{\infty}(\mathbf{R})$ such that

$$u(x) = \begin{cases} 1 & \text{if } \mid x \mid < 1/2 \\ 0 & \text{if } \mid x \mid \geq 1 \end{cases}$$

Letting $f_n(x) = f(x)u(x/n)$, we see that $f_n \to f$ as $n \to \infty$.

Then $\| \hat{f} - \hat{f}_n \|_\infty \leq \| f - f_n \|_1 \leq \int_{|x|>n/2} \mid f(x)dx \mid dx \to 0$ as $n \to \infty$ since $f \in C_{\downarrow}^{\infty}(\mathbf{R})$.

The Fourier transforms $\hat{f}_n \forall n$ and $\hat{f}$ are dominated by a function from $L^1(\mathbf{R}) \cap L^2(\mathbf{R})$:

$$\mid \hat{f}_n \mid \leq \| f_n \|_1 < \infty \text{ for } \mid \omega \mid \leq 1$$

while for

$$\mid \omega \mid \geq 1, \ \mid \hat{f}_n \mid = \omega^{-2} \mid (f_n'')^\wedge \mid \leq \omega^{-2} \mid f_n'' \mid_1$$

$$\leq \omega^{-2} \| f''(x)u(x/n) + 2n^{-1} f'(x)u'(x/n) + n^{-2} f(x)u''(x/n) \|_1$$

$$\leq \omega^{-2} [\| f''(x) \|_1 \| u \|_\infty + 2n^{-1} \| f' \|_1 \| u' \|_\infty + n^{-2} \| f \|_1 \| u'' \|_1] \leq C\omega^{-2} \forall n$$

The domination from $L^1(\mathbf{R})$, yields the Fourier inversion formula:

$$\begin{aligned} f(x) &= \lim_{n \to \infty} f_n(x) \\ &= \lim_{n \to \infty} \frac{1}{\sqrt{2\pi}} \int_{-\infty}^{\infty} \hat{f}_n(\omega)e^{i\omega x} d\omega \\ &= \frac{1}{\sqrt{2\pi}} \int_{-\infty}^{\infty} \hat{f}_n(\omega)e^{i\omega x} d\omega = (\hat{f})^\vee(x), \end{aligned}$$

whereas the domination from $L^2(\mathbf{R})$, yields the Parseval identity:

$$\| f \|_2 = \lim_{n \to \infty} \| f \|_2 = \lim_{n \to \infty} \| \hat{f}_n \|_2 = \| \hat{f}_n \|_2$$

4.3.1   Fourier transforms of functions in $L^2(\mathbf{R})$

If $f \in L^2(\mathbf{R})$, choose $f_n \in C_{\downarrow}^{\infty}(\mathbf{R})$ such that $\lim_{n\to\infty} \| f_n - f \|_2 = 0$. By Parseval's identity for $C_{\downarrow}^{\infty}(\mathbf{R})$,

$$\| \hat{f}_n - \hat{f}_m \|_2 = \| (f_n - f_m)^{\wedge} \|_2 = \| (f_n - f_m) \|_2 \leq \| (f_n - f) \|_2 + \| (f_m - f) \|_2$$

Therefore, $\{\hat{f}_n\}$ is a convergent sequence in the L^2 norm.

Define $\hat{f} = \lim_{n\to\infty} \hat{f}_n$. $\hat{f}$ is well defined and is independent of the choice of the approximating function f_n .

$$\wedge : L^2(\mathbf{R}) \to L^2(\mathbf{R}) \text{ given by } \hat{f}(\omega) = \lim_{n\to\infty} \frac{1}{\sqrt{2\pi}} \int_{-\infty}^{\infty} f_n(x) e^{-i\omega x} dx,$$

where $\{f_n\} \subset C_{\downarrow}^{\infty}(\mathbf{R})$ such that $\lim_{n\to\infty} \| f_n - f \|_2 = 0$ is a linear map.

Parseval's identity for $L^2(\mathbf{R})$ is obvious since:

$$\| f \|_2 = \lim_{n\to\infty} \| f_n \|_2 = \lim_{n\to\infty} \| \hat{f}_n \|_2 = \| \hat{f} \|_2$$

This implies that $\wedge$ is a 1:1 length-preserving map of $L^2(\mathbf{R})$ into itself.

The map $\vee$ is extended in the same manner from $C_{\downarrow}^{\infty}(\mathbf{R})$ to $L^2(\mathbf{R})$.

$$\therefore \vee : L^2(\mathbf{R}) \to L^2(\mathbf{R}) \text{ given by } \check{f}(x) = \lim_{n\to\infty} \frac{1}{\sqrt{2\pi}} \int_{-\infty}^{\infty} f_n(\omega) e^{i\omega x} d\omega$$

where $\{f_n\} \subset C_{\downarrow}^{\infty}(\mathbf{R})$ such that $\lim_{n\to\infty} \| f_n - f \|_2 = 0$, is a linear map.

$$(\hat{f})^{\vee} = (\lim_{n\to\infty} f_n)^{\wedge \circ \vee} = (\lim_{n\to\infty} \hat{f}_n)^{\vee} = \lim_{n\to\infty} \left((\hat{f}_n)^{\vee} \right) = \lim_{n\to\infty} f_n = f = (f^{\vee})^{\wedge} = f^{\vee \circ \wedge}$$

Thus, $\wedge$ and $\vee$ are isometric isomorphisms of $L^2(\mathbf{R})$ onto itself, and $\wedge \circ \vee = \vee \circ \wedge = $id is the identity map.

4.3.2   Fourier transforms of functions in $L^1(\mathbf{R})$

Unlike in the case of $L^2(\mathbf{R})$, the Fourier transform in $L^1(\mathbf{R})$ is not an isometric isomorphism of $L^1(\mathbf{R})$ onto itself, and the inverse Fourier transform may not be defined always. The basic facts of the Fourier transform in $L^1(\mathbf{R})$ are contained in the following theorem.

Theorem 4.18 *The Fourier transform $\hat{f}(\omega) = \frac{1}{\sqrt{2\pi}} \int_{-\infty}^{\infty} f(x)e^{-i\omega x}dx$, for any, $f \in L^1(\mathbf{R})$, exists as an ordinary Lebesgue integral with the following properties*

(a) $\|\hat{f}\|_{\infty} \leq \|f\|_1 (2\pi)^{-1/2}$;

(b) $\hat{f} \in C(\mathbf{R})$: the space of continuous functions ;

(c) $\|f\|_2 = \lim_{|\omega| \to \infty} \hat{f}(\omega) = 0$ (Riemann-Lebesgue lemma) ;

(d) $(f * g)^{\wedge} = \sqrt{2\pi}\hat{f}\hat{g}$;

(e) $\hat{f} = 0 \Leftrightarrow f = 0$.

Proof:

(a) $\hat{f}(\omega) = \frac{1}{\sqrt{2\pi}} \int_{-\infty}^{\infty} f(x)e^{-i\omega x}dx \Rightarrow |\hat{f}(\omega)| \leq (2\pi)^{-1/2} \int_{-\infty}^{\infty} |f| \leq (2\pi)^{-1/2}\|f\|_1$;

(b) $|\hat{f}(\omega) - \hat{(\omega')}| \leq \frac{1}{\sqrt{2\pi}} \int_{-\infty}^{\infty} |e^{-\omega x} - e^{-i\omega' x}||f(x)|dx \to 0$ as $\omega \to \omega'$, by Lebesgue's dominated convergence theorem;

(c) Choose $f_n \in C_{\downarrow}^{\infty}(\mathbf{R}) \ni \| f_n - f \|_1 \leq n^{-1}$. Then $\hat{f}_n$ is rapidly decreasing, and $|\hat{f}_n(\omega) - \hat{f}(\omega)| \leq \| f_n - f \|_1 \leq n^{-1} \forall n \geq 1$;

(d) The interchange of integrals is justified by Fubini's theorem, and

$$(f * g)^{\wedge}(\omega) = \frac{1}{\sqrt{2\pi}} \int_{-\infty}^{\infty} \left[\int_{-\infty}^{\infty} f(x-y)g(y)dy \right] e^{-i\omega x}dx$$

$$= \frac{1}{\sqrt{2\pi}} \int_{-\infty}^{\infty} g(y) \left[\int_{-\infty}^{\infty} f(x-y)e^{-i\omega x}dx \right] dy$$

$$= \frac{1}{\sqrt{2\pi}} \int_{-\infty}^{\infty} g(y) \left[\int_{-\infty}^{\infty} f(x)e^{-i\omega(x+y)}dx \right] dy$$

$$= \sqrt{2\pi} \frac{1}{\sqrt{2\pi}} \int_{-\infty}^{\infty} g(y)e^{-i\omega y}dy \frac{1}{\sqrt{2\pi}} \int_{-\infty}^{\infty} f(x)e^{-i\omega x}dx \sqrt{2\pi}\hat{f}(\omega)\hat{g}(\omega);$$

(e) $\hat{f} = 0 \Rightarrow \int_{-\infty}^{\infty} f(x)g(x)dx = \int_{-\infty}^{\infty} f(x) \left[\frac{1}{\sqrt{2\pi}} \int_{-\infty}^{\infty} g^{\vee}(\omega)e^{-i\omega x}d\omega \right] dx$

$$= \int_{-\infty}^{\infty} \frac{1}{\sqrt{2\pi}} \int_{-\infty}^{\infty} f(x)e^{-i\omega x}dx g^{\vee}(\omega)d\omega$$

$$= \int_{-\infty}^{\infty} \hat{f}(\omega) g(\omega)d\omega = 0$$

$$\forall g \in C_{\downarrow}^{\infty}(\mathbf{R})$$

(by Fubini's theorem).

By choosing g to be the C^∞-approximation of the characteristic function of the interval $[\alpha, \beta]$ for arbitrary $\alpha, \beta \in \mathbf{R}$. One can show that $f(x) = 0 \forall x \in \mathbf{R}$.

The above theorem states that $\wedge$ is a 1 to 1 mapping of $L^1(\mathbf{R})$ into the algebra of functions from $C(\mathbf{R})$ that vanish at infinity.

In general, the inverse transform cannot be used directly on $\hat{f}$ to obtain $f \in L^1(\mathbf{R})$ since $\hat{f}$ may not be integrable.

Example: $\hat{f}(\omega) = \frac{1}{\sqrt{2\pi}} \int_0^1 x^{-1/2} e^{-i\omega x} dx$ is not integrable.

Theorem 4.19 $f \in L^1(\mathbf{R})$ *can be recovered from* $\hat{f}$ *by the formula:*

$f(x) = \lim_{t\to 0} [e^{-\omega^2 t} \hat{f}]^\vee$. The limit being taken in the L^1-norm:

$$\lim_{t\to 0} \left| \left\| [e^{-\omega^2 t}\hat{f}]^\vee - f \right\| \right|_1 = 0.$$

Proof: Consider the Gauss kernel

$$\gamma_t(x) = \frac{e^{-x^2/4t}}{2\sqrt{\pi t}} t > 0, \in C_\downarrow^\infty(\mathbf{R}) \hat{\gamma}_t(\omega) = e^{-t\omega^2}$$

Consider $\gamma_t * f$ and its Fourier transform $(\gamma_t * f)^\wedge = \hat{\gamma}_t \hat{f} = e^{-\omega^2 t} \hat{f}$. Since this is an integrable function, $\vee$ can be applied directly.

$$\begin{aligned}
\sqrt{2\pi}[e^{-\omega^2 t}\hat{f}]^\vee &= \int_{-\infty}^\infty e^{i\omega x} e^{-\omega^2 t} \hat{f}(\omega)d\omega \\
&= \int_{-\infty}^\infty e^{i\omega x} e^{-\omega^2 t} \left(\frac{1}{\sqrt{2\pi}} \int_{-\infty}^\infty f(y) e^{-i\omega x} dy \right) d\omega \\
&= \int_{-\infty}^\infty f(y) \left[\frac{1}{\sqrt{2\pi}} \int_{-\infty}^\infty e^{i\omega(x-y)} e^{-\omega^2 t} d\omega \right] dy \\
&= \gamma_t * f \quad \text{(by Fubini's theorem)}
\end{aligned}$$

Now $\lim_{t\to 0} \| \gamma_t * f - f \|_1 = 0$. Since $\gamma_t * f - f = \int [f(x+y) - f(x)]\gamma_t(y)dy (\because \int \gamma_t = 1)$.

$$\begin{aligned}
\| \gamma_t * f - f \|_1 &\leq \int \| f_y - f \|_1 \gamma_t(y)dy \\
&\leq \int_{|y|<\delta} \| f_y - f \| \gamma_t(y)dy + 4 \| f \|_1 \int_{\delta/\sqrt{t}}^\infty \frac{e^{-y^2/2}dy}{\sqrt{2\pi}}
\end{aligned}$$

The first integral is small for small δ, and the second $\to 0$ as $t \to 0$ for fixed δ.

4.3.3 Poisson summation formula

If $f \in L^1(\mathbf{R})$, then one can construct a periodic function from f by defining $F_f(x) = \sum_{n=-\infty}^{\infty} f(x + 2\pi n)$. The following theorem shows that the function $F_f(x)$ makes sense:

Theorem 4.20 *If $f \in L^1(\mathbf{R})$, then*

$$F_f(x) = \sum_{n=-\infty}^{\infty} f(x + 2\pi n)$$

converges a.e. to a periodic function of period 2π. Also, the convergence is absolute, and $F_f \in L^1([0, 2\pi])$ with $\| F_f \|_{L^1([0,2\pi])} \leq \| f \|_1$.

Proof:

$$\frac{1}{2\pi} \int_{-\infty}^{\infty} | F_f(x) | \, dx \leq \frac{1}{2\pi} \sum_{\pi=-\infty}^{\infty} \int_{-\infty}^{\infty} | f(x + 2\pi n) | \, dx$$

$$\frac{1}{2\pi} \sum_{\pi=-\infty}^{\infty} \int_{n2\pi}^{(n+1)2\pi} | f(x) | \, dx = \frac{1}{2\pi} \int_{-\infty}^{\infty} | f(x) | \, dx < \infty$$

$$\therefore \ \| F_f \|_{L^1[0,2\pi]} \leq \| f_1 \|_1 \, (2\pi)^{-1} < \infty$$

The Fourier series of F_f is given by:

$$F_f(x) = \sum_{n=-\infty}^{\infty} \hat{F}_f(n) e^{inx}$$

where

$$\hat{F}_f(n) = \frac{1}{2\pi} \int_0^{2\pi} F_f(x) e^{-inx} \, dx = \frac{1}{2\pi} \sum_{m=-\infty}^{\infty} \int_0^{2\pi} f(x + 2\pi n) e^{-inx} \, dx$$

$$\sum_{n=-\infty}^{\infty} f(x + 2\pi n) = (2\pi)^{-1/2} \sum_{n=-\infty}^{\infty} \hat{f}(n) e^{inx}$$

This equation is called the Poisson summation formula for f.

Theorem 4.21 *If $f \in C_{\downarrow}^{\infty}(\mathbf{R})$, then the Poisson summation formula holds for f.*

Proof: $\hat{F}_f(n) = (2\pi)^{-1/2}\hat{f}(n)$. Since $f \in C_{\downarrow}^{\infty}(\mathbf{R})$ so is f.

$$\therefore n^p\hat{f}(n) \to 0 \text{ as } |n| \to \infty \forall p < \infty, \text{ i.e.,}$$

$$|\hat{f}(n)| \le \frac{C}{|n|^p}\forall n > n_0. \; \forall p < \infty \; \therefore \; \sum_{n=-\infty}^{\infty} |\hat{f}(n)| < \infty \text{ (for } p > 1)$$

$$\therefore \; |\sum_{n=-\infty}^{\infty} \hat{F}_f(n)e^{inx}| \le (2\pi)^{-1/2} \sum_{n=-\infty}^{\infty} |\hat{f}(n)| < \infty$$

$\therefore$ The Fourier series of $F_f(x)$ converges $\forall x \in \mathbf{R}$. Hence,

$$F_f(x) = \sum_{n=-\infty}^{\infty} \hat{F}_f(n)e^{inx} = \frac{1}{\sqrt{2\pi}} \sum_{n=-\infty}^{\infty} \hat{f}(n)e^{inx} = \sum_{n=-\infty}^{\infty} f(x + 2\pi n)$$

The following theorems give the conditions on $f \in L^1(\mathbf{R})$ for the Poisson summation formula for f to hold.

Theorem 4.22 *If $f \in L^1(\mathbf{R})$ is such that*

$$\sum_{n=-\infty}^{\infty} f(x + 2\pi n)$$

converges everywhere to some continuous function, then the Poisson series

$$(2\pi)^{-1/2} \sum_{n=-\infty}^{\infty} \hat{f}(n)e^{inx}$$

converges everywhere, and the Poisson summation formula holds for f.

Theorem 4.23 *If f is a measurable function such that*

$$|f(x)| \le \frac{C}{(1+|x|^\gamma)}, |\hat{f}(x)| \le \frac{C'}{(1+|x|^\gamma)}$$

for constants $C, C' > 0 \forall x \in \mathbf{R}$ and for some $\gamma > 1$, then Poisson summation formula holds for f.

Proof: Indeed,

$$|\hat{f}(x)| \le \frac{C'}{(1+|x|^\gamma)} \Rightarrow f$$

is continuous, and the hypotheses of the previous theorem are satisfied.

Chapter 5

Wavelet Analysis

5.1 Time-Frequency Analysis and the Windowed Fourier Transform

If $f \in L^2(\mathbf{R})$ is an analog signal with finite energy $\|f\|_2$, then its Fourier transform $\hat{f}$ gives the spectrum of this signal. Practically speaking, however, the Fourier transform is not suited for computing the spectral information of a signal because it requires all previous as well as future information about the signal (i.e., its value over the entire time domain) to evaluate its spectral density at a single frequency ω. Since the Fourier transform of a time-varying signal is a function independent of time, it does not register frequencies varying with time. For spectral information to be useful, it should be possible to identify regions in time corresponding to desired spectral characteristics at all frequencies. In order to achieve time-localization of spectral characteristics of a time varying signal, a "window function" is introduced into the Fourier transform.

A window function $w(t)$ is a function in $L^2(\mathbf{R})$ such that both w and $\hat{w}$ have rapid decay (i.e., w is well-localized in time, while $\hat{w}$ is well-localized in frequency). Multiplying a signal by a window function before taking its Fourier transform has the effect of restricting the spectral information of the signal to the domain of influence of the window function. Using translates of the window function on the time axis to cover the entire time domain, the signal is analyzed for spectral information in localized neighborhoods in time. We make these statements more precise below:

Definition 5.1 *A nontrivial function $w \in L^2(\mathbf{R})$ is called a window function if $tw(t) \in L^2(\mathbf{R})$.*

Remark 5.2 $tw(t) \in L^2(\mathbf{R}) \Rightarrow |t|^{1/2} w(t) L^2(\mathbf{R}).$

Therefore, writing

$$w(t) = (1+|t|)^{-1}(1+|t|)w(t)$$

and applying Schwartz inequality, we have

$$\|w\|_1 \leq \|(1+|t|)^{-1}\|_2 \|(1+|t|)\|_2 < \infty$$

Hence, $w \in L^1(\mathbf{R})$. This implies that $\hat{w}$ is continuous. By Parseval's identity, $\hat{w} \in L^2(\mathbf{R})$. However, in general it is not true that $w\hat{w}(\omega) \in L^2(\mathbf{R})$. In other words, it is possible that while w is a window function, $\hat{w}$ is not. An example of such a window function is the Haar function $\psi_{\mathbf{H}}$:

$$\psi_{\mathbf{H}}(x) = \begin{cases} 1 & \text{if } 0 \leq x < \frac{1}{2} \\ -1 & \text{if } \frac{1}{2} \leq x < 1 \\ 0 & \text{otherwise} \end{cases}$$

Definition 5.3 *The **center** t^* and **radius** Δ_w of a window function w are defined by*

$$t^* = \|w\|_2^{-2} \int_{-\infty}^{\infty} t |w(t)|^2 \, dt,$$

and

$$\Delta_w = \|w\|_2^{-1} \left[\int_{-\infty}^{\infty} (t-t^*)^2 |w(t)|^2 \, dt \right]^{1/2}$$

respectively. $2\Delta_w$ is the width of the window function wi.

For a window function w to be useful in time-frequency analysis , it is necessary that both w and $\hat{w}$ are window functions. In this case we can define the center ω^* and radius $\Delta_{\hat{w}}$ of $\hat{w}$ as defined above.

Henceforth we will assume that both w and $\hat{w}$ are window functions (and hence continuous) with rapid decay in time and frequency, respectively.

If w is any window function *windowed Fourier transform* ,such that $t^* = 0$, then define the $T^w(\tau,\omega)(f)$ of any analog signal $f \in L^2(\mathbf{R})$ by

$$T^w(\tau,\omega)(f) = \frac{1}{\sqrt{2\pi}} \int_{-\infty}^{\infty} f(t)\overline{w(t-\tau)}e^{-i\omega t} dt \; \forall \tau \in \mathbf{R}, \; \forall \omega \in \mathbf{R},$$

Letting $\mathcal{W}_{\tau,\omega}(t) = w(t-\tau)e^{i\omega t}$, we get

$$T^w(\tau,\omega)(f) = \frac{1}{\sqrt{2\pi}} \int_{-\infty}^{\infty} f(t)\overline{\mathcal{W}_{\tau,\omega}(t)}dt = \langle f, \mathcal{W}_{\tau,\omega}\rangle$$

Hence, $T^w(\tau,\omega)(f)$ gives the localized spectral information of f in the time window

$$[t^* + \tau - \Delta_w, t^* + \tau + \Delta_w]$$

By Parseval's identity,

$$\langle f, \mathcal{W}_{\tau,\omega}\rangle = \langle \hat{f}, \hat{\mathcal{W}}_{\tau,\omega}\rangle, \text{ where } \hat{\mathcal{W}}_{\tau,\omega}(\xi) = e^{i\tau\omega}\hat{w}(\xi-\omega)e^{-i\tau\xi}$$

is a frequency window function with center $\omega + \omega^*$ and radius $\Delta_{\hat{w}}$. Hence $T^w(\tau,\omega)(f)$ also gives the localized spectral information of f in the frequency window $[\omega^* + \omega - \Delta_{\hat{w}}, \omega^* + \omega + \Delta_{\hat{w}}]$. Thus we have a time-frequency window :

$$[t^* + \tau - \Delta_w, t^* + \tau + \Delta_w] \times [\omega^* + \omega - \Delta_{\hat{w}}],$$

centered at $(t^* + \tau, \omega + \omega^*)$ in the $t - \omega$ plane, with width $2\Delta_w$ and height $2\Delta_{\hat{w}}$. The width (width of the time window) and height (width of the frequency window) of the time frequency window are constant for all time and frequency values and has constant area $4\Delta_w\Delta_{\hat{w}}$.

The above considerations indicate that in order to achieve a high degree of localization in time and frequency we need to choose a window function with sufficiently narrow time and frequency windows. However, the following important inequality called Heisenberg's Uncertainty Principle imposes a theoretical lower bound on the area of the time frequency window of any window function. This lower bound is independent of the function in question.

5.1.1 Heisenberg's Uncertainty Principle

Heisenberg's Uncertainty Principle is an important principle with far reaching consequences in quantum mechanics. It imposes a lower bound on the product of the mean square errors incurred in the simultaneous measurement of the two complementary parameters of a function with respect to the Fourier transform. Informally stated, this principle says that a function's feature (frequency component) and the feature's location (position at which that frequency component is found) cannot both be measured to an arbitrary degree of precision simultaneously. The proof of this for $f \in C_{\downarrow}^{\infty}(\mathbf{R})$ is easy :

Theorem 5.4 *If*
$$f \in C_\downarrow^\infty(\mathbf{R}),$$

then
$$\frac{\| xf \|_2}{\| f \|_2} \cdot \frac{\| \omega \hat{f} \|_2}{\| \hat{f} \|_2} \geq \frac{1}{2}$$

Proof If $f \in C_\downarrow^\infty(\mathbf{R})$, then

$$\| xf \|_2 \cdot \| \omega \hat{f} \|_2 = \| xf \|_2 \cdot \| (i\omega)\hat{f} \|_2 = \| xf \|_2 \cdot \| f' \|_2$$

(by Parseval's identity) .

$$\geq \| xf'\bar{f} \|_1 \geq \frac{1}{2} \left| \int_{-\infty}^\infty x(f'\bar{f} + \bar{f}'f)dx \right| = \frac{1}{2} \left| \int_{-\infty}^\infty x(| f |^2)'dx \right|$$

(by the Schwartz inequality)

$$\frac{1}{2} \left| \int_{-\infty}^\infty x(| f |^2)'dx \right| = \left| | f(x) |^2 x \Big|_{-\infty}^\infty - \int_{-\infty}^\infty | f |^2 \, dx \right| = \left| \int_{-\infty}^\infty | f |^2 \, dx \right| = \| f \|_2^2$$

(integration by parts)

$$\therefore \ \| xf \|_2 \cdot \| \omega \hat{f} \|_2 \geq \frac{1}{2} \| f \|_2^2 = \frac{1}{2} \| f \|_2 \cdot \| \hat{f} \|_2$$

(by Parseval's identity).

$$\frac{\| xf \|_2}{\| f \|_2} \cdot \frac{\| \omega \hat{f} \|_2}{\| \hat{f} \|_2} \geq \frac{1}{2}$$

This can be extended to all those $f \in L^2(\mathbf{R})$ such that $(1+ | x |)f(x)$ and $f'(x) \in L^2(\mathbf{R})$, using the fact that $C_\downarrow^\infty(\mathbf{R})$ is dense in $L^2(\mathbf{R})$.

Let w be a window function such that $\hat{w}$ is also a window function. Moreover, assume w.l.g. that both w and $\hat{w}$ are centered at zero. Then we have

$$4\Delta_w \Delta_{\hat{w}} = 4 \cdot \frac{\| xw \|_2}{\| w \|_2} \cdot \frac{\| \omega\hat{w} \|_2}{\| \hat{w} \|_2} \geq 2$$

It can be shown that equality holds iff

$$w(t) = \frac{Ce^{i\alpha t}}{2\sqrt{\pi\beta}} e^{\frac{-t^2}{4\beta}}$$

for constants $\beta > 0$, α and $C \neq 0$.

When the window function is a Gaussian as above, the windowed Fourier transform $T^w(\tau,\omega)(f)$ is called a Gabor transform. Thus Gabor transforms have the tightest time-frequency windows of all windowed Fourier transforms.

Since a sinusoid's frequency is the number of cycles per unit time, it is necessary to be able to localize high frequency transient spectral information to a relatively narrow time interval while allowing a relatively wider time interval to identify low frequency characteristics in order to capture complete information. In other words, it is desirable to be able to zoom in on the signal to identify short duration transients corresponding to high frequency bursts (small scale features), and zoom out from the signal to completely capture more gradual variations corresponding to low frequency components (large scale features). We have already seen that we cannot reduce the size of the time-frequency window beyond that of the Gabor transform. On the other hand, since the Gabor transform is a windowed Fourier transform, its time frequency window is rigid and does not vary over time or frequency.

We, therefore, have to modify the windowed Fourier transform in a fundamentally different way to achieve varying time and frequency windows. The only way we can vary the size of the time window for different degrees of localization is by reciprocally varying the size of the frequency window at the same time, so as to keep the area of the window constant. This means a trade-off between time and frequency localization. This is achieved by directly windowing the signal instead of its Fourier transform, and its Fourier transform instead of its inverse Fourier transform, and by scaling the window function appropriately to change its time window width. We thus introduce the Integral wavelet transform.

5.2   The Integral Wavelet Transform

Define an integral transform on $L^2(\mathbf{R})$ by:

$$(\mathcal{W}_\psi f)(a,b) = \int_{-\infty}^{\infty} f(t)\overline{\Psi_{a,b}(t)}\,dt$$

where $\Psi_{a,b}(t) = \mid a \mid^{-1/2} \Psi(\frac{t-b}{a})$, with $a, b \in \mathbf{R}$, $a \neq 0$ and $\Psi \in L^2(\mathbf{R})$,
Further Ψ satisfies the condition

$$\int_{-\infty}^{\infty} \Psi(t)\,dt = 0$$

Ψ is called the *wavelet function* and the $\Psi_{a,b}$ are called *wavelets* . $\mathcal{W}_\Psi$ is called the *integral wavelet transform* w.r.t. the *wavelet* Ψ.

If both the *wavelet function* Ψ and its Fourier transform $\hat{\Psi}$ are window functions with centers t^* and ω^*, and radii Δ_Ψ and $\Delta_{\hat{\psi}}$, respectively, then the integral wavelet transform

$$(\mathcal{W}_\Psi f)(a,b) =|\, a\, |^{-1/2} \int_{-\infty}^{\infty} f(t)\overline{\Psi_{a,b}(t)}dt$$

of a signal $f \in L^2(\mathbf{R})$ localizes it within a time window given by $[at^* + b - a\Delta_\psi, at^* + b + a\Delta_\psi]$. If $\vartheta(\omega) = \hat{\Psi}(\omega + \omega^*)$, then clearly, ϑ is also a window function, with center 0 and radius $\Delta_{\hat{\psi}}$. Moreover,

$$(\mathcal{W}_\psi f)(a,b) = a\,|\, a\, |^{-1/2} \int_{-\infty}^{\infty} \hat{f}(\omega)\overline{\vartheta(\omega - a^{-1}\omega^*)}e^{ib\omega}\, d\omega$$

(by Parseval's identity). This means that $\mathcal{W}_\psi$ localizes the spectrum of f to a *frequency window* given by $[a^{-1}\omega^* - a^{-1}\Delta_{\hat{\psi}}, a^{-1}\omega^* + a^{-1}\Delta_{\hat{\psi}}]$. Thus, we have a *time-frequency* window for analyzing finite energy analog signals with the help of the integral wavelet transform $\mathcal{W}_\psi$. Since only positive frequencies are of interest, the wavelet function Ψ is chosen with $\omega^* > 0$.

The width (time window size) $2a\Delta_\psi$ of the time-frequency window is inversely proportional to the center frequency $a^{-1}\omega^*$, while the height (frequency window size) $2a^{-1}\Delta_{\hat{\psi}}$ is directly proportional to the center frequency. Thus, the area of the time-frequency windows is constant and is given by $4\Delta_\psi\Delta_{\hat{\psi}}$, The ratio of the center frequency to the width of the frequency band is also a constant, given by

$$\frac{a^{-1}\omega^*}{2a^{-1}\Delta_{\hat{\psi}}} = \frac{\omega^*}{2\Delta_{\hat{\psi}}}$$

Henceforth we assume that both Ψ and $\hat{\Psi}$ are window functions. We also use the variables t and x, and ω and ξ interchangeably. We continue to interpret t and ω as the time and frequency variables, respectively, and x and ξ as the space and scale variables, respectively. Thus, f is appropriately a time-varying signal or a function defined over space.

We now turn our attention briefly to the question of the recovery of a finite energy analog signal from its integral wavelet transform, *i.e.,* the question of the inversion of the integral wavelet transform.

For the wavelet transform to be invertible, the basic wavelet Ψ must

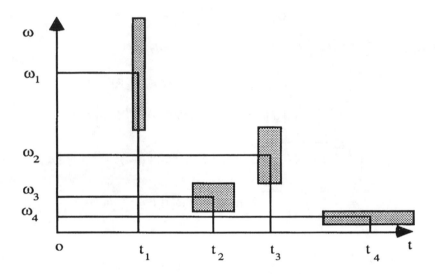

Figure 5.1: Wavelet time frequency windows.

satisfy the admissibility criterion

$$\int_{-\infty}^{\infty} \frac{|\hat{\Psi}(\xi)|^2}{|\xi|} d\xi < \infty$$

The following theorem justifies this requirement.

Theorem 5.5 *If $f, g \in L^2(\mathbf{R})$, then the following resolution of the identity formula holds:*

$$\int_{-\infty}^{\infty} \int_{-\infty}^{\infty} (\mathcal{W}_\psi f)(a,b)\overline{(\mathcal{W}_\psi g)(a,b)} a^{-2} da\, db = C_\psi \langle f, g \rangle$$

where

$$C_\psi = \int_{-\infty}^{\infty} \frac{|\hat{\psi}(\xi)|^2}{|\xi|} d\xi$$

Proof : $\int_{-\infty}^{\infty} \int_{-\infty}^{\infty} (\mathcal{W}_\psi f)(a,b)\overline{(\mathcal{W}_\psi g)(a,b)} a^{-2} da\, db$

$$= \int_{-\infty}^{\infty} \int_{-\infty}^{\infty} \left[a \mid a \mid^{-1/2} \int_{-\infty}^{\infty} \hat{f}(\omega)\overline{\hat{\Psi}(a\omega)} e^{ib\omega} d\omega \right]$$
$$\left[a \mid a \mid^{-1/2} \int_{-\infty}^{\infty} \overline{\hat{g}(\omega)} \hat{\Psi}(a\omega) e^{-ib\omega} d\omega \right] a^{-2} da\, db$$

$$= \int_{-\infty}^{\infty} \int_{-\infty}^{\infty} \left[\overline{\int_{-\infty}^{\infty} \overline{\hat{f}(\omega)} \hat{\Psi}(a\omega) e^{ib\omega} d\omega} \right] \left[\int_{-\infty}^{\infty} \overline{\hat{g}(\omega)} \hat{\Psi}(a\omega) e^{-ib\omega} d\omega \right] \mid a \mid^{-1} da\, db$$

$$= \int_{-\infty}^{\infty} \int_{-\infty}^{\infty} \overline{\hat{f}(\omega)\hat{\Psi}(a\omega)} \hat{g}(\omega)\hat{\Psi}(a\omega) \mid a \mid^{-1} da\, d\omega$$

(by Parseval's identity)

$$= \int_{-\infty}^{\infty} \int_{-\infty}^{\infty} \hat{f}(\omega)\overline{\hat{g}(\omega)} \mid \hat{\Psi}(a\omega) \mid^2 \mid a \mid^{-1} da\, d\omega$$

$$= \int_{-\infty}^{\infty} \hat{f}(\omega)\overline{\hat{g}(\omega)} \int_{-\infty}^{\infty} \frac{|\hat{\Psi}(\zeta)|^2}{|\zeta|} d\zeta\, d\omega$$

(by Fubini's theorem, and by letting $\zeta = a\omega$)

$$= C_\psi \langle \hat{f}, \hat{g} \rangle = C_\psi \langle f, g \rangle$$

(by Parseval's Theorem)

Corollary 5.6. $f = C_\psi \int_{-\infty}^{\infty} \int_{-\infty}^{\infty} (\mathcal{W}_\psi f)(a,b)\Psi_{a,b} a^{-2} da\, db$

where the equality is meant in the sense that taking the inner products on both sides of the above equation with any function in $L^2(\mathbf{R})$ yields an identity.

Proof : $\langle f, g \rangle = C_\psi^{-1} \int_{-\infty}^{\infty} \int_{-\infty}^{\infty} (\mathcal{W}_\psi f)(a,b) \overline{\mathcal{W}_\psi g)(a,b)} a^{-2} dadb$

$$= C_\psi^{-1} \int_{-\infty}^{\infty} \int_{-\infty}^{\infty} (\mathcal{W}_\psi f)(a,b) \langle \psi_{a,b}, g \rangle \rangle a^{-2} dadb$$

$$= \left\langle C_\psi^{-1} \int_{-\infty}^{\infty} \int_{-\infty}^{\infty} (\mathcal{W}_\psi f)(a,b) \psi_{a,b} a^{-2} dadb, g \right\rangle$$

The equality in the above corollary is valid in a stronger sense:

Corollary 5.7

$$\lim_{\substack{\alpha \to 0 \\ \beta,\gamma \to \infty}} \| f - C_\psi^{-1} \int_{-\gamma}^{\gamma} \int_{-\alpha}^{\beta} (\mathcal{W}_\psi f)(a,b) \Psi_{a,b} a^{-2} dadb \|_2 = 0$$

Proof : The integral

$$C_\psi^{-1} \int_{-\gamma}^{\gamma} \int_{-\alpha}^{\beta} (\mathcal{W}_\psi, f)(a,b) \Psi_{a,b} a^{-2} dadb$$

can be treated as a bounded operator on $L^2(\mathbf{R})$:

$$C_\psi^{-1} \int_{-\gamma}^{\gamma} \int_{-\alpha}^{\beta} (\mathcal{W}_\psi f)(a,b) \Psi_{a,b} a^{-2} dadb : L^2(\mathbf{R}) \to L^2(\mathbf{R})$$

given by

$$\left(C_\psi^{-1} \int_{-\gamma}^{\gamma} \int_{-\alpha}^{\beta} (\mathcal{W}_\Psi f)(a,b) \Psi_{a,b} a^{-2} dadb \right)(g)$$

$$= \left\langle C_\psi^{-1} \int_{-\gamma}^{\gamma} \int_{-\alpha}^{\beta} (\mathcal{W}_\Psi f)(a,b) \Psi_{a,b} a^{-2} dadb, g \right\rangle \forall g \in L^2(\mathbf{R})$$

$$= C_\psi^{-1} \int_{-\gamma}^{\gamma} \int_{-\alpha}^{\beta} (\mathcal{W}_\Psi f)(a,b) \langle \Psi_{a,b}, g \rangle a^{-2} dadb$$

$$\left\| C_\psi^{-1} \int_{-\gamma}^{\gamma} \int_{-\alpha}^{\beta} (\mathcal{W}_\Psi f)(a,b) \langle \Psi_{a,b}, g \rangle a^{-2} dadb \right\|_2 = 4\gamma (\frac{1}{\alpha} - \frac{1}{\beta}) \| f \|_2 \| g \|_2$$

$$\left\| f - C_\psi^{-1} \int_{-\gamma}^{\gamma} \int_{-\alpha}^{\beta} (\mathcal{W}_\Psi f)(a,b) \Psi_{a,b} a^{-2} dadb \right\|_2$$

$$= \sup_{\|g\|=1} \left| \left\langle f - C_\psi^{-1} \int_{-\gamma}^{\gamma} \int_{-\alpha}^{\beta} (\mathcal{W}_\Psi f)(a,b) \Psi_{a,b} a^{-2} dadb, g \right\rangle \right|$$

$$\leq sup_{||g||=1} \mid C_\psi^{-1} \int \int_{\substack{|a|\geq\beta \\ or|a|\leq\alpha \\ or|b|\geq\gamma}} (\mathcal{W}_\psi f)(a,b)\overline{(\mathcal{W}_\psi g)(a,b)}a^{-2}dadb \mid$$

$$\leq \sup_{||g||=1} \left[C_\psi^{-1} \int\int_{\substack{|a|\geq\beta \\ or|a|\leq\alpha \\ or|b|\geq\gamma}} \mid (\mathcal{W}_\psi f)(a,b)\mid^2 a^{-2}dadb \right]^{1/2}$$

$$\cdot \left[C_\psi^{-1} \int_{-\infty}^{\infty}\int_{-\infty}^{\infty} \mid (\mathcal{W}_\psi g)(a,b)\mid^2 a^{-2}dadb \right]^{1/2}$$

The expression in the first brackets approaches zero as $\alpha \to 0$ and $\beta,\gamma \to \infty$, and the expression in the second brackets $=|| g ||^2 = 1$

Remark 5.8 If Ψ is an admissible wavelet , then

$$\int_{-\infty}^{\infty} \Psi(x)dx = 0$$

Indeed, since Ψ is a window function, $\hat{\Psi}$ is continuous. Since Ψ is admissible, $C_\psi < \infty$. Therefore, $\hat{\Psi}(0) = 0$, *i.e.*,

$$\int_{-\infty}^{\infty} \Psi(x)dx = 0$$

We now present a heuristic approach to the integral wavelet transform that helps illuminate the purpose and nature of it better. Consider a window function ϕ that is smooth (*i.e.*, $\phi \in C_{\downarrow}^{\infty}(\mathbf{R})$) and compactly supported such that $\hat{\phi}$ is also a window function. Define a linear operator $\mathcal{A}_\phi : L^2(\mathbf{R}) \to L^2(\mathbf{R})$ by

$$\mathcal{A}_\phi(f)(t) = \frac{\int_{-\infty}^{\infty} f(\tau)\phi(t-\tau)d\tau}{\int_{-\infty}^{\infty} \phi(\tau)d\tau},$$

$\forall f \in L^2(\mathbf{R})$. This operator is a moving average or a smoothing operator on the function f with respect to the weighting function $\frac{\phi(t)}{\int_{-\infty}^{\infty}\phi(\tau)d\tau}$, provided $\int_{-\infty}^{\infty} \phi(\tau)d\tau < \infty$. Indeed, without a loss of generality, we may assume that $\int_{-\infty}^{\infty} \phi(\tau)d\tau = 1$.

Thus, $\mathcal{A}_\phi$ is an operator which blurs the features of f at scales less than the support of the weighting function ϕ. In other words, $\mathcal{A}_\phi(f)(t) = (f * \phi)(t)$ is a low-pass filtered version of f with ϕ as the impulse response of the low-pass filter.

By dilating and contracting the weighting function ϕ, we can smooth f to a greater of lesser degree, respectively. Accordingly, define a family of operators parameterized by a scaling variable $a > 0$ as follows:

$$A_\phi(f)(t, a) = \int_{-\infty}^{\infty} f(\tau) \frac{1}{a} \phi(\frac{t - \tau}{a}) d\tau = (f * \phi_a)(t)$$

Note that

$$\int_{-\infty}^{\infty} \phi_a(t) d\tau = \int_{-\infty}^{\infty} \phi(\frac{t}{a}) \frac{dt}{a} = \int_{-\infty}^{\infty} \phi(t) dt = 1$$

The magnitude of a is proportional to the amount of blurring of f. Thus, this parameter a may be thought of as the minimum relative scale with respect to the support of ϕ at which the details (features) of f are preserved in $A_\phi(f)(t, a)$. Indeed, as $a \to 0$, $\phi_a(t) \to \delta(t)$, the Dirac-delta function at the origin, and hence, $A_\phi(f)(t, a) \to f(t)$ as $a \to 0$.

The "loss" of information or detail in going from scale a to scale $a + \Delta a$ is captured in the difference:

$$\Delta_\phi(f)(t, a) = A_\phi(f)(t, a) - A_\phi(f)(t, a + \Delta a)$$

This difference in detail for a fixed change Δa in the scale parameter is lesser for larger values of a and larger for smaller values a. Thus, the "instantaneous" loss of detail, or the information at scale a is given by

$$\mathcal{D}_\phi(f)(t, a) = \lim_{\Delta a \to 0} \frac{A_\phi(f)(t, a) - A_\phi(f)(t, a + \Delta a)}{\frac{\Delta a}{a}}$$

$$= -a \frac{\partial}{\partial a}(A_\phi(t, a))$$

Since ϕ is compactly supported, we may interchange the order of integration and differentiation. Therefore,

$$-a \frac{\partial}{\partial a}(A_\phi(f)(t, a)) = -a \frac{\partial}{\partial a}((f * \phi_a)(t)) = -(a \frac{\partial}{\partial a} \phi_a * f)(t)$$

Now,

$$-a \frac{\partial}{\partial a}(\phi_a(t)) = -a \frac{\partial}{\partial a}\left(\frac{1}{a}\phi(\frac{t}{a})\right) = \frac{1}{a}\left(\phi(\frac{t}{a}) + (\frac{t}{a})\phi'(\frac{t}{a})\right)$$

Define $\Psi(t) = \phi(t) + t\phi'(t)$. Then,

$$-a \frac{\partial}{\partial a}(\phi_a(t)) = \Psi_a(t) = \frac{1}{a}\Psi(\frac{t}{a})$$

and

$$\int_{-\infty}^{\infty} \Psi(t)dt = \int_{-\infty}^{\infty} \phi(t)dt + \int_{-\infty}^{\infty} t\phi'(t)dt = 0$$

Thus,

$$\mathcal{D}_\phi(f)(t,a) = (\Psi_a * f) \triangleq \mathcal{W}_\Psi(f)(t,a)$$

where

$$\mathcal{W}_\Psi(f)(t,a) = \int_{-\infty}^{\infty} f(\tau)\frac{1}{a}\Psi\left(\frac{t-\tau}{a}\right)d\tau$$

is the wavelet transform of f at time (or position) t and scale a. $\mathcal{W}_\Psi(f)(t,a)$ is the coefficient that measures the strength of features at scale a at time (position) t.

For a more detailed exposition of this approach, see M. Holschneider [20].

5.3   The Discrete Wavelet Transform

The continuous (integral) wavelet transform involved wavelets of the form $\Psi_{a,b}(x) = \mid a \mid^{-1/2} \Psi(\frac{x-b}{a})$ with $a,b \in \mathbf{R}, a \neq 0$ and Ψ is admissible $(\int_{-\infty}^{\infty} \frac{|\hat{\Psi}(\zeta)|^2}{\zeta}d\zeta < \infty$.

As we have already seen, the integral wavelet transform, unlike the Fourier transform, has very good time-frequency localization properties. Moreover, the functions $\Psi_{a,b}$ belong to $L^2(\mathbf{R})$.

We would like to explore the possibility of extracting a discrete subset of the set of functions

$$\{\Psi_{a,b}\}_{(a,b)\in\mathbf{R}^* \times \mathbf{R}'}$$

which forms a basis of $L^2(\mathbf{R})$ and inherits the time frequency localization property of the continuous family. This involves the singling out of an appropriate lattice of the set $\mathbf{R}^* \times \mathbf{R}(\mathbf{R}^* = \mathbf{R}\backslash\{0\})$ and selecting wavelets parametrized by elements of this lattice.

In signal processing applications, one is interested in nonnegative frequencies only. Since the concept of frequency corresponds to that of feature scale in image processing, we would like to consider only positive values of the frequency (scale) parameter a . With this restriction, the admissibility criterion becomes

$$\int_0^{\infty} \frac{\mid \hat{\Psi}(\zeta) \mid^2}{\mid \zeta \mid}d\zeta = \int_{-\infty}^0 \frac{\mid \hat{\Psi}(\zeta) \mid^2}{\mid \zeta \mid}d\zeta < \infty$$

We have seen that the wavelet $\Psi_{a,b}$ is centered at $at^* + b$ in time and has its time window width equal to $2a\Delta_\psi$, and $\hat{\Psi}_{a,b}$ is centered at $a^{-1}\omega^*$ in frequency and has its frequency window width equal to $2a^{-1}\Delta_{\hat\psi}$. We let frequency ω be equal to $a^{-1}\omega^*$. Since $a > 0$ and $\omega^* > 0$, $\omega \in (0, \infty)$.

In order to discretize the continuous family of wavelets $\{\Psi_{a,b}\}$, we first partition the frequency domain $(0, \infty)$ into dyadic intervals as follows:

$$(0, \infty) = \bigcup_{j=-\infty}^{\infty} (2^{-j}\Delta_{\hat\psi}, 2^{-j+1}\Delta_{\hat\psi})$$

Since shifting the phase of Ψ by θ is equivalent to shifting $\hat{\Psi}$ by θ on the frequency axis, *i.e.*, $\Psi'(t) = e^{i\theta t}\Psi(t) \Rightarrow \hat{\Psi}'(\omega) = \hat{\Psi}(\omega - \theta)$, while leaving the time and frequency radii, w.l.g. we can assume that $\omega^* = 3\Delta_{\hat\psi}$. This yields

$$(a^{-1}\omega^* - a^{-1}\Delta_{\hat\psi}, a^{-1}\omega^* + a^{-1}\Delta_{\hat\psi}] = (2a^{-1}\Delta_{\hat\psi}, 4a^{-1}\Delta_{\hat\psi})$$

If we discretize the dilation parameter a by letting it take the values $2^j, \forall j \in \mathbf{Z}$, then we have

$$(a^{-1}\omega^* - a^{-1}\Delta_{\hat\psi}, a^{-1}\omega^* + a^{-1}\Delta_{\hat\psi}] = (2^{-j+1}\Delta_{\hat\psi}, 2^{-j+2}\Delta_{\hat\psi}), j \in \mathbf{Z}.$$

Thus, by setting $\omega^* = 3\Delta_{\hat\psi}$, and discretizing the dilation parameter a as above, the frequency domains of the resulting set of wavelets form the dyadic partition of the frequency domain we constructed.

We now discretize the translation parameter b by observing that at each scale 2^j the width of the time window is $2^{j+1}\Delta_\psi$. Hence, we may discretize the time domain at each scale separately by setting the distance between two consecutive wavelets' centers equal to $2^{j+1}\Delta_\psi$. This will ensure that the discrete set of wavelets will cover the entire time domain at each scale. Denoting the width $2\Delta_\psi$ of the time window of ψ by b_0, we have b taking the discrete range of values $k2^j b_0, k \in \mathbf{Z}$.

Introducing the notation $\Psi_{j,k}(t) = 2^{-j/2}\Psi(2^{-j}t - kb_0), \forall j, k \in Z$, we have a discrete set of wavelets whose time-frequency windows cover the whole of the ωt plane. As will be seen later, the choice of b_0 is arbitrary, and it can be replaced by any other real number by appropriately rescaling the basic wavelet Ψ. For now, however, we will consider b_0 to be an arbitrary fixed positive real number.

The wavelet transform $\mathcal{W}_\psi$ of any function $f \in L^2(\mathbf{R})$, restricted to this discrete set of wavelets, is given by

$$(\mathcal{W}_\psi f)(\frac{1}{2^j}, \frac{k}{2^j}) = 2^{-j/2} \int_{-\infty}^{\infty} f(x)\overline{\Psi_{j,k}(x)}dx = \langle f, \Psi_{j,k}\rangle \ \forall j, k \in \mathbf{Z}$$

Thus the *discrete wavelet coefficients* of any $f \in L^2(\mathbf{R})$ are given by

$$\langle f, \Psi_{j,k} \rangle, j, k \in \mathbf{Z}.$$

The motivation behind the discretization of the continuous wavelet transform is two-fold: Not only is any function $f \in L^2(\mathbf{R})$ completely characterized by its discrete wavelet coefficients $\langle f, \Psi_{j,k} \rangle, j, k \in \mathbf{Z}$, but it is also possible to recover any function in $L^2(\mathbf{R})$ from its discrete wavelet coefficients in a numerically stable way.

In other words, $L^2(\mathbf{R})$ is "spanned" by the discrete set of wavelets $\{\Psi_{j,k}\}_{j,k \in \mathbf{Z}}$ for appropriate choice of Ψ and b_0, and there exists an algorithm to determine the discrete wavelet coefficients of a function $f \in L^2(\mathbf{R})$.

The discrete set of wavelets $\{\Psi_{j,k}\}_{j,k \in \mathbf{Z}}$ characterize a function $f \in L^2(\mathbf{R})$ by means of the discrete wavelet coefficients $\langle f, \Psi_{j,k} \rangle j, k \in \mathbf{Z}$ if

$$\langle f, \Psi_{j,k} \rangle = \langle g, \Psi_{j,k} \rangle \forall j, k \in \mathbf{Z} \Leftrightarrow f = g,$$

or equivalently,

$$\langle f, \Psi_{j,k} \rangle = 0 \ \forall j, k \in \mathbf{Z} \Leftrightarrow f = 0.$$

The characterization is *numerically stable* if small perturbations in the wavelet coefficients $\langle f, \Psi_{j,k} \rangle$ of f correspond to small perturbations of the function f in the L^2-norm.

For any basic wavelet Ψ that is admissible, has good decay in time and frequency, and satisfies $\hat{\Psi}(0) = 0$,

$$\sum_{j,k \in \mathbf{Z}} |\langle f, \Psi_{j,k} \rangle|^2 \le C \parallel f \parallel_2^2$$

for some constant C. This implies that the sequence

$$\{\langle f, \Psi_{j,k} \rangle\}_{j,k \in \mathbf{Z}}$$

of discrete wavelet coefficients of $f \in L^2(\mathbf{R})$ w.r.t. the discretized family of wavelets $\{\Psi_{j,k}\}_{j,k \in \mathbf{Z}}$ belongs to

$$l^2(\mathbf{Z}^2) = \left\{ \{c_{i,j}\}_{i,j \in \mathbf{Z}} : \sum_{i,j \in \mathbf{Z}} |c_{i,j}|^2 < \infty \right\}$$

Thus, $f \mapsto \{\langle f, \Psi_{j,k} \rangle\}_{j,k \in \mathbf{Z}}$ is a mapping from $L^2(\mathbf{R})$ into $l^2(\mathbf{Z}^2)$.

Now $L^2(\mathbf{R})$ is equipped with the metric topology induced by the norm $\parallel \ \parallel_2$ on $L^2(\mathbf{R})$ while $l^2(\mathbf{Z}^2)$ has the metric topology induced by the norm

$$\parallel c \parallel^2 = \sum_{i,j \in \mathbf{Z}} |c_{i,j}|^2 \ \forall c \in l^2(\mathbf{Z}^2)$$

Therefore, we can take the numerical stability of the characterization of functions in $L^2(\mathbf{R})$ by their sequences of discrete wavelet coefficients to mean: whenever the discrete wavelet coefficient sequences of two functions are close in $l^2(\mathbf{Z}^2)$, the functions themselves are close in $L^2(\mathbf{R})$. That is, if

$$\sum_{i,j\in\mathbf{Z}} |\langle f, \Psi_{i,k}\rangle|^2$$

is small, then $\| f \|_2^2$ is small.

In particular, $\exists \alpha < \infty \ni \sum_{i,j\in\mathbf{Z}} |\langle f, \Psi_{i,j}\rangle|^2 \le 1 \Rightarrow \| f \|^2 \le \alpha$

For $f \in L^2(\mathbf{R})$, define

$$\tilde{f} = f \left[\sum_{i,j\in\mathbf{Z}} |\langle f, \Psi_{i,j}\rangle|^2 \right]^{-1/2}$$

Then,

$$\sum_{i,j\in\mathbf{Z}} |\langle \tilde{f}, \Psi_{i,k}\rangle|^2 \le 1 \text{ and } \| f \|^2 < \alpha$$

$$\therefore \sum_{i,j\in\mathbf{Z}} |\langle f, \Psi_{i,k}\rangle|^2 \ge \alpha^{-1} \| f \|^2 .$$

i.e.,

$$A \| f \|^2 \le \sum_{i,j\in\mathbf{Z}} |\langle f, \Psi_{i,j}\rangle|^2,$$

for $A > 0$. Thus, the condition for numerical stability of the characterization of $f \in L^2(\mathbf{R})$ by its discrete wavelet coefficients is that there exist constants $0 < A \le B < \infty$ such that

$$A \| f \|^2 \le \sum_{i,j\in\mathbf{Z}} |\langle f, \Psi_{i,j}\rangle|^2 \le B \| f \|^2, \ \forall f \in L^2(\mathbf{R})$$

We direct the interested reader to Daubechies [9] for necessary and sufficient conditions on a wavelet Ψ under which the discrete family $\{\Psi_{i,j}\}_{i,j\in\mathcal{Z}}$ satisfies the above inequalities for all functions in $L^2(\mathbf{R})$.

We will restrict our attention to a large subclass of wavelets that arise from certain structures on $L^2(\mathbf{R})$ called *multiresolution analyses*. These wavelets yield discrete families of dilations and translations that are orthonormal bases for $L^2(\mathbf{R})$. This elegant framework for understanding and constructing wavelet bases was formulated by Y. Meyer and S. Mallat [27, 28, 29].

5.4   Multiresolution Analysis (MRA) of $L^2(\mathbf{R})$

Definition 5.9 *An MRA of $L^2(\mathbf{R})$ is a sequence of closed subspaces $\{V_j\}_{j \in \mathbf{Z}}$ of $L^2(\mathbf{R})$ satisfying the following properties :*

(a) $V_j \subset V_{j-1} \; \forall j \in \mathbf{Z};$ *(nesting property)*

(b) $Closure_{L^2(\mathbf{R})} \left(\bigcup_{j \in \mathbf{Z}} V_j \right) = L^2(\mathbf{R})$; *(density of the union in $L^2(\mathbf{R})$))*

(c) $\bigcap_{j \in \mathbf{Z}} V_j = \{0\}$;

(d) $f(x) \in V_j \Leftrightarrow f(2x) \in V_{j-1} \; \forall j \in \mathbf{Z}$; *(scaling property)*

(e) $f(x) \in V_0 \Rightarrow f(x - n) \in V_0 \; \forall n \in \mathbf{Z};$ *(invariance under integral translations)*

(f) $\exists \phi \in V_0 \ni \{\phi_{0,n}\}_{n \in \mathbf{Z}}$ *is an orthonormal basis of V_0, where $\phi_{j,k}(x) = 2^{-j/2}\phi(2^{-j}x - k) \; \forall j, k \in \mathbf{Z}$; (existence of a scaling function).*

Condition (d) implies that each of the subspaces V_j is a scaled version of the central subspace V_0, and together with (e) and (f) it implies that $\{\phi_{j,k}\{_{k \in \mathbf{Z}}$ is an orthonormal basis for $V_j \; \forall j \in \mathbf{Z}$.

$\forall j \in \mathbf{Z}$, denote by W_j the orthogonal complement of V_j in V_{j-1}. Then, $V_{j-1} = V_j \oplus W_j$ where $\oplus$ is the direct sum, and $\forall u \in V_{j-1}, u = u + w$, where $v \in V_j$ and $w \in W_j$ with $\langle v, w \rangle = 0$; also the decomposition of u is unique:

Indeed, if $u = v + w = v' + w'$ for $v' \in V_j$ and $w' \in W_j$, then $v - v' = w - w'$. This is possible iff $v = v'$ and $w = w'$ since $x \in V_j$ and $x \in W_j \Rightarrow \langle x, x \rangle = 0$, *i.e.*, $x = 0$.

$W_j \perp W_{j'} \; \forall j \neq j'$, since if $j > j'$, then $W_{j'} \subset V_j \perp W_j$. Thus, if $j < i$, $\forall V_j = V_j \oplus \bigoplus_{k=0}^{i-j-1} W_{j-k'}$.

As $Closure_{L^2(\mathbf{R})}(\bigcup_{j \in \mathbf{Z}} V_j) = L^2(\mathbf{R})$ and $\bigcap_{j \in \mathbf{Z}} = \{0\}$, $L^2(\mathbf{R}) = \bigoplus_{j \in \mathbf{Z}} W_{j'}$, where the W_j are mutually orthogonal closed subspaces of $L^2(\mathbf{R})$.

It is clear that the W_j are mutually orthogonal. If $\{w_n\}_{n=1}^{\infty}$ is a sequence of vectors in W_j with $\lim_{n \to \infty} w_n = w$, then $w \in v_{j-1}$, since $W_j \subset V_{j-1}$ and V_{j-1} is closed; also, $\forall v \in V_j$ and

$$\forall n \geq 1 \langle v, w_n \rangle = 0 \quad \therefore \quad \lim_{n \to \infty} \langle v, w_n \rangle = \langle v, \lim_{n \to \infty} w_n \rangle = \langle v, w \rangle = 0$$

by continuity of $\langle , \rangle$. As $v \in \mathbf{V}_j$ and $w \in \mathbf{V}_{j-1}$, $w \in \mathbf{W}_j$. Hence $\mathbf{W}_j$ is closed.

Denote by $\mathbf{P}_j$ the orthogonal projection operator of $L^2(\mathbf{R})$ onto $\mathbf{V}_j$ and by $\mathbf{Q}_j$ the orthogonal projection operator of $L^2(\mathbf{R})$ onto $\mathbf{W}_j$. Then $\mathbf{Q}_j = \mathbf{P}_{j-1} - \mathbf{P}_j$.

The subspaces $\mathbf{W}_j$ inherit the scaling property from the $\mathbf{V}_j$: $f(x) \in \mathbf{W}_j \Leftrightarrow f(2x) \in \mathbf{W}_{j-1}$.

Given an MRA of $L^2(\mathbf{R})$, one can construct a basic wavelet ψ such that $\{\psi_{j,k}\}_{k \in \mathbf{Z}}$ is an orthonormal basis for $\mathbf{W}_j$ $\forall j \in \mathbf{Z}$, and $\{\psi_{j,k}\}_{j,k \in \mathbf{Z}}$ is an orthonormal basis of $L^2(\mathbf{R})$ where $\psi_{j,k}(x) = 2^{-j/2}\psi(2^{-j}x - k)$ $\forall j, k \in \mathbf{Z}$. The following theorem demonstrates this :

Theorem 5.10 *If the sequence of closed subspaces $\{\mathbf{V}_j\}_{j \in \mathbf{Z}}$ of $L^2(\mathbf{R})$ is an MRA of $L^2(\mathbf{R})$, then*

$$\exists \psi \in L^2(\mathbf{R}) \ni \left\{ \psi_{j,k} : \psi_{j,k}(x) = 2^{-j/2}\psi(2^{-j}x - k) \, \forall j, k \in \mathbf{Z} \right\}$$

is an orthonormal basis of $L^2(\mathbf{R})$ such that

$$\mathbf{P}_{j-1} = \mathbf{P}_j + \sum_{k \in \mathbf{Z}} \langle \bullet, \psi_{j,k} \rangle \psi_{j,k} \quad \forall j \in \mathbf{Z}$$

Proof : If $\{\psi_{j,k}\}_{j,k \in \mathbf{Z}}$ is an orthonormal basis of $L^2(\mathbf{R})$ as in the theorem, then since $\mathbf{P}_j : L^2(\mathbf{R}) \to \mathbf{V}_j$ is surjective, $\mathbf{Q}_j = \mathbf{P}_{j-1} - \mathbf{P}_j$ is also surjective and

$$\mathbf{Q}_j f = \sum_{k \in \mathbf{Z}} \langle f, \psi_{j,k} \rangle \psi_{j,k} \quad \forall f \in L^2(\mathbf{R})$$

$$\mathbf{Q}_j f = \sum_{k \in \mathbf{Z}} \langle f, \psi_{j,k} \rangle \psi_{j,k} \quad \forall f \in L^2(\mathbf{R})$$

Hence, $\{\psi_{j,k}\}_{k \in \mathbf{Z}}$ is an orthonormal basis for $\mathbf{W}_j$ $\forall j \in \mathbf{Z}$.

Since the subspaces $\mathbf{W}_j$ have the scaling property $f(x) \in \mathbf{W}_j \Leftrightarrow f(2x) \in \mathbf{W}_{j-1}$, $\{\psi_{0,k}\}_{k \in \mathbf{Z}}$ is an orthonormal basis of $\mathbf{W}_0$ iff $\{\psi_{j,k}\}_{k \in \mathbf{Z}}$ is an orthonormal basis of $\mathbf{W}_j$ $\forall j \in \mathbf{Z}$.

The above observations imply that it is sufficient to find a function $\psi \in \mathbf{W}_0 \ni \{\psi_{0,k}\}_{k \in \mathbf{Z}}$ is an orthonormal basis of $\mathbf{W}_0$.

Construction of ψ : Now $\psi \in \mathbf{V}_0 \subset \mathbf{V}_{-1}$ and $\{\phi_{-1,k}\}_{k \in \mathbf{Z}}$ is an orthonormal basis for $\mathbf{V}_{-1}$. Therefore,

$$\phi = \sum_{k \in \mathbf{Z}} h_k \phi_{-1,k}$$

where

$$h_k = \langle \phi, \phi_{-1,k} \rangle \text{ and } \sum_{k \in \mathbf{Z}} |\, h_k \,|^2 = 1$$

since $\langle \phi, \phi \rangle = 1$.

$$\phi_{-1,k}(x) = \sqrt{2}\phi(2x - k) \Rightarrow \phi(x) = \sqrt{2} \sum_{k \in \mathbf{Z}} h_k \phi(2x - k)$$

$$\therefore \hat{\phi}(\xi) = \frac{1}{\sqrt{2}} \sum_{k \in \mathbf{Z}} h_k \hat{\phi}\left(\frac{\xi}{2}\right) e^{-ik\xi/2} = m_0\left(\frac{\xi}{2}\right)\hat{\phi}\left(\frac{\xi}{2}\right)$$

with

$$m_0(\xi) = \frac{1}{\sqrt{2}} \sum_{k \in \mathbf{Z}} h_k e^{-ik\xi}$$

$$\therefore m_0(\xi + 2\pi) = m_0(\xi)$$

Since $\{\phi_{0,k}\}_{k \in \mathbf{Z}}$ is an orthonormal set of functions in $L^2(\mathbf{R})$, we have

$$\int_{-\infty}^{\infty} \phi(x)\overline{\phi(x - k)}dx = \int_{-\infty}^{\infty} |\, \hat{\phi}(\xi) \,|^2 \, e^{ik\xi} d\xi$$

$$= \int_{0}^{2\pi} \sum_{n \in \mathbf{Z}} |\, \hat{\phi}(\xi + 2\pi n) \,|^2 \, e^{ik\xi} d\xi = \delta_{k,0} \quad \forall k \in \mathbf{Z}$$

$$i.e., \sum_{n \in \mathbf{Z}} |\, \hat{\phi}(\xi + 2\pi n) \,|^2 = (2\pi)^{-1} \text{ a.e.}$$

$$\because \hat{\phi}(\xi) = m_0\left(\frac{\xi}{2}\right)\hat{\phi}\left(\frac{\xi}{2}\right),$$

$$\sum_{k \in \mathbf{Z}} |\, \hat{\phi}(\xi + 2\pi k) \,|^2 = \sum_{k \in \mathbf{Z}} |\, m_0\left(\frac{\xi}{2} + \pi k\right)\hat{\phi}\left(\frac{\xi}{2} + \pi k\right) \,|^2 = (2\pi)^{-1} \text{ a.e.}$$

Letting $\zeta = \frac{\xi}{2}$,

$$\sum_{k \in \mathbf{Z}} |\, m_0(\zeta + \pi k)\hat{\phi}(\zeta + \pi k) \,|^2 = (2\pi)^{-1} \text{ a.e.}$$

i.e.,

$$\sum_{k \in \mathbf{Z}} |\, m_0(\zeta + 2\pi k) \,|^2 |\, \hat{\phi}(\zeta + 2\pi k) \,|^2$$

$$+ \sum_{k \in \mathbf{Z}} |\, m_0(\zeta + \pi + 2\pi k) \,|^2 |\, \hat{\phi}(\zeta + \pi + 2\pi k) \,|^2$$

$$= (2\pi)^{-1}$$

Using $m_0(\zeta + 2\pi) = m_0(\zeta)$ and $\sum_{k \in \mathbf{Z}} |\hat{\phi}(\zeta + 2\pi k)|^2 = (2\pi)^{-1}$ a.e., we have

$$|m_0(\zeta)|^2 + |m_0(\zeta + \pi)|^2 = 1 \text{ a.e.}$$

$$f \in \mathbf{W}_0 \Rightarrow f \in \mathbf{V}_{-1} \text{ and } f \perp \mathbf{V}_0$$

$$f \in \mathbf{V}_{-1} \Rightarrow f = \sum_{n \in \mathbf{Z}} f_n \phi_{-1,n'}$$

where

$$f_n = \langle f, \phi_{-1,n} \rangle \quad \forall n \in \mathbf{Z}$$

$$\therefore \hat{f}(\xi) = \frac{1}{\sqrt{2}} \sum_{n \in \mathbf{Z}} f_n \hat{\phi}(\frac{\xi}{z}) e^{-in\xi/2} = m_f(\frac{\xi}{2}) \hat{\phi}(\frac{\xi}{2}),$$

where

$$m_f(\xi) = \frac{1}{\sqrt{2}} \sum_{n \in \mathbf{Z}} f_n e^{-in\xi}$$

$$\therefore (\xi + 2\pi) = m_f(\xi)$$

$$f \perp \mathbf{V}_0 \Rightarrow \langle f, \phi_{0,n} \rangle = 0 \quad \forall k \in \mathbf{Z}.$$

i.e.,

$$\int_{-\infty}^{\infty} \hat{f}(\xi) \overline{\hat{\phi}(\xi)} e^{ik\xi} d\xi = 0 \quad \forall k \in \mathbf{Z}$$

$$\therefore \int_{0}^{2\pi} \sum_{n \in \mathbf{Z}} \hat{f}(\xi + 2\pi n) \overline{\hat{\phi}(\xi + 2\pi n)} e^{ik\xi} d\xi = 0 \quad \forall k \in \mathbf{Z}$$

Thus,

$$\sum_{n \in \mathbf{Z}} \hat{f}(\xi + 2\pi n) \overline{\hat{\phi}(\xi + 2\pi n)} = 0 \quad \text{a.e.}$$

where this series converges absolutely in $L^1([-\pi, \pi])$. *i.e.,*

$$\sum_{n \in \mathbf{Z}} m_f(\frac{\xi}{2} + \pi n) \hat{\phi}(\frac{\xi}{2} + \pi n) \overline{m_0(\frac{\xi}{2} + \pi n) \hat{\phi}(\frac{\xi}{2} + \pi n)} = 0 \quad \text{a.e.}$$

Letting $\frac{\xi}{2} = \zeta$,

$$\sum_{n \in \mathbf{Z}} m_f(\zeta + 2\pi n) |\hat{\phi}(\zeta + 2\pi n)|^2 \overline{m_0(\zeta + 2\pi n)}$$

$$+ \sum_{n \in \mathbf{Z}} m_f(\zeta + \pi + 2\pi n) |\hat{\phi}(\zeta + \pi + 2\pi n)|^2 \overline{m_0(\frac{\xi}{2} + \pi + 2\pi n)} = 0 \quad \text{a.e.}$$

i.e., $m_f(\zeta) \overline{m_0(\zeta)} + m_f(\zeta + \pi n) \overline{m_f(\zeta + \pi n)} = 0 \quad \text{a.e.}$

Since $\mid m_0(\zeta)\mid^2 + \mid m_0(\zeta + \pi)\mid^2 = 1$, a.e., $m_0(\zeta)$ and $m_0(\zeta + \pi)$ cannot vanish simultaneously on a set of nonzero measure.

Let $m_f(\zeta) = \mu(\zeta)\overline{m_0(\zeta + \pi)}$ a.e., where $\mu(\zeta)$ is a rational function of $e^{i\zeta}$, then $\mu(\zeta) + \mu(\zeta + \pi) = 0$ a.e. and $\mu(\zeta + 2\pi) = \mu(\zeta)$. $\therefore$ $\mu(\zeta) = e^{i\zeta}\lambda(2\zeta)$ where λ is 2π-periodic, and $m_f(\zeta) = e^{i\zeta}\lambda(2\zeta)\overline{m_0(\zeta + \pi)}$ a.e. Thus,

$$\hat{f}(\xi) = m_f(\tfrac{\xi}{2})\hat{\phi}(\tfrac{\xi}{2}) = e^{i\frac{\xi}{2}}\lambda(\xi)\overline{m_0(\tfrac{\xi}{2} + \pi)}\hat{\phi}(\tfrac{\xi}{2})$$

Rewriting this, we have

$$\hat{f}(\xi) = [e^{i\frac{\xi}{2}}\overline{m_0(\tfrac{\xi}{2} + \pi)}(\tfrac{\xi}{2})] \cdot \lambda(\xi),$$

where $\lambda(\xi)$ is the only factor dependent on f. Choosing

$$\hat{\psi}(\xi) = e^{i\frac{\xi}{2}}\overline{m_0(\tfrac{\xi}{2} + \pi)}\hat{\phi}(\tfrac{\xi}{2}),$$

we get $\hat{f}(\xi) = \lambda(\xi)\hat{\psi}(\xi)$. Letting

$$\lambda(\xi) = \sum_{k \in \mathbf{Z}}\lambda_k e^{-ik\xi} (\lambda \in L^2([0, 2\pi]),$$

we have

$$f(x) = \sum_{k \in \mathbf{Z}}\lambda_k \psi(x - k).$$

This suggests that the $\psi(x - k)$, $k \in \mathbf{Z}$ are a basis $\mathbf{W}_0$. Also $\psi \in L^2(\mathbf{R})$:

$$\hat{\psi}(\xi) = e^{i\frac{\xi}{2}}\overline{m_0(\tfrac{\xi}{2} + \pi)}\hat{\phi}(\tfrac{\xi}{2})$$

$$= \sqrt{2}\sum_{n \in \mathbf{Z}}\bar{h}_n(-1)^n \int_{-\infty}^{\infty}\hat{\phi}(\xi)e^{i\xi(2x+n+1)}d\xi$$

$$= \sqrt{2}\sum_{n \in \mathbf{Z}}(-1)^{n-1}\overline{h_{-n-1}}\phi(2x - n) \in \mathbf{V}_{-1}$$

$\psi \in \mathbf{W}_0$:

Indeed,

$$\hat{\phi}_{0,k}(\xi) = \hat{\phi}(\xi)e^{-ik\xi} = m_0(\tfrac{\xi}{2})\hat{\phi}(\tfrac{\xi}{2})e^{-ik\xi},$$

and

$$\hat{\psi}(\xi) = e^{i\frac{\xi}{2}}\overline{m_0(\tfrac{\xi}{2} + \pi)}\hat{\phi}(\tfrac{\xi}{2}).$$

$$\langle \hat{\phi}_{0,k}, \hat{\psi} \rangle = \int_{-\infty}^{\infty} m_0(\tfrac{\xi}{2}) m_0(\tfrac{\xi}{2} + \pi) \mid \hat{\phi}(\xi) \mid^2 e^{-i\frac{\xi}{2}(2k+1)} d\xi$$

$$= 2 \sum_{n \in \mathbf{Z}} \int_0^{2\pi} m_0(\zeta) m_0(\zeta + \pi) \mid \hat{\phi}(\zeta + 2\pi n) \mid^2 e^{-i\zeta(2k+1)} d\zeta$$

$$\frac{1}{\pi} \int_0^{2\pi} m_0(\zeta) m_0(\zeta + \pi) e^{-i\zeta(2k+1)} d\zeta$$

(Interchanging the summation and the integral and using $\sum_{n \in \mathbf{Z}} \mid \hat{\phi}(\zeta + 2\pi n) \mid^2 = (2\pi)^{-1}$)

$$= \frac{1}{2\pi} \int_0^{2\pi} \sum_{n \in \mathbf{Z}} \sum_{n \in \mathbf{Z}} (-1)^m h_n h_m e^{-i\zeta(n+m+2k+1)} d\zeta$$

(using $m_0(\zeta) = \frac{1}{2\pi} \sum_{n \in \mathbf{Z}} h_n e^{-in\zeta}$)

$$= \sum_{n+m=-(2k+1)} (-1)^m + h_m h_n$$

$$= \frac{1}{2} \sum_{n+m=-(2k+1)} (-1)^m h_n h_m + (-1)^m h_m h_n$$

$$= \frac{1}{2} \sum_{n+m=-(2k+1)} ((-1)^{n+m} + 1) h_m h_n = 0$$

Therefore, $\psi \perp \phi_{0,k}$ $\forall k \in \mathbf{Z}$, *i.e.*, $\psi \perp \mathbf{V}_0$ and along with $\psi \in \mathbf{V}_{-1}$ we have $\psi \in \mathbf{W}_0$.

$\{\psi_{0,k}\}_{k \in \mathbf{Z}}$ is an orthonormal system :

$$\int_{-\infty}^{\infty} \psi(x) \overline{\psi(x-k)} dx = \int_{-\infty}^{\infty} \mid \hat{\psi}(\xi) \mid^2 e^{ik\xi} d\xi$$

$$= \int_0^{2\pi} \sum_{n \in \mathbf{Z}} \mid \hat{\psi}(\xi + 2\pi n) \mid^2 e^{ik\xi} d\xi.$$

$$\sum_{n \in \mathbf{Z}} \mid \hat{\psi}(\xi + 2\pi n) \mid^2 = \sum_{n \in \mathbf{Z}} \mid m_0(\tfrac{\xi}{2} + n\pi + \pi) \mid^2 \mid \hat{\phi}(\tfrac{\xi}{2} + n\pi) \mid^2$$

$$= \sum_{n \in \mathbf{Z}} \mid m_0(\tfrac{\xi}{2} + 2(n+1)\pi) \mid^2 \mid \hat{\phi}(\tfrac{\xi}{2} + (2n+1)\pi) \mid^2$$

$$+ \sum_{n \in \mathbf{Z}} \mid m_0(\tfrac{\xi}{2} + 2n\pi + \pi) \mid^2 \mid \hat{\phi}(\tfrac{\xi}{2} + 2n\pi) \mid^2$$

$$=| m_0(\tfrac{\xi}{2}) |^2 \sum_{n\in\mathbf{Z}} | \hat{\phi}(\tfrac{\xi}{2} + (2n+1)\pi) |^2 + | m_0(\tfrac{\xi}{2} + \pi) |^2 \sum_{n\in\mathbf{Z}} | \hat{\phi}(\tfrac{\xi}{2} + 2n\pi) |^2$$

$$=| m_0(\tfrac{\xi}{2}) |^2 \sum_{n\in\mathbf{Z}} | \hat{\phi}(\tfrac{\xi}{2} + \pi + 2n\pi) |^2 + | m_0(\tfrac{\xi}{2} + \pi) |^2 \sum_{n\in\mathbf{Z}} | \hat{\phi}(\tfrac{\xi}{2} + 2n\pi) |^2$$

$$= (2\pi)^{-1} \left\{ | m_0(\tfrac{\xi}{2}) |^2 + | m_0(\tfrac{\xi}{2} + \pi) |^2 \right\} = (2\pi)^{-1} \text{ a.e.}$$

$$\therefore \langle \psi, \psi_{0,k} \rangle = \delta_{0,k}$$

$$\langle \hat{\phi}_{0,k}, \hat{\psi}_{0,j} \rangle = \int_{-\infty}^{\infty} m_0(\tfrac{\xi}{2}) m_0(\tfrac{\xi}{2}+\pi) | \hat{\phi}(\xi) |^2 \, e^{-i\frac{\xi}{2}(2(k-j)+1)} d\xi = 0 \quad \forall j,k \in \mathbf{Z}$$

Therefore, $\{\psi_{0,k}\}_{k\in\mathbf{Z}} \subset \mathbf{W}_0$.

To verify that $\{\psi_{0,k}\}_{k\in\mathbf{Z}}$ is an orthonormal basis for $\mathbf{W}_0$, it is sufficient to show that $\hat{f}(\xi) = \lambda(\xi)\hat{\psi}(\xi)$, $\lambda \in L^2([0, 2\pi])$. This then proves that $\forall f \in \mathbf{W}_0$,

$$f = \sum_{k\in\mathbf{Z}} \lambda_k \psi_{0,k} \text{ with } \sum_{k\in\mathbf{Z}} | \lambda_k |^2 < \infty$$

$$\int_0^{2\pi} | \lambda(\xi) |^2 \, d\xi = 2 \int_0^{\pi} | \mu(\zeta) |^2 \, d\zeta$$

Now,

$$\int_0^{2\pi} | m_f(\xi) |^2 \, d\xi = \int_0^{2\pi} | m_0(\xi+\pi) |^2 | \mu(\xi) |^2 \, d\xi$$

$$= \int_0^{2\pi} | \mu(\xi) |^2 \left[| \mu(\xi) |^2 + | m_0(\xi+\pi) |^2 \right] d\xi$$

$$= \int_0^{\pi} | \mu(\xi) |^2 \, d\xi = \frac{1}{2} \int_0^{2\pi} | \lambda(\xi) |^2 \, d\xi$$

$$\int_0^{2\pi} | \lambda(\xi) |^2 \, d\xi = 2\pi \, || f ||^2 < \infty$$

This proves that $\{\psi_{0,k}\}_{k\in\mathbf{Z}}$ is an orthonormal basis for $\mathcal{W}_0$.

In the definition of an MRA, the requirement that $\{\phi_{0,k}\}_{k\in\mathbf{Z}}$ forms an orthonormal basis for $\mathbf{V}_0$ can be replaced by a weaker requirement.

Definition 5.11 *In a Hilbert space* **H**, *a basis* $\{\varphi_i\}_{i\in\mathbf{x}}$ *is called a* Riesz basis *if*

$$\exists \, 0 < \alpha \le \beta < \infty \quad \ni \alpha \, || v ||^2 \le \sum_{i\in\mathbf{I}} | \langle v, \varphi_i \rangle |^2 \le \beta \, || v ||^2 \qquad \forall v \in \mathbf{H}$$

Remark 5.12 Any bounded operator **A** with a bounded inverse maps any orthonormal basis to a Riesz basis. Also all Riesz bases can be obtained as images of orthonormal bases under such operators.

The weaker condition on $\{\phi_{0,k}\}_{k\in\mathbf{Z}}$ is then : The $\{\phi_{0,k}\}_{k\in\mathbf{Z}}$ form a Riesz basis for $\mathbf{V}_0$.

One can construct an orthonormal basis $\{\phi_{0,k}^{\#}\}_{k\in\mathbf{Z}}$ from a Riesz basis $\{\phi_{0,k}\}_{k\in\mathbf{Z}}$ of $\mathbf{V}_0$. Indeed, $\{\phi_{0,k}\}_{k\in\mathbf{Z}}$ is a Riesz basis of $\mathbf{V}_0 \Leftrightarrow span\langle\{\phi_{0,k}\}_{k\in\mathbf{Z}}\rangle = \mathbf{V}_0$ and $\exists 0 < A \leq B < \infty \ni \forall c = (c_k)_{k\in\mathbf{Z}} \in l^2(\mathbf{Z})$,

$$A \sum_{k\in\mathbf{Z}} | c_k |^2 \leq \left\| \sum_{k\in\mathbf{Z}} c_k \phi_{0,k} \right\|^2 \leq B \sum_{k\in\mathbf{Z}} | c_k |^2 .$$

$$\left\| \sum_{k\in\mathbf{Z}} c_k \phi_{0,k} \right\|^2 = \int_{-\infty}^{\infty} | \sum_{k\in\mathbf{Z}} c_k e^{-ik\xi} \hat{\phi}(\xi) |^2 \, d\xi$$

$$\int_0^{2\pi} | \sum_{k\in\mathbf{Z}} c_k e^{-ik\xi} |^2 \sum_{k\in\mathbf{Z}} | \hat{\phi}(\xi + 2\pi n) |^2 \, d\xi.$$

$$\sum_{k\in\mathbf{Z}} | c_k |^2 = \frac{1}{2\pi} \int_0^{2\pi} | \sum_{k\in\mathbf{Z}} c_k e^{-ik\xi} |^2 \, d\xi$$

$$\therefore 0 < A(2\pi)^{-1} \leq \sum_{n\in\mathbf{Z}} | \hat{\phi}(\xi + 2n\pi) |^2 \leq B(2\pi)^{-1} < \infty \quad \text{a.e.}$$

Defining

$$\hat{\phi}^0(\xi) = (2\pi)^{-1/2} \left[\sum_{n\in\mathbf{Z}} | \hat{\phi}(\xi + 2n\pi) |^2 \right]^{-1/2} \hat{\phi}(\xi),$$

it is easy to see that

(a) $\phi^0 \in L^2(\mathbf{R})$, and

(b) $\sum_{n\in\mathbf{Z}} | \hat{\phi}^0(\xi + 2n\pi) |^2 = (2\pi)^{-1}$ a.e.

Therefore, $\{\phi_{0,k}^0\}_{k\in\mathbf{Z}}$ is an orthonormal set in $L^2(\mathbf{R})$.

The space $\mathbf{V}_0^0$ spanned by the $\{\phi_{0,k}^0\}_{k\in\mathbf{Z}}$ is given by

$$\mathbf{V}_0^0 = \left\{ f : f = \sum_{n\in\mathbf{Z}} f_n^0 \phi_{0,n}^0, \ (f_n^0)_{n\in\mathbf{Z}} \in l^2(\mathbf{Z}) \right\}$$

$$= \{f : \hat{f} = v'\hat{\phi}^0 \text{ with } v2\pi \text{ - periodic, } v \in L^2([0, 2\pi])\}$$
$$= \{f : \hat{f} = v'\hat{\phi} \text{ with } v'2\pi \text{ - periodic, } v' \in L^2([0, 2\pi])\}$$
$$\{f : f = \sum_{n\in\mathbf{Z}} f_n\phi_{0,n}, \ (f_n)_{n\in bfZ} \in l^2(\mathbf{Z})\} = \mathbf{V}_0$$

Thus, we may redefine an MRA as follows :

Definition 5.13 *An MRA of $L^2(\mathbf{R})$ is a sequence of closed subspaces $\{\mathbf{V}_j\}_{j\in\mathbf{Z}}$ of, $L^2(\mathbf{R})$ satisfying the following properties :*

(a) $\mathbf{V}_j \subset \mathbf{V}_{j-1} \ \forall j \in \mathbf{Z}$; *(nesting property)*

(b) $Closure_{L^2(\mathbf{R})}\left(\bigcup_{j\in\mathbf{Z}} \mathbf{V}_j\right) = L^2(\mathbf{R})$; *(density of the union in $L^2(\mathbf{R})$)*

(c) $\bigcap_{j\in\mathbf{Z}} \mathbf{V}_j = \{0\}$;

(d) $f(x) \in \mathbf{V}_j \Leftrightarrow f(2x) \in \mathbf{V}_{j-1} \ \forall j \in \mathbf{Z}$; *(scaling property)*

(e) $f(x) \in \mathbf{V}_0 \Rightarrow f(x - n) \in \mathbf{V}_0 \ \forall n \in \mathbf{Z}$; *(invariance under integral translations)*

(f) $\exists \phi \in \mathbf{V}_0 \ni \{\phi_{0,n}\}_{n\in\mathbf{Z}}$, *is a Riesz basis of $\mathbf{V}_0$, where* $\phi_{j,k}(x) = 2^{-j/2}\phi(2^{-j}x - k) \ \forall j, k \in \mathbf{Z}$; *(existence of a scaling function)*.

It is also possible to construct an MRA by first choosing an appropriate scaling function ϕ_0 and obtaining $\mathbf{V}_0$ by taking the linear span of integer translates of ϕ. The other subspaces $\mathbf{V}_j$ can be generated as scaled versions of $\mathbf{V}_0$.

5.4.1 Constructing an MRA from a scaling function

Let $\phi \in L^2(\mathbf{R})$ such that the following two requirements hold :

(a) $\phi(x) = \sum_{k\in\mathbf{Z}} c_k\phi(2x - k)$ where $\sum_{k\in\mathbf{Z}} | c_k |^2 < \infty$.

(b) $\exists 0 < \alpha \le \beta < \infty \ni 0 < \alpha \le \sum_{n\in\mathbf{Z}} | \hat{\phi}(\xi + 2n\pi) |^2 \le \beta < \infty$ a.e.

Define $\mathbf{V}_j$ to be the closed subspace spanned by the set of vectors $\{\phi_{j,k}\}_{k\in\mathbf{Z}}$ given by

$$\phi_{j,k}(x) = 2^{-j/2}\phi(2^{-j}x - k) \ \forall j, k \in \mathbf{Z}$$

Then conditions (a) and (b) are necessary and sufficient to guarantee that:

1. $\{\phi_{j,k}\}_{k\in\mathbf{Z}}$ is a Riesz basis in each $\mathbf{V}_j$, and the ladder property $\mathbf{V}_j \subset \mathbf{V}_{j-1}$ $\forall j \in \mathbf{Z}$;

2. $f(x) \in \mathbf{V}_j \Leftrightarrow f(2x) \in \mathbf{V}_{j-1}$ $\forall f \in L^2(\mathbf{R})$;

3. $f(x) \in \mathbf{V}_0 \Leftrightarrow f(x-n) \in \mathbf{V}_0$ $\forall n \in \mathbf{Z}$.

To verify that the ladder of spaces generated by ϕ forms an MRA, it is sufficient to show that the following properties also hold:

4. $\bigcap_{j\in\mathbf{Z}} \mathbf{V}_j = \{0\}$.

5. $closure_{L^2(\mathbf{R})}(\bigcup_{j\in\mathbf{Z}} \mathbf{V}_j) = L^2(\mathbf{R})$.

The following two theorems verify this.

Theorem 5.14 *If $\phi \in L^2(\mathbf{R})$ is such that*

$$0 < A \leq \sum_{n\in\mathbf{Z}} |\hat{\phi}(\xi + 2n\pi)|^2 \leq B < \infty \quad a.e.$$

for constants A and B and

$$\mathbf{V}_j = closure_{L^2(\mathbf{R})}(span\langle\{\phi_{j,k}\}_{k\in\mathbf{Z}}\rangle),$$

then

$$\bigcap_{j\in\mathbf{Z}} \mathbf{V}_j = \{0\}$$

Proof : Let $f = \sum_{k\in\mathbf{Z}} f_k\phi_{0,k} \in \mathbf{V}_0$. Then

$$\| f \|^2 = \int_{-\infty}^{\infty} \sum_{jkI\in\mathbf{Z}} f_j\bar{f}_k e^{-i\xi(j-k)} |\hat{\phi}(\xi)|^2 d\xi < \infty \quad (\because f \in \mathbf{V}_0 \subset L^2(\mathbf{R}))$$

$$\therefore \| f \|^2 = \sum_{jk\in\mathbf{Z}} f_j\bar{f}_k \int_{-\infty}^{\infty} e^{-i\xi(j-k)} |\hat{\phi}(\xi)|^2 d\xi$$

$$= \sum_{jk\in\mathbf{Z}} f_j\bar{f}_k \int_{-\infty}^{\infty} e^{-i\xi(j-k)} \sum_{n\in\mathbf{Z}} |\hat{\phi}(\xi + 2n\pi)|^2 d\xi$$

Since

$$0 < A \leq \sum_{n\in\mathbf{Z}} |\hat{\phi}(\xi + 2n\pi)|^2 \leq B < \infty \text{ a.e.}$$

$$\therefore A\sum_{k\in\mathbf{Z}} |f_k|^2 \leq \| f \|^2 \leq B\sum_{k\in\mathbf{Z}} |f_k|^2 \leq \quad \therefore \sum_{k\in\mathbf{Z}} |f_k|^2 < \infty$$

Also,

$$B^{-1} \| f \|^2 \leq \sum_{k \in \mathbf{Z}} | f_k |^2 \leq A^{-1} \| f \|^2$$

Now,

$$| \langle f, \phi_{0,k} \rangle | = \left| \int_{-\infty}^{\infty} \sum_{j,k \in \mathbf{Z}} f_j e^{-i\xi(j-k)} | \hat{\phi}(\xi) |^2 \, d\xi_0 \right|^2$$

$$\therefore A^2 | f_k |^2 \leq | \langle f, \phi_{0,k} \rangle |^2 \leq B^2 | f_k |^2$$

$$\therefore A^2 \sum_{k \in \mathbf{Z}} | f_k |^2 \leq \sum_{k \in \mathbf{Z}} | \langle f, \phi_{0,k} \rangle | \leq B^2 \sum_{k \in \mathbf{Z}} | f_k |^2$$

i.e.,

$$B^{-1} A^2 \| f \|^2 \leq \sum_{k \in \mathbf{Z}} | \langle f, \phi_{0,k} \rangle |^2 \leq A^{-1} B^2 \| f \|^2$$

Therefore, $\{\phi_{0,k}\}_{k \in \mathbf{Z}}$ is a Riesz basis of $\mathbf{V}_0$.

The map $f \mapsto \mathbf{D}^j f$ given by $(\mathbf{D}^j f)(x) = 2^{-j/2} f(2^{-j} x)$ is unitary (*i.e.,* $\| \mathbf{D}^j f \| = \| f \|$). Since $Image(\mathbf{D}^j \mathbf{V}_0) = \mathbf{V}_j$, and $\mathbf{D}^j \phi_{0,k} = \phi_{0,k}$, $\forall f \in \mathbf{V}_j$, we have

$$B^{-1} A^2 \| f \|^2 \leq \sum_{k \in \mathbf{Z}} | \langle f, \phi_{j,k} \rangle |^2 \leq A^{-1} B^2 \| f \|^2 \, .$$

Hence, $\{\phi_{j,k}\}_{k \in \mathbf{Z}}$ is a Riesz basis of $\mathbf{V}_j$ $\forall j \in \mathbf{Z}$.

Let

$$f \in \bigcap_{j \in \mathbf{Z}} \mathbf{V}_j$$

Then, $\forall \varepsilon > 0$ there exists a compactly supported and continuous function $f_\varepsilon \ni \| f - f_\varepsilon \| \leq \varepsilon$. If $\mathbf{P}_j$ is the orthogonal projection operator on $L^2(\mathbf{R})$ onto $\mathbf{V}_j$, then

$$\| f - \mathbf{P}_j f_\varepsilon \| = \| \mathbf{P}_j (f - f_\varepsilon) \| \leq \| f - f_\varepsilon \| \leq \varepsilon$$

$$\therefore \| f \| = \varepsilon + \| \mathbf{P}_j f_\varepsilon \| \quad \forall j \in \mathbf{Z}$$

Now,

$$\| \mathbf{P}_j f_\varepsilon \| \leq B^{-1/2} A^{-1} \left[\sum_{k \in \mathbf{Z}} | \langle f_\varepsilon, \phi_{j,k} \rangle |^2 \right]^{1/2}$$

and

$$\sum_{k \in \mathbf{Z}} | \langle f_\varepsilon, \phi_{0,k} \rangle |^2 \leq 2^{-j} \sum_{k \in \mathbf{Z}} \left[\int_{|x| \leq M} | f_\varepsilon(x) | \, | \phi(2^{-j} x - k) | \, dx \right]^2$$

(where sup $f_\epsilon \subset [-M, M]$)

$$\leq 2^{-j} \sup_{x \in \mathbf{R}} \mid f_\epsilon \mid^2 \cdot 2M \sum_{k \in \mathbf{Z}} \left[\int_{|x| \leq M} \mid \phi(2^{-j} x - k) \mid dx \right]^2$$

$$= \sup_{x \in \mathbf{R}} \mid f_\epsilon \mid^2 \cdot 2M \cdot \int_{R(j)} \mid \phi(y) \mid^2 dy$$

where

$$\mathbf{R}(j) = \bigcup_{k \in \mathbf{Z}} [k - 2^{-j} M, k + 2^{-j} M]$$

with j chosen such that $2^{-j} M \leq 1/2$. Therefore,

$$\sum_{k \in \mathbf{Z}} \mid \langle f_\epsilon, \phi_{0,k} \rangle \mid^2 \leq 2M \sup_{x \in \mathbf{R}} \mid f_\epsilon \mid^2 \int_{-\infty}^{\infty} \chi_{\mathbf{R}(j)}(x) \mid \phi(x) \mid^2 dx,$$

where $\chi_{\mathbf{R}(j)}(x)$ is the characteristic function of the set $\mathbf{R}(j)$. As $j \to \infty$, the sequence

$$\{\chi_{\mathbf{R}(j)}(x)\}_{j \geq \frac{\log M}{\log 2} + 1} \to 0 \quad \forall x \notin \mathbf{Z}$$

Therefore, choosing j large enough so that

$$\sum_{k \in \mathbf{Z}} \mid \langle f, \phi_{j,k} \rangle \mid^2 \leq B^{-1} A^{-2} \epsilon^2,$$

we have $\| \mathbf{P}_j f_\epsilon \| \leq \epsilon$. *i.e.*, $\| f \| \leq 2\epsilon$. Since $\epsilon > 0$ is arbitrary, $\| f \| = 0$ $\therefore f = 0$. Hence, $\bigcap_{j \in \mathbf{Z}} \mathbf{V}_j = \{0\}$. To verify that

$$Closure_{L^2(\mathbf{R})} \left(\bigcup_{j \in \mathbf{Z}} \mathbf{V}_j \right) = L^2(\mathbf{R})$$

it is necessary to assume that $\hat{\phi}(\xi)$ is bounded and that $\hat{\phi}(0) \neq 0$

Theorem 5.15 *If $\phi \in L^2(\mathbf{R})$ is such that*

(a) $0 < A \leq \sum_{n \in \mathbf{Z}} \mid \hat{\phi}(\xi + 2n\pi) \mid^2 \leq B < \infty$ *for constants A and B,*

(b) $\hat{\phi}(\xi)$ *is bounded $\forall \xi \in \mathbf{R}$,*

(c) $\hat{\phi}(\xi)$ *continuous at $\xi = 0$ with $\hat{\phi}(0) \neq 0$,*

(d) $\mathbf{V}_j = closure_{l^2(\mathbf{R})} (span\langle \{\phi_{j,k}\}_{k \in \mathbf{Z}} \rangle)$, *then*

$$Closure_{L^2(\mathbf{R})} \left(\bigcup_{j \in \mathbf{Z}} V_j \right) = L^2(\mathbf{R})$$

Proof : $\forall j \in \mathbf{Z}$ and $\forall j \in \mathbf{V}_j$, we have

$$B^{-1}A^2 \parallel f \parallel^2 \leq \sum_{k \in \mathbf{Z}} \mid \langle f, \phi_{j,k} \rangle \mid^2 \leq A^{-1}B^2 \parallel f \parallel^2$$

Let

$$f \in (\bigcup_{j \in \mathbf{Z}} \mathbf{V}_j)^{\perp}.$$

Given an arbitrary $\varepsilon > 0$, there exists a C^{∞}-function f_{ε} with compact support such that $\parallel f - f_{\varepsilon} \parallel \leq \varepsilon$. ($C_c^{\infty}(\mathbf{R})$ is dense in $L^2(\mathbf{R})$) Therefore, $\forall j \in \mathbf{Z}$

$$\parallel \mathbf{P}_j f_{\varepsilon} \parallel = \parallel \mathbf{P}_j(f_{\varepsilon} - f) \parallel \leq \varepsilon \quad (\because \ \mathbf{P}_j f = 0)$$

Also, since

$$0 < B^{-1}A^2 \parallel f \parallel^2 \leq \sum_{k \in \mathbf{Z}} \mid \langle f, \phi_{j,k} \rangle \mid^2 \leq A^{-1}B^2 \parallel f \parallel^2 < \infty,$$

$$\parallel \mathbf{P}_j f_{\varepsilon} \parallel^2 \geq B^{-1}A^{-2}[\sum_{k \in \mathbf{Z}} \mid \langle f_{\varepsilon}, \phi_{j,k} \rangle \mid^2]$$

Now,

$$\sum_{k \in \mathbf{Z}} \mid \langle f_{\varepsilon}, \phi_{j,k} \rangle \mid^2 = \sum_{k \in \mathbf{Z}} \mid \int_{-\infty}^{\infty} \hat{f}_{\varepsilon}(\xi) 2^{j/2} \overline{\hat{\phi}(2^j \xi)} e^{ik2^j \xi} d\xi \mid^2$$

$$\sum_{k \in \mathbf{Z}} 2^j \mid \int_0^{2\pi 2^{-j}} \sum_{n \in \mathbf{Z}} \hat{f}_{\varepsilon}(\xi + 2\pi n 2^{-j}) \overline{\hat{\phi}(2^j \xi + 2\pi n)} e^{ik2^j \xi} d\xi \mid^2$$

$$2\pi \int_0^{2\pi 2^{-j}} \mid \sum_{n \in \mathbf{Z}} \hat{f}_{\varepsilon}(\xi + 2\pi n 2^{-j}) \overline{\hat{\phi}(2^j \xi + 2\pi n)} e^{ik2^j \xi} d\xi \mid^2$$

(Parseval's theorem for periodic functions)

$$= 2\pi \sum_{m \in \mathbf{Z}} \int_{-\infty}^{\infty} \hat{f}_{\varepsilon}(\xi) \overline{\hat{f}(\xi + 2\pi m 2^{-j})} \overline{\hat{\phi}(2^j \xi)} \hat{\phi}(2^{j^J} \xi + 2\pi m) d\xi$$

$$= 2\pi \int_{-\infty}^{\infty} \mid \hat{f}_{\varepsilon}(\xi) \mid^2 \overline{\mid \hat{\phi}(2^j \xi) \mid}^2 d\xi + \mathbf{E}(f)$$

where

$$| \mathbf{E}(f) | = 2\pi \left| \sum_{0 \neq m \in \mathbf{Z}} \int_{-\infty}^{\infty} \hat{f}_\varepsilon(\xi) \overline{\hat{f}(\xi + 2\pi m 2^{-j})} \hat{\phi}(2^j \xi) \overline{\hat{\phi}(2^j \xi + 2\pi m)} d\xi \right|$$

Since $f_\varepsilon \in C_c^\infty(\mathbf{R})$, there exists a constant $C \ni | \hat{f}_\varepsilon(\xi) | \leq C(1 + | \xi |^2)^{-3/2}$. Therefore,

$$| \mathbf{E}(f) | \leq C^2 \sup_{\xi \in \mathbf{R}} | \hat{\phi}(\xi) |^2 \sum_{0 \neq m \in \mathbf{Z}} \int_{-\infty}^{\infty} (1 + | \xi + 2^{-j} \pi m |^2)^{-3/2}$$
$$(1 + | \xi - 2^{-j} \pi m |^2)^{-3/2} d\xi$$
$$\leq C^2 \sup_{\xi \in \mathbf{R}} | \hat{\phi}(\xi) |^2 \sum_{0 \neq m \in \mathbf{Z}} (1 + 2^{-2j} \pi^2 m^2)^{-1/2} \int_{-\infty}^{\infty} (1 + | \zeta |^2)^{-1} d\zeta$$

(applying twice the bound $\sup_{x,y \in \mathbf{R}} (1 + y^2) [1 + (x-y)^2]^{-1} [1 + (x+y)^2]^{-1} < \infty$)

$$\leq C' 2^j$$
$$\therefore 2\pi \int_{-\infty}^{\infty} | \hat{f}_\varepsilon(\xi) |^2 | \hat{\phi}(2^j \xi) |^2 d\xi \leq B\varepsilon^2 + C'' 2^j$$

As $\hat{\phi}(\xi)$ is uniformly bounded and continuous at $\xi = 0$,

$$\int_{-\infty}^{\infty} | \hat{f}_\varepsilon(\xi) |^2 | \hat{\phi}(2^j \xi) |^2 d\xi \rightarrow | \hat{\phi}(0) |^2 || \hat{f}_\varepsilon ||^2 \quad \text{as } j \rightarrow -\infty$$

$$\therefore || \hat{f}_\varepsilon ||^2 \leq C | \hat{\phi}(0) |^{-1} \varepsilon$$

where C is independent of ε. Since $\varepsilon > 0$ is arbitrary, $|| f || = 0$.

If conditions $\hat{\phi}(\xi)$ are uniformly bounded and $\hat{\phi}(\xi)$ is continuous at $\xi = 0$ with $\hat{\phi}(0) \neq 0$, then impose restrictions on the coefficients c_n in the two-scale relation

$$\phi(x) = \sum_{n \in \mathbf{Z}} c_n \phi(2x - n)$$

Indeed,

$$\phi(x) = \sum_{n \in \mathbf{Z}} c_n \phi(2x - n) \Rightarrow \hat{\phi}(\xi) = m_0(\frac{\xi}{2}) \hat{\phi}(\frac{\xi}{2})$$

where

$$m_0(\xi) = \frac{1}{2} \sum_{n \in \mathbf{Z}} c_n e^{-in\xi}$$

i.e., $\hat{\phi}(0) = m_0(0) \hat{\phi}(0)$. Therefore,

$$m_0(0) = \frac{1}{2} \sum_{n \in \mathbf{Z}} c_n = 1$$

Therefore,

$$\sum_{n \in \mathbf{Z}} c_n = 2$$

$\hat{\phi}(\xi) = m_0(\frac{\xi}{2})\hat{\phi}(\frac{\xi}{2}) \Rightarrow m_0$ is continuous except possibly at the zeros of $\hat{\phi}$. Therefore, m_0 is continuous at zero. Further, if $\hat{\phi}(\xi) \leq C(1+ \mid \xi \mid)^{-1/2-\varepsilon}$, then $\hat{\phi}$ is continuous; therefore,

$$\sum_{n \in \mathbf{Z}} \mid \hat{\phi}(\xi + 2\pi n) \mid^2$$

is continuous. Therefore,

$$\hat{\phi}^0(\xi) = (2\pi)^{-1/2} \left[\sum_{n \in \mathbf{Z}} \mid \hat{\phi}(\xi + 2\pi n) \mid^2 \right]^{-1/2} \hat{\phi}(\xi)$$

is continuous. Thus,

$$m_0^0(\xi) = \frac{\hat{\phi}^0(2\xi)}{\hat{\phi}^0(\xi)} \quad \text{satisfies} \quad m_0^0(0) = 1$$

Now, since $\mid m_0^0(\xi) \mid^2 + \mid m_0^0(\xi + \pi) \mid^2 = 1$, $m_0^0(\pi) = 0$. *i.e.*, $m_0(\pi) = 0$.

$$\left(\because m_0^0(\xi) = m_0(\xi) \left[\sum_{n \in \mathbf{Z}} \mid \hat{\phi}(\xi + 2\pi n) \mid^2 \right]^{1/2} \left[\sum_{n \in \mathbf{Z}} \mid \hat{\phi}(\xi + 2\pi n) \mid^2 \right]^{-1/2} \right),$$

i.e., $\sum_{n \in \mathbf{Z}} c_n(-1)^n = 0$. Therefore

$$\sum_{n \in \mathbf{Z}} c_n = 2 \quad \text{and} \quad \sum_{n \in \mathbf{Z}} (-1)^n c_n = 0 \quad \Rightarrow \quad \sum_{n \in \mathbf{Z}} c_{2n} = 1 \quad = \quad \sum_{n \in \mathbf{Z}} c_{2n+1}$$

Thus, we have a recipe for constructing orthonormal wavelet bases :

1. Choose $\phi \in L^2(\mathbf{R})$ with the following properties :

(a) $\hat{\phi}$ is bounded and continuous at 0;

(b) ϕ and $\hat{\phi}$ exhibit satisfactory decay as $x, \xi \to \pm\infty$;

(c) $\int_{-\infty}^{\infty} \phi(x)dx \neq 0$ (*i.e.*, $\hat{\phi}(0) = 0$);

(d) $\exists 0 < \alpha \leq \beta < \infty \ni \alpha \leq \sum_{n \in \mathbf{Z}} \mid \hat{\phi}(\xi + 2\pi n) \mid^2 \leq \beta$ a.e.;

(e) $\phi(x) = \sum_{n \in \mathbf{Z}} c_n \phi(2x - n)$

2. If $\{\phi_{j,k}\}_{k\in\mathbf{Z}}$ is not an orthonormal set, then set

$$\hat{\phi}^0(\xi) = (2\pi)^{-1/2} \left[\sum_{n\in\mathbf{Z}} | \hat{\phi}(\xi + 2\pi n) |^2 \right]^{-1/2} \hat{\phi}(\xi)$$

3. Define $\hat{\psi}(\xi) = e^{i\frac{\xi}{2}} \overline{m_0^0(\frac{\xi}{2} + \pi)} \hat{\phi}^0(\frac{\xi}{2})$, where

$$m_0^0(\xi) = m_0(\xi) \left[\sum_{n\in\mathbf{Z}} | \hat{\phi}(\xi + 2\pi n) |^2 \right]^{1/2} \left[\sum_{n\in\mathbf{Z}} | \hat{\phi}(\xi + 2\pi n) |^2 \right]^{-1/2} ;$$

i.e., $\psi = \sum_{n\in\mathbf{Z}} (-1)^n h^0_{-n+1} \phi^0_{0,n}$ and $m_0^0(\xi) = \frac{1}{\sqrt{2}} \sum_{n\in\mathbf{Z}} h^0_n e^{-in\xi}.$

5.5 Wavelet Decomposition and Reconstruction of Functions

5.5.1 Multiresolution decomposition and reconstruction of functions in $\mathbf{L}^2(\mathbf{R})$

Given $\{\langle f, \phi_{j,k}\rangle\}_{k\in\mathbf{Z}}$ for some $f \in L^2(\mathbf{R})$, and some $j \in \mathbf{Z}$, one can rescale units and assume that $j = 0$, *i.e.*, $\langle f, \phi_{j,k}\rangle \mapsto \langle f, \phi_{0,k}\rangle$ $\forall k \in \mathbf{Z}$. We now compute the wavelet coefficients $\{\langle f, \psi_{j,k}\rangle\}_{k\in\mathbf{Z}}$ $\forall j \geq 1$:

Recall that

$$\psi = \sum_{n\in\mathbf{Z}} g_n \phi_{-1,n}, \quad g_n = \langle \psi, \phi_{-1,n}\rangle = (-1)^n h_{-n+1} \quad \forall n \in \mathbf{Z}.$$

Therefore,

$$\psi_{j,k}(x) = 2^{-j/2} \psi(2^{-j}x - k) = 2^{-j/2} \sum_{n\in\mathbf{Z}} g_n 2^{1/2} \phi(2^{-j+1}x - 2k - n)$$

$$= \sum_{n\in\mathbf{Z}} g_n \phi_{j-1,2k+n}(x) = \sum_{n\in\mathbf{Z}} g_{n-2k} \phi_{j-1,n}(x)$$

$$\therefore \langle f, \psi_{1,k}\rangle = \sum_{n\in\mathbf{Z}} \overline{g_{n-2k}} \langle f, \phi_{0,n}\rangle \quad \forall k \in \mathbf{Z}$$

In other words, the sequence

$$\{\langle f, \psi_{1,k}\rangle\}_{k\in\mathbf{Z}}$$

is obtained by convolution of the sequence

$$\{\langle f, \phi_{0,n}\rangle\}_{n\in\mathbf{Z}}$$

with the sequence

$$\bar{\mathbf{g}} = \{\bar{g}_{-n}\rangle_{n\in\mathbf{Z}}$$

and the retention of only the even-indexed elements of the resulting sequence. Also,

$$\forall j \in \mathbf{Z} \quad \langle f, \psi_{j,k}\rangle = \sum_{n\in\mathbf{Z}} \overline{g_{n-2k}}\langle f, \phi_{j-1,n}\rangle \quad \forall k \in \mathbf{Z};$$

i.e., if the sequence

$$\{\langle f, \phi_{j-1,n}\rangle\}_{n\in\mathbf{Z}}$$

is known, then the sequence

$$\{\langle f, \psi_{j,k}\rangle\}_{k\in\mathbf{Z}}$$

is obtained by convolving the sequence

$$\{\langle f, \phi_{j-1,n}\rangle\}_{n\in\mathbf{Z}}$$

with the sequence

$$\bar{\mathbf{g}} = \{\bar{g}_{-n}\}_{n\in\mathbf{Z}}$$

and retaining only the even-indexed terms of the resulting sequence. Moreover, since

$$\phi_{j,k}(x) = 2^{-j/2}\phi(2^{-j}x - k) = \sum_{n\in\mathbf{Z}} h_{n-2k}\phi j - 1, n(x) \quad \forall j, k \in \mathbf{Z},$$

we have $\forall j \in \mathbf{Z}$

$$\langle f, \phi_{j,k}\rangle = \sum_{n\in\mathbf{Z}} \overline{h_{n-2k}}\langle f, \phi_{j-1,n}\rangle \quad \forall k \in \mathbf{Z}.$$

The above two formulas give rise to the following procedure of obtaining the wavelet coefficients

$$\{\langle f, \psi_{j,k}\rangle\}_{j,k\in\mathbf{Z}}$$

starting from the sampled sequence

$$\{\langle f, \phi_{0,n}\rangle\}_{n\in\mathbf{Z}}$$

of $f \in L^2(\mathbf{R})$.

For $j \geq 1$, the sequence

$$\{\langle f, \phi_{j,n} \rangle\}_{n \in \mathbf{Z}}$$

can be obtained from the sequence

$$\{\langle f, \phi_{j-1,n} \rangle\}_{n \in \mathbf{Z}}$$

by convolving it with the sequence

$$\bar{\mathbf{g}} = \{\bar{g}_{-n}\}_{n \in \mathbf{Z}}$$

and retaining only the even terms of the resulting sequence.

The following diagram illustrates the above procedure:

$$
\begin{array}{ccccccc}
\{\langle f, \phi_{0,n} \rangle\}_{n \in \mathbf{Z}} & \rightarrow & \{\langle f, \phi_{1,n} \rangle\}_{n \in \mathbf{Z}} & \rightarrow & \cdots & \{\langle f, \phi_{j-1,n} \rangle\}_{n \in \mathbf{Z}} & \rightarrow & \cdots \\
\downarrow & & \downarrow & & & \downarrow & & \\
\{\langle f, \psi_{1,k} \rangle\}_{k \in \mathbf{Z}} & & \{\langle f, \psi_{2,k} \rangle\}_{k \in \mathbf{Z}} & & & \{\langle f, \psi_{j,k} \rangle\}_{k \in \mathbf{Z}} & &
\end{array}
$$

Thus, one can compute successively coarser approximations of f along with the difference in information between successive levels of approximation.

If $f \in L^2(\mathbf{R})$, denote by f^j the projection $\mathbf{P}_j f$ of f onto the closed subspace $\mathbf{V}_j$ of $L^2(\mathbf{R})$, and denote by g^j the projection $\mathbf{Q}_j f$ of f onto the closed subspace $\mathbf{W}_j$ of $L^2(\mathbf{R})$. Then,

$$f^{j-1} = f^j + g^j$$

since

$$\mathbf{V}_{j-1} = \mathbf{V}_j \oplus \mathbf{W}_j$$

and

$$f^j = g^j + g^{j+1} + L + g^{j+m} + f^{j+m}$$

since

$$\mathbf{V}_j = \bigoplus_{k=0}^{m} \mathbf{W}_{j+k} \oplus \mathbf{V}_{j+m}$$

If $a = \{a_n\}_{n \in \mathbf{Z}}$ is any sequence, then the sequence $\{a_{2n}\}_{n \in \mathbf{Z}}$ is denoted by $a \downarrow_2$. The mapping $a \downarrow_2$ is called *downsampling* by factor 2. Thus,

$$(..., a_{-4}, a_{-3}, a_{-2}, a_{-1}, a_0, a_1, a_2, a_3, a_4, ...) \xrightarrow{\downarrow_2} (..., a_{-4}, a_{-2}, a_0, a_2, a_4, ...)$$

If $a = \{a_n\}_{n\in\mathbf{Z}}$ is any sequence, then the sequence $\{b_n\}_{n\in\mathbf{Z}}$ defined by

$$b_n = \begin{cases} a_n/2 & \text{if } n \text{ is even} \\ 0 & \text{if } n \text{ is odd} \end{cases}$$

is denoted by $a \uparrow_2$. The mapping a a $a \uparrow_2$ is called *upsampling* by factor 2. Thus,

$$(\ldots, a_{-3}, a_{-2}, a_{-1}, a_0, a_1, a_2, a_3, \ldots) \xrightarrow{\uparrow_2}$$

$$\xrightarrow{\uparrow_2} (\ldots, 0, a_{-3}, 0, a_{-2}, 0, a_{-1}, 0, a_0, 0, a_1, 0, a_2, 0, a_3, 0, \ldots)$$

If $a = \{a_n\}_{n\in\mathbf{Z}}$ is any sequence, denote by $\bar{a}$ the sequence $\{\bar{a}_n\}_{n\in\mathbf{Z}}$, and for any other sequence $b = \{b_n\}_{n\in\mathbf{Z}}$, define the action of the operator $a\#$ on the sequence b, denoted by $a\#b$, to be the sequence

$$\left\{ \sum_{m\in\mathbf{Z}} a_{m-2n} b_m \right\}_{n\in\mathbf{Z}}$$

i.e.,

$$(a\#b)_n = \sum_{m\in\mathbf{Z}} a_{m-2n} b_m$$

Denote $c_k^j = \langle f, \phi_{j,k} \rangle$ and $d_k^j = \langle f, \psi_{j,k} \rangle$. Thus

$$c^j = \{c_n^j\} n \in \mathbf{Z} = \{\langle f, \phi_{j,n} \rangle\}_{n\in\mathbf{Z}}$$

and

$$d^j = \{d_n^j\} n \in \mathbf{Z} = \{\langle f, \psi_{j,n} \rangle\}_{n\in\mathbf{Z}}$$

The formulas $\forall j \in \mathbf{Z}$

$$\langle f, \psi_{j,k} \rangle = \sum_{n\in\mathbf{Z}} \overline{g_{n-2k}} \langle f, \phi_{j-1,n} \rangle \quad \forall k \in \mathbf{Z}$$

and $\forall j \in \mathbf{Z}$

$$\langle f, \phi_{j,k} \rangle = \sum_{n\in\mathbf{Z}} \overline{h_{n-2k}} \langle f, \phi_{j-1,n} \rangle \quad \forall k \in \mathbf{Z}$$

can then be written as

$$c_k^j = \sum_{n\in\mathbf{Z}} \overline{h_{n-2k}} c_n^{j-1} = (\bar{h}\#c^{j-1})_k$$

and

$$d_k^j = \sum_{n\in\mathbf{Z}} \overline{g_{n-2k}} c_n^{j-1} = (\bar{g}\#c^{j-1})_k,$$

i.e., $c^j = \bar{h}\#c^{j-1}$ and $d^j = \bar{g}\#c^{j-1}$.

These formulas give the effect of the orthonormal basis transformation

$$\{\phi_{j,n}\}_{n\in\mathbf{Z}} \;\mapsto\; (\{\phi_{j+1,n}\}_{n\in\mathbf{Z}}, (\{\psi_{j+1,n}\}_{n\in\mathbf{Z}})$$

in $\mathbf{V}_j$ on the coefficients of the projection of $f \in L^2(\mathbf{R})$ onto $\mathbf{V}_j$. This fine-to-coarse decomposition of the projection f^0 of f onto the subspaces $\mathbf{V}_j$ and $\mathbf{W}_j$ is schematically represented by the following diagram :

$$
\begin{array}{ccccccccc}
f^0 & & f^1 & & f^2 & & f^3 & & \\
c^0 & \xrightarrow{\ \overline{H}\ } & c^1 & \xrightarrow{\ \overline{H}\ } & c^2 & \xrightarrow{\ \overline{H}\ } & c^3 & \xrightarrow{\ \overline{H}\ } & \cdots \\
& \overline{G}\searrow & & \overline{G}\searrow & & \overline{G}\searrow & & \overline{G}\searrow & \cdots \\
& & d^1 & & d^2 & & d^3 & & \\
& & g^1 & & g^2 & & g^3 & &
\end{array}
$$

In practical applications, this decomposition process is stopped after finitely many iterations giving a finite sequence of sequences :

$$c^0 \mapsto (d^1, d^2, d^3, K, d^j, c^j)$$

where c^0 is the coefficient sequence of f^0, the d^j $1 \le j \le J$ are the wavelet coefficient sequences representing the difference in information between two consecutive levels of resolution of f^0, and c^J is the coefficient sequence of the projection of f^0 in $\mathbf{V}_J$.

As the above decomposition operation is equivalent to a cascade of orthonormal basis transformations, the reconstruction operation is the adjoint of the decomposition operation. That is, since $f^{j-1} = f^j + g^j$,

$$
\begin{aligned}
c_n^{j-1} &= \langle f^{j-1}, \phi_{j-1,n}\rangle = \langle f^j + g^j, \phi_{j-1,n}\rangle \\[2mm]
&= \Big\langle \sum_{k\in\mathbf{Z}} c_k^j \phi_{j,k} + \sum_{k\in\mathbf{Z}} d_k^j \psi_{j,k},\, \phi_{j-1,n}\Big\rangle \\[2mm]
&= \sum_{k\in\mathbf{Z}} c_k^j \langle \phi_{j,k}, \phi_{j-1,n}\rangle + \sum_{k\in\mathbf{Z}} d_k^j \langle \psi_{j,k}, \phi_{j-1,n}\rangle
\end{aligned}
$$

Also, since

$$\phi_{j,k} = \sum_{n\in\mathbf{Z}} h_{n-2k}\,\phi_{j-1,n}$$

and

$$\psi_{j,k} = \sum_{n\in\mathbf{Z}} g_{n-2k}\,\phi_{j-1,n'}$$

we have

$$c_n^{j-1} = \sum_{k\in\mathbf{Z}} [h_{n-2k} c_k^j + g_{n-2k} d_k^j]$$

That is, $c^{j-1} = h^* \# c^j + g^* \# d^j$, where

$$(a^* \# b)_n = \sum_{m \in \mathbf{Z}} a_{n-2m} b_m$$

and $a^* \#$ is the adjoint of the operator $a\#$.

The above formula suggests a reconstruction algorithm to construct finer resolutions of f from its coarser resolutions and the differences in information between successive resolutions. That is, the coefficient sequence c^{j-1} of f^{j-1} in $\mathbf{V}_{j-1}$ is obtained as follows :

The coefficient sequence c^j of f^{j-1} in $\mathbf{V}_{j-1}$ and the coefficient sequence c^j of f^j in $\mathbf{V}_j$ (representing the difference in information between the $(j-1)$th and jth levels of resolution) are both upsampled by factor 2. The resulting sequences are then convolved with the sequences h and g, respectively, and the resulting convolution products are added up term-by-term to yield the sequence c^{j-1}. Thus, we have the following fast wavelet algorithm.

5.6   The Fast Wavelet Algorithm

This algorithm consists of two sub-algorithms.

1. Wavelet Decomposition Algorithm :

Given by the formulas

$$c_k^j = \sum_{n \in \mathbf{Z}} \overline{h_{n-2k}} c_n^{j-1} \text{ and } d_k^j = \sum_{n \in \mathbf{Z}} \overline{g_{n-2k}} c_n^{j-1}$$

or $c^j = \bar{h} \# c^{j-1}$ and $d^j = \bar{g} \# d^{j-1}$ and represented schematically by

$$
\begin{array}{ccccccccc}
f^0 & & f^1 & & f^2 & & f^3 & & \\
c^0 & \xrightarrow{\overline{H}} & c^1 & \xrightarrow{\overline{H}} & c^2 & \xrightarrow{\overline{H}} & c^3 & \xrightarrow{\overline{H}} & \cdots \\
& \overline{G} \searrow & & \overline{G} \searrow & & \overline{G} \searrow & & \overline{G} \searrow & \cdots \\
& & d^1 & & d^2 & & d^3 & & \\
& & g^1 & & g^2 & & g^3 & &
\end{array}
$$

2. Wavelet Reconstruction Algorithm :

Given by the formula

$$c_n^{j-1} = \sum_{k \in \mathbf{Z}} [h_{n-2k} c_k^j + g_{n-2k} d_k^j]$$

or $c^j = \bar{h}\#c^j + \bar{g}\#d^j$ and represented schematically by the following diagram:

$$
\begin{array}{ccccccccc}
f^j & & f^{j-1} & & \cdots & & f^1 & & f^0 \\
c^j & \xrightarrow{\overline{H}} & c^{j-1} & \xrightarrow{\overline{H}} & \cdots & \xrightarrow{\overline{H}} & c^1 & \xrightarrow{\overline{H}} & c^0 \\
& \overline{G}\searrow & & \overline{G}\searrow & & & & \overline{G}\searrow & \\
d^j & & d^{j-1} & & & & d^1 & & \\
g^j & & g^{j-1} & & & & g^1 & &
\end{array}
$$

We now briefly introduce the Z-transform of a sequence along with a few of its basic properties. This will enable us to rewrite the decomposition and reconstruction formulas in a form more conducive for discussion of the FWT from the standpoint of subband filtering schemes later.

If $a = \{a_n\}_{n \in \mathbf{Z}}$ is any sequence, define by

$$
\mathbf{Z}(a) = A(z) = \sum_{n \in \mathbf{Z}} a_n z^n
$$

its Z-transform, which is a formal series (i.e., without regard to convergence issues). If $b = \{b_n\}_{n \in \mathbf{Z}}$ is any other sequence, then it is easily verified that

$$
\mathbf{Z}(aob) = \sum_{n \in \mathbf{Z}} (\sum_{m \in \mathbf{Z}} a_{n-m} b_m) z^n = (\sum_{n \in \mathbf{Z}} a_n z^n)(\sum_{n \in \mathbf{Z}} b_n z^n) = A(z)B(z)
$$

where the product is evaluated term-by-term in the formal algebraic sense.

For any sequence $a = \{a_n\}_{n \in \mathbf{Z}}$, defined

$$
\bar{A}(z) = \sum_{n \in \mathbf{Z}} \overline{a_{-n}} z^n
$$

if $A(z)$ is the Z-transform of the sequence $a = \{a_n\}_{n \in \mathbf{Z}}$ then $\frac{1}{2}[A(z) + A(-z)]$ is the Z-transform of the downsampled sequence $a \downarrow_2 = \{a_{2n}\}_{n \in \mathbf{Z}}$, and $A(z^2)$ is the Z-transform of the upsampled sequence

$$
a \uparrow_2 = \{0 \text{ if } n \equiv 1 \ (\text{mod } 2), a_{n/2} \text{ if } n \equiv 0 \ (\text{mod})\}_{n \in \mathbf{Z}}
$$

The decomposition formulas can be rewritten as

$$
C^{j-1}(z^2) = \frac{1}{2}[\bar{H}(z)C^j(z) + \bar{H}(-z)C^j(-z)]
$$

and

$$
D^{j-1}(z^2) = \frac{1}{2}[\bar{G}(z)C^j(z) + \bar{G}(-z)C^j(-z)]
$$

while the reconstruction formula can be rewritten as

$$
C^j(z) = [H(z)C^{j-1}(z^2) + G(z)D^{j-1}(z^2)]
$$

where $C^j(z)$ and $D^j(z)$ are the Z-transforms of c^j and d^j, respectively.

Both the decomposition and the reconstruction formulas can be combined into a single identity by substituting the former in the latter as follows

$$C^j(z) = \frac{1}{2}[H(z)\bar{H}(z)+G(z)\bar{G}(z)]C^j(z)+\frac{1}{2}[H(z)\bar{H}(-z)+G(z)\bar{G}(-z)]C^j(-z)$$

This method of decomposing and reconstructing a function (signal) is called a subband filtering scheme and illustrated by the following diagram:

$$
\begin{array}{ccccccccc}
& & \bar{H} & \to & \downarrow_2 & \to & c^j \downarrow_2 & \to & \uparrow_2 & H \\
& \nearrow & & & & & & & & & \searrow \\
c^{j-1} & & & & & & & & & & & \oplus & = c^{j-1} \\
& \searrow & & & & & & & & & \nearrow \\
& & \bar{G} & \to & \downarrow_2 & \to & d^j \downarrow_2 & \to & \uparrow_2 & G
\end{array}
$$

The fast wavelet algorithm can be used to modify a signal's characteristics locally in frequency and time without affecting the signal globally. This is extremely important for purposes of filtering and compression of signals.

In practical situations f is given in a sampled form, say $\{f(n)\}_{n\in\mathbf{Z}}$. Then, $c^0 = \{\langle f, \phi_{0,n}\rangle\}_{n\in\mathbf{Z}}$ can be computed by assuming that $f \in \mathbf{V}_0$.

That is,

$$f = \sum_{k\in\mathbf{Z}}\langle f, \phi_{0,k}\rangle\phi_{0,k} \ f(n) = \sum_{k\in\mathbf{Z}}\langle f, \phi_{0,k}\rangle\phi(n-k)$$

If $f = \{f(n)\}_{n\in\mathbf{Z}}$ and $\Phi = \{\phi(n)\}_{n\in\mathbf{Z}}$, then since

$$f(n) = \sum_{k\in\mathbf{Z}}\langle f, \phi_{0,k}\rangle\phi(n-k)$$

$$\mathbf{Z}(f) = \mathbf{Z}(c^0)\mathbf{Z}(\Phi)$$

$$\sum_{n\in\mathbf{Z}}f(n)z^{-n} = (\sum_{j\in\mathbf{Z}}\langle f, \phi_{0,j}\rangle z^{-j})(\sum_{k\in\mathbf{Z}}\phi(k)z^{-k})$$

in the sense that the coefficients of like powers of z on both sides are equal. Since $f = \{f(n)\}_{n\in\mathbf{Z}}$, $\{\langle f, \phi_{0,n}\rangle\}_{n\in\mathbf{Z}}$ and $\Phi = \{\phi(n)\}_{n\in\mathbf{Z}}$ are all in $l^2(\mathbf{Z})$, we can let $z = e^{i\xi}$ to get

$$\sum_{n\in\mathbf{Z}}f(n)e^{-in\xi} = (\sum_{j\in\mathbf{Z}}\langle f, \phi_{0,j}\rangle e^{-ij\xi})(\sum_{k\in\mathbf{Z}}\phi(k)e^{-ik\xi})$$

where each sum is in $L^2([0, 2\pi])$. Thus,

$$\sum_{j\in\mathbf{Z}}\langle f\phi_{0,j}\rangle e^{-ik\xi} = (\sum_{n\in\mathbf{Z}}f(n)e^{-in\xi})(\sum_{m\in\mathbf{Z}}\phi(k)e^{-im\xi})^{-1}$$

and, therefore,

$$\langle f\phi_{0,j}\rangle e^{-in\xi} = \frac{1}{2\pi} \int_0^{2\pi} \left(\sum_{n\in\mathbf{Z}} f(n)e^{-in\xi}\right)\left(\sum_{m\in\mathbf{Z}} \phi(k)e^{-im\xi}\right)^{-1} e^{ik\xi} d\xi$$

$$= \sum_{n\in\mathbf{Z}} f(n)s_{k-n'}$$

where

$$s_n = \frac{1}{2\pi} \int_0^{2\pi} \left(\sum_{k\in\mathbf{Z}} \phi(k)e^{-ik\xi}\right)^{-1} e^{in\xi} d\xi$$

In the following chapter we will consider some important examples of wavelets and develop some simple and effective methods for constructing wavelet functions.

Chapter 6

Construction of Wavelets

So far, we have discussed wavelets abstractly through their properties. The question that naturally arises is whether it is possible to systematically construct wavelet bases with certain desired properties. In this chapter, we will discuss an important method of construction of wavelets of arbitrary smoothness. As we have noted before, there is quite a bit of freedom that we may exercise in choosing the wavelet function generating a wavelet basis. Therefore, the specific choice and the method of construction of a wavelet function depends on the requirements and motivation for its construction. We first explore a rich arsenal of functions giving rise to wavelets, namely spline functions. Next, we take a signal processing perspective and study the fast wavelet algorithm discussed in the previous chapter from this point of view. We then construct two important classes of compactly supported wavelet bases, namely the compactly supported orthonormal and the biorthogonal wavelet bases. These wavelet bases give rise to FIR and linear phase FIR subband filtering schemes, respectively. We obtain characterizations of such subband filtering schemes arising from these wavelet bases.

6.1   The Battle-Lemarié Family of Wavelets

An elegant and simple method of construction of orthonormal wavelet bases employs cardinal B-spline functions . The resulting class of wavelets are known as the Battle-Lemarié wavelets , after G. Battle and P.G. Lemarié who constructed these wavelets independently using different techniques. Cardinal B-splines are piecewise polynomial functions with compact supports that are easy to realize and implement and can be made as smooth as desired. We describe the construction of the Battle-Lemarié B-spline wavelets below.

6.1.1 Cardinal B-splines

For each $n \in N$ let $C^n(\mathbf{R})$ be the space of all functions that are n times continuously differentiable everywhere on $\mathbf{R}$, $C^0(\mathbf{R}) = C(\mathbf{R})$ be the space of all functions that are continuous everywhere on $\mathbf{R}$, and $C^{-1}(\mathbf{R})$ be the space of all functions that are piecewise continuous everywhere on $\mathbf{R}$. Also, for each $n \in N$ let $P^n[x]$ be the space of all polynomials of degree at most n in x.

Definition 6.1 *For every $n \in N$, the space C^n of **cardinal splines of order n with knot sequence $\mathbf{Z}$** is the set*

$$\{f : f \in C^{n-2}(\mathbf{R}) \ , \text{ and } f \mid_{[k,k+1]} \in P^{n-1}[x], k \in \mathbf{Z}\}$$

For instance, the space C^1 consists of all piecewise constant functions. Clearly, the set $\{\chi(x - k) : k \in \mathbf{Z}\}$ is a basis for C^1, where χ is the characteristic function of the half open unit interval $[0, 1]$.

To obtain a basis for C^n $n \in \mathbf{N}$, consider the set C^n_M given by

$$C^n_M = \{g : g = f \mid_{[-M,M]} \text{ for some } f \in C^n\}$$

where M is a positive integer. If f is any function in C^n_M, let $f_{n,k} = f\mid_{[k,k+1]}$ for $j = -M, ..., M - 1$. Since $f \in C^{n-2}(\mathbf{R})$,

$$(f^{(i)}_{n,k} - f^{(i)}_{n,k-1})(k) = 0 \ \ \forall 0 \le i \le n - 2, \ n \ge 2$$

The difference between the right limit and the left limit of f^{n-1} at the knot points $k \in \mathbf{Z}$ is given by

$$c_k = f^{(n-1)}_{n,k}(k + 0) - f^{(n-1)}_{n,k-1}(k - 0) = \lim_{\varepsilon \to 0+} [f^{(n-1)}(k + \varepsilon) - f^{(n-1)}(k - \varepsilon)]$$

Hence, the relation between two neighboring segments of f is given by

$$f_{n,k}(x) = f_{n,k-1}(x) + \frac{c_k}{(n-1)!}(x - k)^{n-1}$$

and if $x_+ = \max(x, 0)$, and $x_+^n = (x_+)^n$ $\forall n \geq 1$, then

$$f(x) = f \mid_{[-M,-M+1]} (x) + \sum_{k=-M+1}^{M-1} \frac{c_k}{(n-1)!}(x-k)_+^{n-1}$$

$\forall x \in [-M, M]$ and $\forall f \in C_M^n$

Thus, the set

$$\{1, x, ..., x^{n-1}, (x+M-1)_+^{n-1}, ..., (x-M+1)_+^{n-1}\}$$

is a basis of C_M^n . Hence, C_M^n has dimension $2M + n - 1$.

Let $(\delta f)(x) = f(x+1) - f(x)$ and $(\delta^n f)(x) = (\delta^{n-1}(\delta f))(x) \forall n \geq 2$. If $p(x)$ is a polynomial of degree d in x, then $(\delta p)(x) = p(x+1) - p(x)$ is a polynomial of degree $d-1$ in x, and $(\delta^j p)(x) = (\delta^{j-1}(\delta p))(x)$ is a polynomial of degree $d-j$ $\forall 0 \leq j \leq d$. Hence, we may replace the monomials $1, x, ..., x^{n-1}$ in the above basis for C_M^n by the monomials $(x+n-1)^{n-1}, ..., x^{n-1}$. Also, since we are interested in values of x in the interval $[-M, M]$, we can further replace these monomials by the truncated monomials $(x+M+n-1)_+^{n-1}, ..., (x+M)_+^{n-1}$. Thus, the set $\{(x-j)_+^{n-1} : -M-n+1 \leq j \leq M-1\}$ is also a basis of C_M^n. Now, since

$$C^n = \bigcup_{M=1}^{\infty} C_M^n$$

the union of the bases of the C_M^n is a basis of C^n. That is, the set $\{(x-j)_+^{n-1} : j \in \mathbf{Z}\}$ is a basis of C^n.

While the elements of this basis have a simple and uniform structure, namely, that they are all obtained by taking integer translates of the single function x_+^{n-1}, for our purposes they suffer from the serious shortcoming that none of the basis elements are in $L^2(\mathbf{R})$. It is possible, however, to transform this basis into another by means of vector space operations alone such that the resultant basis is a subset of $L^2(\mathbf{R})$. This is done by letting

$$(\Delta f)(x) = f(x) - f(x-1)$$

and

$$(\Delta^n f)(x) = (\Delta^{n-1}(\Delta f))(x) \quad \forall n \geq 2$$

As in the case of the δ operator, if $p(x)$ is a polynomial of degree d in x, then $(\Delta p)(x)$ is a polynomial of degree $d-1$ in x, and $(\Delta^{d+1} p)(x) = 0$, $\forall x$. Also, let $M_1 = \chi_{[0,1]}$, the characteristic function of $[0,1]$, and

$$M_n(x) = \frac{1}{(n-1)!} \Delta^n x_+^{n-1} \quad \forall n \geq 2$$

Clearly, M_n is a finite linear combination of the basis functions $\{(x - j)_+^{n-1} : j \in \mathbf{Z}\}$. More precisely,

$$M_n(x) = \frac{1}{(n-1)!} \sum_{j=0}^{n} (-1)^j \binom{n}{j} (x - j)_+^{n-1} \quad \forall n \geq 2$$

Since $\forall j \geq 0$, $(x - j)_+^{n-1} = 0$ $\forall x < 0$, $M_n(x) = 0$ $\forall x < 0$. Also, since $(\Delta^{d+1} p)(x) = 0$ for any polynomial $p(x)$ of degree $\leq d$, $M_n(x) = 0$ $\forall x \geq n$. Hence, $\sup M_n \subseteq [0, n]$ (in fact $M_n = [0, n]$). Therefore, M_n is in $L^2(\mathbf{R})$. Since, for each $j \in \mathbf{Z}$, $M_n(x-j)$ has support, $[j, m+j][j, m+j] \cap [-M, M] \neq \Phi \Leftrightarrow -M - n + 1 \leq j \leq M - 1$. It can be verified that the set

$$\{M_n(x - j) : \ -M - n + 1 \leq j \leq M - 1\}$$

is linearly independent and, therefore, is a basis for C^n.

As before, we take the union of the bases for the spaces C_M^n to obtain a basis for C^n; *i.e.*, the set

$$\mathcal{B} = \{M_n(x - j) : \ j \in \mathbf{Z}\}$$

is a basis for C^n. The series

$$\sum_{j=-\infty}^{\infty} c_j \mathbf{M}_n(x - j)$$

converges pointwise everywhere since for each x, only finitely many of the $M_n(x - j)$ are nonzero.

The Battle-Lemarié orthonormal wavelet bases are constructed by generating an MRA using scaled versions of the space $\mathbf{C}^n$ as described below.

6.1.2   Cardinal B-spline MRA of $\mathbf{L}^2(\mathbf{R})$

Let

$$V_0^n = \operatorname*{closure}_{L^2(\mathbf{R})}(\mathbf{C}^n \cap L^2(\mathbf{R}))$$

Clearly, $\mathcal{B} \subset V_0^n$. It will be later shown that $\mathcal{B}$ is a Riesz basis of V_0^n.

If $C^{n;j}$ is the space of cardinal splines with knot sequence

$$2^j \mathbf{Z} = \{k2^j\}_{k \in \mathbf{Z}}, \quad j \in \mathbf{Z}$$

then clearly $C^{n;j} \subset C^{n;j'}$, hence we have a bi-infinite nested sequence of spline spaces corresponding to the dyadic knot sequences $2^{-j}\mathbf{Z}$, $\forall j \in \mathbf{Z}$:

$$\ldots \subset C^{n;2} \subset C^{n;1} \subset C^{n;0} \subset C^{n;-1} \subset C^{n;-2} \subset \ldots$$

where $C^{n;0} = \mathbf{C}^n$. Letting $\mathcal{V}_j^n = \underset{L^2(\mathbf{R})}{\mathrm{closure}}(C^{n;j} \cap L^2(\mathbf{R}))$, we have

$$\ldots \subset \mathcal{V}_1^n \subset \mathcal{V}_0^n \subset \mathcal{V}_{-1}^n \subset \ldots$$

If $\mathcal{B} = \mathcal{B}_0\{M_n(x-j) : j \in \mathbf{Z}\}$ is a Riesz basis of $\mathcal{V}_0^n$, then

$$\mathcal{B}_j = \{2^{-j/2}M_n(2^{-j}x - k) : k \in \mathbf{Z}\}$$

is a Riesz basis of $\mathcal{V}_j^n$, the Riesz bounds being independent of the scale j. It will be shown that

$$\underset{L^2(\mathbf{R})}{\mathrm{closure}}\left(\bigcup_{j \in \mathbf{Z}} \mathcal{V}_j^n\right) = L^2(\mathbf{R})$$

and

$$\bigcap \mathcal{V}_j^n = \{0\}$$

We now introduce a family of functions $\{N_n\}_{n \geq 1}$ which will be shown to coincide with the family of B-spline functions $\{M_n\}_{n \geq 1}$ defined above. This redefinition helps unravel some important properties of B-splines .

Let $N_1 = \chi_{[0,1]}$, and

$$N_n(x) = (N_{n-1} * N_1)(x) = \int_0^1 N_{n-1}(x-y)dy \quad \forall n \geq 2$$

We then have the following theorem.

Theorem 6.2 *If for $n \geq 1$ N_n is as defined above, then the following identities hold:*

(a) $\int_{-\infty}^{\infty} f(x)N_n(x)dx = \int_0^1 \ldots \int_0^1 f(x_1 + \ldots + x_n)dx_1\ldots dx_n \quad \forall f \in C(\mathbf{R})$

(b) $\int_{-\infty}^{\infty} f^n(x)N_n(x)dx = \sum_{j=0}^n (-1)^{n-j}\binom{n}{j}f(j) \quad \forall f \in C^n(\mathbf{R})$.

(c) $N_n(x) = M_n(x), \quad \forall x \in \mathbf{R}$

(d) $\mathrm{supp}\, N_n = [0, n]$

(e) $N_n(x) > 0$, for $0 < x < n$

(f) $\sum_{j=-\infty}^{\infty} N_n(x - j) = 1, \forall x \in \mathbf{R}$

(g) $N_n'(x) = (\Delta N_{n-1})(x) = N_{n-1}(x) - N_{n-1}(x - 1)$

(h) $N_n(x) = \frac{x}{n-1} N_{n-1}(x) + \frac{n-x}{n-1} N_{n-1}(x - 1)$

(i) $N_n(\frac{n}{2} + x) = N_n(\frac{n}{2} - x)$ $\forall x \in \mathbf{R}$

Proof : Property (a): Clearly, (a) is true for $n = 1$. Suppose it is true for $n = m - 1$. Then,

$$\int_{-\infty}^{\infty} f(x) N_m(x) dx = \int_{-\infty}^{\infty} f(x) \left\{ \int_0^1 N_{m-1}(x - y) dy \right\} dx$$

and by Fubini's theorem,

$$= \int_0^1 \left\{ \int_{-\infty}^{\infty} f(x) N_{m-1}(x - y) dx \right\} dy$$

Next, letting $x - y = u$,

$$= \int_0^1 \left\{ \int_{-\infty}^{\infty} f(u + y) N_{m-1}(u) du \right\} dy$$

and by hypothesis,

$$= \int_0^1 \int_0^1 \cdots \int_0^1 f(x_1 + \ldots + x_{m-1} + y) dx_1 \ldots dx_{m-1} dy.$$

Then, letting $y = x_m$,

$$\int_0^1 \cdots \int_0^1 f(x_1 + \ldots + x_m) dx_1 \ldots dx_m,$$

i.e., it is true for $n = m$. Thus, by induction (a) is true for all $n \geq 1$.

Property (b): Let $f \in \mathbf{C}^n(\mathbf{R})$. Clearly (b) is true for $n = 1$. Suppose it is true for $n = m - 1$. Now,

$$\int_{-\infty}^{\infty} f^{(m)}(x) N_m(x) dx = \int_0^1 \cdots \int_0^1 f^{(m)}(x_1 \ldots + x_m) dx_1 \ldots dx_m$$

$$= \int_0^1 \cdots \int_0^1 [f^{(m-1)}(1 + x_2 + \ldots + x_m) - f^{(m-1)}(x_2 + \ldots + x_m)] dx_2 \ldots dx_m$$

$$= \sum_{k=0}^{m-1} (-1)^{m-1-k} \binom{m-1}{k} f(1 + k) - \sum_{k=0}^{m-1} (-1)^{m-1-k} \binom{m-1}{k} f(k)$$

(the above by hypothesis)

$$= \sum_{k=1}^{m}(-1)^{m-k}\binom{m-1}{k-1}f(k) + \sum_{k=0}^{m-1}(-1)^{m-k}\binom{m-1}{k}f(k)$$

$$= f(m) + \sum_{k=1}^{m}(-1)^{m-k}\binom{m}{k}f(k) + (-1)^m f(0)$$

$$= \sum_{k=0}^{m}(-1)^{m-k}\binom{m}{k}f(k)$$

Hence, (b) is true for $n = m$ and, hence by induction, it is true for all $n \geq 1$.

Property (c): Observe that (c) is true for $n = 1$, since $M_1 = \chi_{[0,1]} = N_1$. Suppose it is true for $n = m - 1$. Then, $\forall f \in C(\mathbf{R})$

$$\int_{-\infty}^{\infty} f(x)N_m(x)dx = \int_{-\infty}^{\infty} f(x)\left\{\int_0^1 N_{m-1}(x-y)dy\right\}dx$$

$$= \int_{-\infty}^{\infty} f(x)\left\{\int_0^1 N_{m-1}(x-y)dy\right\}dx$$

(the above by hypothesis)

$$= \int_{-\infty}^{\infty} f(x)\left\{\int_0^1 \frac{1}{(m-2)!}\sum_{k=0}^{m-1}(-1)^k\binom{m-1}{k}(x-y-k)_+^{m-2}dy\right\}dx$$

$$= \int_{-\infty}^{\infty} f(x)\left\{\frac{1}{(m-2)!}\sum_{k=0}^{m-1}(-1)^k\binom{m-1}{k}\int_0^1(x-y-k)_+^{m-2}dy\right\}dx$$

$$= \int_{-\infty}^{\infty} f(x)\left\{\frac{1}{(m-1)!}\sum_{k=0}^{m-1}(-1)^{k+1}\binom{m-1}{k}[(x-k)_+^{m-1}-(x-1-k)_+^{m-1}]\right\}dx$$

$$= \int_{-\infty}^{\infty} f(x)\{\frac{1}{(m-1)!}\sum_{k=0}^{m-1}(-1)^k\binom{m-1}{k}(x-k)_+^{m-1}$$

$$+\frac{1}{(m-1)!}\sum_{k=1}^{m}(-1)^k\binom{m-1}{k-1}(x-k)_+^{m-1}\}dx$$

$$= \int_{-\infty}^{\infty} f(x)\left\{\frac{1}{(m-1)!}\sum_{k=0}^{m}(-1)^k\binom{m}{k}(x-k)_+^{m-1}\right\}dx$$

$$= \int_{-\infty}^{\infty} f(x)M_m(x)dx$$

Hence, (c) is true for $n = m$ Thus, by induction it is true $\forall n \geq 1$.

Properties (d), (e), (f) and (i): These follow readily from the definition of N_n and induction.

Property (g): Property (g) can be obtained by differentiating both sides of the identity

$$N_n(x) = \int_0^1 N_{n-1}(x-y)dy$$

Property (h): Property (h) can be derived by applying the Leibnitz rule for the n-fold difference of products of functions:

$$(\Delta^n fg)(x) = \sum_{k=0}^n \binom{n}{k}(\Delta^k f)(x)(\Delta^{n-k}g)(x-k)$$

which in turn may be derived by induction.

Letting $f(x) = x$ and $g(x) = x_+^{n-2}$ in the above formula, and noting that $x_+^{n-1} = x.x_+^{n-2}$, we have

$$
\begin{aligned}
N_n(x) &= M_n(x) = \frac{1}{(n-1)!}\Delta^n x_+^{n-1} \\
&= \frac{1}{(n-1)!}\left\{x\Delta^n x_+^{n-2} + n\Delta^{n-1}(x-1)_+^{n-2}\right\} \quad (\Delta^j f = 0 \ \forall j \ge 2) \\
&= \frac{1}{(n-1)!}\left\{x[\Delta^n x_+^{n-2} + \Delta^{n-1}(x-1)_+^{n-2}] + n\Delta^{n-1}(x-1)_+^{n-2}\right\}
\end{aligned}
$$

(by property vii)

$$
\begin{aligned}
&= \frac{x}{n-1}M_{n-1}(x) + \frac{n-x}{n-1}M_{n-1}(x-1) \\
&= \frac{x}{n-1}N_{n-1}(x) + \frac{n-x}{n-1}N_{n-1}(x-1)
\end{aligned}
$$

Theorem 6.3 $B_0 = \{N_n(x-j) : j \in \mathbf{Z}\}$ *is a Riesz basis of* $\mathbf{V}_0^n$.

Proof : By the remarks at the end of Chapter 4,

$$\sum_{k=-\infty}^{\infty} |\hat{N}_n(\xi + 2\pi k)|^2 = \sum_{k=-\infty}^{\infty}\left\{\int_{-\infty}^{\infty} N_n(x+k)\overline{N_n(x)}dx\right\}e^{-ik\xi}$$

Now, it can be easily shown from the definition of N_n that

$$\int_{-\infty}^{\infty} N_n(x+k)\overline{N_n(x)}dx = N_{2n}(n+k)$$

Using

$$N_n(x) = \frac{x}{n-1}N_{n-1}(x) + \frac{n-x}{n-1}N_{n-1}(x-1),$$

N_{2n} can be computed recursively. Also, since N_{2n} has compact support, only finitely many of the $N_{2n}(n+k)$, $k \in \mathbf{Z}$ are nonzero. Thus,

$$\sum_{k=-\infty}^{\infty} |\hat{N}_n(\xi + 2\pi k)|^2 = \sum_{k=-n+1}^{n-1} N_{2n}(n+k)e^{ik\xi}$$

Now using the properties $N_n(x) > 0$, $0 < x < n$, and

$$\sum_{j=-\infty}^{\infty} N_n(x-j) = 1 \quad \forall x \in \mathbf{R},$$

we conclude that

$$\sum_{k=-\infty}^{\infty} |\hat{N}_n(\xi + 2\pi k)|^2 \leq 1,$$

Thus, we have an upper Riesz bound for the spline basis generated by N_n. The lower Riesz bound is obtained by introducing a class of polynomials called Euler-Frobenius polynomials of degree $2n - 2$, $n \geq 1$ given by

$$E_{2n-1}(z) = (2n-1)! z^{n-1} \sum_{k=-n+1}^{n-1} N_{2n}(n+k)z^k$$

All the $2n-2$ roots $\lambda_1, ..., \lambda_{2n-2}$ of E_{2n-1} are real, negative, and simple, and if they are taken (w.l.g.) to be arranged monotonically, $\lambda_1 > ... > \lambda_{2n-2}$, then they satisfy

$$\lambda_1 \lambda_{2n-2} = ... = \lambda_{n-1} \lambda_n = 1$$

Hence,

$$\sum_{k=-\infty}^{\infty} |\hat{N}_n(\xi + 2\pi k)|^2 = \frac{1}{(2n-1)!} \prod_{k=1}^{2n-2} |e^{i\xi} - \lambda_k|$$

$$= \frac{1}{(2n-1)!} \prod_{k=1}^{n-1} |e^{i\xi} - \lambda_k| |e^{i\xi} - \lambda_{2n-1-k}|$$

$$= \frac{1}{(2n-1)!} \prod_{k=1}^{n-1} |e^{i\xi} - \lambda_k| |e^{i\xi} - \lambda_k^{-1}|$$

$$= \frac{1}{(2n-1)!} \prod_{k=1}^{n-1} \frac{1 - 2\lambda_k \cos\xi + \lambda_k^2}{|\lambda_k|}$$

$$= \frac{1}{(2n-1)!} \prod_{k=1}^{n-1} \frac{(1+\lambda_k^2)^2}{|\lambda_k|}$$

(since the λ_k are all negative). Thus,

$$\frac{1}{(2n-1)!}\prod_{k=1}^{n-1}\frac{(1+\lambda_k^2)^2}{|\lambda_k|} \leq \sum_{k=-\infty}^{\infty}|\hat{N}_n(\xi+2\pi k)|^2 \leq 1$$

which proves the theorem. The following corollary is immediate :

Corollary 6.4 For all $n \geq 2$ and any $j \in \mathbf{Z}$,

$$\mathcal{B}_j = \left\{2^{-j/2}N_n(2^{-j}x-k) : \ldots \in \mathbf{Z}\right\}$$

is a Riesz basis of $\mathcal{V}_j^n$ with the same Riesz bounds as those for $\mathcal{V}_0^n$.

Proof : The map $f \mapsto \mathcal{D}^j$ given by $(\mathcal{D}^{-j}f)(x) = 2^{-j/2}f(2^{-j}x)$ is unitary (*i.e.,* $\|\mathcal{D}^j f\| = \|f\|$). Now, since

$$\text{Image}(\mathcal{D}^j\mathcal{V}_0^n) = \mathcal{V}_j^n$$

and

$$\mathcal{D}^j N_n(x-k) = N_n^{j;k}(x) = 2^{-j/2}\mathbf{N}_n(2^{-j}x-k)$$

and $\mathcal{B}_0$ is a Riesz basis for $\mathcal{V}_0^n$ with bounds A and B, we have

$$\forall f \in \mathcal{V}_0^n, \quad A\|f\|^2 \leq \sum_{k\in\mathbf{Z}}|\langle f, \mathbf{N}_n^{0;k}\rangle|^2 \leq B\|f\|^2$$

The unitarity of the map $f \mapsto \mathcal{D}^j f$ implies

$$\forall f \in \mathcal{V}_j^n, \quad A\|f\|^2 \leq \sum_{k\in\mathbf{Z}}|\langle f, \mathbf{N}_n^{j;k}\rangle|^2 \leq B\|f\|^2$$

Hence, $\mathcal{B}_j = \{\mathbf{N}_n^{j;k} : k \in \mathbf{Z}\}$ is a Riesz basis of $\mathcal{V}_j^n$ with the same Riesz bounds A and B.

By Theorem 5.22, we have for $n \geq 2$,

$$\bigcap_{j\in\mathbf{Z}} \mathcal{V}_j^n = \{0\}$$

Also, since $\mathbf{N}_n(x) > 0$, for $0 < x < n$ and supp $\mathbf{N}_n = [0,1]$,

$$\hat{\mathbf{N}}_n(0) = \int_{-\infty}^{\infty} \mathbf{N}_n(x)dx \neq 0$$

Furthermore, $\hat{\mathbf{N}}_n$ is bounded and continuous at 0 for $n \geq 2$. Thus, by Theorem 5.23, we have for $n \geq 2$,

$$\text{closure}_{L^2(\mathbf{R})}(\bigcup_{j\in\mathbf{Z}} \mathcal{V}_j^n) = L^2(\mathbf{R})$$

We now derive the two-scale relation for cardinal B-splines of order n.

We first note that since $\mathbf{N}_n = N_{n-1} * N_1, \hat{\mathbf{N}}_n = \hat{N_{n-1}}\hat{N}_1$, and since

$$\hat{N}_1 = \frac{1 - e^{-i\xi}}{i\xi},$$

$$\hat{\mathbf{N}}_n = \left(\frac{1 - e^{-i\xi}}{i\xi}\right)^n$$

As $\mathcal{V}_{j+1}^n \subset \mathcal{V}_j^n$ $j \in \mathbf{Z}$, we have

$$N_n^{j+1;0} = \sum_{k=-\infty}^{\infty} c_k^n \mathbf{N}_n^{j;k},$$

where $\{c_k^n\}_{k\in\mathbf{Z}} \in l^2(\mathbf{Z})$. That is,

$$2^{-(j+1)/2}\mathbf{N}_n(2^{-j-1}x) = \sum_{k=-\infty}^{\infty} c_k^n 2^{-j/2}\mathbf{N}_n(2^{-j}x - k)$$

Letting $2^{-j}x = y$ and simplifying, we have

$$\mathbf{N}_n(y) = \sqrt{2} \sum_{k=-\infty}^{\infty} c_k^n \mathbf{N}_n(2y - k)$$

Taking the Fourier transform of both sides we have

$$\hat{N}n(\xi) = 2^{-1/2} \sum_{k=-\infty}^{\infty} c_k^n \hat{N}_n\left(\frac{\xi}{2}\right) e^{-ik\frac{\xi}{2}}$$

Using $\hat{\mathbf{N}}_n = (\frac{1-e^{-i\xi}}{i\xi})^n$, we have

$$2^{-1/2} \sum_{k=-\infty}^{\infty} c_k^n e^{-ik\frac{\xi}{2}} = \left(\frac{1 - e^{-i\xi}}{i\xi}\right)^n \left(\frac{i\xi/2}{1 - e^{-i\xi/2}}\right)^n$$

$$= \left(\frac{1 + e^{-i\xi/2}}{2}\right)^n$$

$$= 2^{-n} \sum_{k=0}^{n} \binom{n}{k} e^{-ik\xi/2}$$

This gives

$$c_k^n = \begin{cases} 2^{-n+1/2} \binom{n}{k} & \text{for } 0 \le k \le n \\ 0 & \text{otherwise} \end{cases}$$

Thus, we have the desired two-scale relation

$$N_n(x) = \frac{1}{\sqrt{2}} \sum_{k=0}^{n} 2^{-n} \binom{n}{k} N_n(2x - k)$$

$\mathcal{B}_0$ is not an orthonormal basis for V_0^n. We may, however, use the orthonormalization technique discussed in Chapter 5 to obtain an orthonormal basis for V_0^n and thus obtain an orthonormal wavelet basis for $L^2(\mathbf{R})$. This operation does not, however, preserve the compactness of support of the functions in $\mathcal{B}_0$ and of the associated wavelet basis of $L^2(\mathbf{R})$, and as a result all the Battle-Lemarié wavelets have the whole real line as their supports.

We have obtained an important technique for generating wavelet bases of desired smoothness by means of multiresolution analysis.

6.2   Subband Filtering Schemes

In this section, we study the structural similarity between the fast wavelet algorithm of Mallat introduced in the previous chapter and a well-known filtering scheme from digital signal processing. This comparison will prompt us to look at an important class of wavelets. We first introduce the concept of a subband filtering scheme.

6.2.1   Bandlimited functions

A function $f \in L^2(\mathbf{R})$ is said to be bandlimited if its Fourier transform is compactly supported. That is, $\hat{f}(\xi) = 0$ for $\mid \xi \mid > T$, for some $T > 0$. By rescaling and shifting, we may assume that $T = \pi$. Thus, $\hat{f} \in L^2[-\pi, \pi]$ and, therefore,

$$\hat{f}(\xi) = \sum_{n \in \mathbf{Z}} a_n e^{-in\xi},$$

where

$$a_n = \frac{1}{2\pi} \int_{-\pi}^{\pi} \hat{f}(\xi) e^{in\xi} d\xi = \frac{1}{2\pi} \int_{-\infty}^{\infty} \hat{f}(\xi) e^{in\xi} d\xi = \frac{1}{\sqrt{2\pi}} f(n)$$

Now,

$$f(x) = \frac{1}{\sqrt{2\pi}} \int_{-\infty}^{\infty} \hat{f}(\xi)e^{ix\xi}d\xi$$

$$f(x) = \frac{1}{\sqrt{2\pi}} \int_{-\pi}^{\pi} \hat{f}(\xi)e^{ix\xi}d\xi$$

$$= \frac{1}{\sqrt{2\pi}} \int_{-\pi}^{\pi} \left[\sum_{n\in\mathbf{Z}} a_n e^{-in\xi}\right] e^{ix\xi}d\xi$$

$$= \frac{1}{\sqrt{2\pi}} \sum_{n\in\mathbf{Z}} a_n \int_{-\pi}^{\pi} e^{-i(x-n)\xi}d\xi$$

$$\sum_{n\in\mathbf{Z}} f(n)\frac{\sin \pi(x-n)}{\pi(x-n)}.$$

Since

$$\sum_{n\in\mathbf{Z}} |f(n)|^2 = 2\pi \sum_{n\in\mathbf{Z}} |a_n|^2 < \infty,$$

the interchanging of the summation and the integral above can be justified. In general if we let $\hat{f} \in L^2[-T,T]$ for a general $T > 0$, then

$$f(x) = \sum_{n\in\mathbf{Z}} f(n\frac{\pi}{T})\frac{\sin \pi(Tx-n\pi)}{\pi(Tx-n\pi)}.$$

This expansion is known as Shannon's theorem . Here,

$$\frac{T}{\pi} = \frac{|\operatorname{supp}\hat{f}|}{2\pi}$$

is the sampling frequency, called the Nyquist frequency of f. Sampling at a lower frequency results in the phenomenon of *aliasing* , wherein the function is treated as if its bandwidth is narrower than it actually is. This causes the overlapping of shifted versions of its Fourier transform in its periodized form, as the shifts are smaller than the actual bandwidth, thus resulting in a distorted representation.

Conversely, any square summable sequence $\{a_n\}_{n\in\mathbf{Z}}$ can be thought of as the discrete sampling of a bandlimited function with $\operatorname{supp}\hat{f} \subset [-\pi,\pi]$, where

$$f(x) = \sum_{n\in\mathbf{Z}} a_n \frac{\sin \pi(x-n)}{\pi(x-n)},$$

or equivalently,

$$\hat{f}(\xi) = \frac{1}{\sqrt{2\pi}} \sum_{n\in\mathbf{Z}} a_n e^{-in\xi}.$$

6.2.2   Discrete filtering

Filtering a discretely sampled function given by a sequence $a = \{a_n\}_{n \in \mathbf{Z}}$ is equivalent to convolving the sequence with a *filter* sequence $h = \{h_n\}_{n \in \mathbf{Z}}$ to obtain the filtered sequence

$$b = \{b_n\}_{n \in \mathbf{Z}} = \left\{ \sum_{m \in \mathbf{Z}} h_{n-m} a_m \right\}_{n \in \mathbf{Z}}$$

This corresponds to multiplying

$$\hat{f}(\xi) = \frac{1}{\sqrt{2\pi}} \sum_{n \in \mathbf{Z}} a_n e^{-in\xi}$$

by a 2π periodic function

$$H(\xi) = \sum_{n \in \mathbf{Z}} h_n e^{-in\xi}$$

called the *frequency response* of the filter:

$$\hat{f}(\xi) H(\xi) = \frac{1}{\sqrt{2\pi}} \sum_{n \in \mathbf{Z}} \left(\sum_{m \in \mathbf{Z}} h_{n-m} a_m \right) e^{-in\xi},$$

or equivalently,

$$(f * h)(x) = \sum_{n \in \mathbf{Z}} \left(\sum_{m \in \mathbf{Z}} h_{n-m} a_m \right) \frac{\sin \pi (x - n)}{\pi (x - n)}$$

The filter is said to be *low-pass* if $H \mid_{[-\pi, \pi]}$ is supported mostly on $\{\xi : \mid \xi \mid \le \pi/2\}$, and is said to be *high-pass* if it is supported mostly on $\{\xi : \mid \xi \mid \ge \pi/2\}$. The ideal low-pass filter H^l and the ideal high-pass filter H^h are defined by

$$H^l(\xi) = \begin{cases} 1 & \text{if } \mid \xi \mid < \pi/2 \\ 0 & \text{if } \pi/2 < \mid \xi \mid < \pi \end{cases}$$

and

$$H^h(\xi) = \begin{cases} 0 & \text{if } \mid \xi \mid < \pi/2 \\ 1 & \text{if } \pi/2 < \mid \xi \mid < \pi \\ \end{cases},$$

respectively. The corresponding sequences

$$h^l = \{h_n^l\}_{n \in \mathbf{Z}}$$

and

$$h^h = \{h_n^h\}'_{n \in \mathbf{Z}},$$

respectively, are given by

$$h^l_n = \begin{cases} \frac{1}{2} & \text{for } n = 0 \\ 0 & \text{for } n = 2k, \, k \neq 0 \\ \frac{(-1)^k}{(2k+1)\pi} & \text{for } n = 2k + 1 \end{cases}, \quad h^h_n = \begin{cases} \frac{1}{2} & \text{for } n = 0 \\ 0 & \text{for } n = 2k, \, k \neq 0 \\ \frac{(-1)^k}{(2k+1)\pi} & \text{for } n = 2k + 1 \end{cases}$$

The action of the ideal low-pass filter H^l on f results in a bandlimited function whose support is contained in $[-\pi/2, \pi/2]$ and is uniquely characterized by its sampled values in $2\mathbf{Z}$, *i.e.*,

$$(h^l * f)(x) = \sum_{n \in \mathbf{Z}} \left(\sum_{m \subset \mathbf{Z}} h^l_{2n-m} a_m \right) \frac{\sin[\pi(x - 2n)/2]}{\pi(x - 2n)/2}$$

(where $T = \pi/2$), and similarly, the result of applying the ideal high-pass filter H^l on f results in a frequency-shifted version of a bandlimited function whose support is contained in $[-\pi/2, \pi/2]$ and is also uniquely characterized by its sampled values in $2\mathbf{Z}$, *i.e.*,

$$(h^l * f)(x) = \int_{\pi/2 < |\xi| \leq \pi} \sum_{n \in \mathbf{Z}} \left(\sum_{m \in \mathbf{Z}} h^l_{2n-m} a_m \right) e^{-2in\xi} e^{ix\xi} d\xi$$

$$= \sum_{n \in \mathbf{Z}} \left(\sum_{m \in \mathbf{Z}} h^l_{2n-m} a_m \right) \frac{\sin[\pi(x - 2n)/2]}{\pi(x - 2n)/2} \{2 \cos[\pi(x - 2n)/2] - 1\}.$$

The complete characterization of the filtered functions $h^l * f$ and $h^h * f$ by the even-indexed elements of the convolved sequences

$$h^l * a = \left\{ \sum_{m \in \mathbf{Z}} h^l_{n-m} a_m \right\}_{n \in \mathbf{Z}}$$

and

$$h^h * a = \left\{ \sum_{m \in \mathbf{Z}} h^h_{n-m} a_m \right\}_{n \in \mathbf{Z}}$$

suggests the retention of only the even-indexed elements of these sequences. This is exactly the operation of downsampling by a factor of 2.

Reconstruction of the original sequence from the filtered and downsampled sequences

$$a^l = (h^l * a) \downarrow_2 = \left\{ \sum_{m \in \mathbf{Z}} h^l_{2n-m} a_m \right\}_{n \in \mathbf{Z}}$$

and

$$a^h = (h^h * a) \downarrow_2 = \left\{ \sum_{m \in \mathbf{Z}} h^h_{2n-m} a_m \right\}_{n \in \mathbf{Z}}$$

is carried out as follows:

Since

$$f(n) = f * h^l(x) + f * h^h(n) \quad (\because \ h^l + h^h = 1)$$

we have

$$a_n = \sum_{k \in \mathbf{Z}} \left\{ a^l_k + a^h_k (2 \cos[\pi(n - 2k)/2] - 1) \right\} \frac{\sin[\pi(x - 2n)/2]}{\pi(x - 2n)/2},$$

i.e.,

$$a_{2n} = a^l_m + a^h_m \text{ and } a_{2n+1} = \sum_{k \in \mathbf{Z}} \frac{2(-1)^k}{\pi(2k+1)} (a^l_{n-k} - a^h_{n-k})$$

These two equations can be combined into a single equation

$$a_n = 2 \sum_{k \in \mathbf{Z}} (h^l_{n-2} a^l_k + h^h_{n-2k} a^h_k)$$

which can be interpreted in terms of a sequence operations with the sequences a^l, a^h, h^l, and h^h as follows:

(a) Upsampling the sequences a^l and a^h,

(b) Convolving these upsampled sequences with the filters h^l and h^h, respectively, and

(c) Adding the resulting two convolved sequences.

This method of decomposition and reconstruction of a discretely sampled function (signal) using low-pass and high-pass filters is called a *subband filtering scheme* with exact reconstruction.

The elements of the filter sequences h^l and h^h for the ideal low-pass and high-pass filters decay too slowly to be of any use. In practice, however, filters h^0 and h^l with much faster decaying coefficients are used which is possible only if the corresponding frequency responses H^0 and H^l are smoother than H^l and H^h. This implies that their supports are larger than those of the ideal filters and, hence, overlap causing aliasing . To eliminate aliasing, appropriate "matching" filters $\tilde{h}^0$ and $\tilde{h}^l$ must be used during the reconstruction stage. The conditions on such filters are best expressed using the notion of Z-transforms introduced in Section 5.4.3.

For $z = e^{-i\xi}$, the Z-transform is a Fourier series :

$$A(e^{-i\xi}) = A(\xi) = \sum_{n \in \mathbf{Z}} a_n e^{-in\xi}.$$

The decomposition stage of the subband filtering scheme is then given by

$$A^0(Z^2) = \frac{1}{2}[H^0(z)A(z) + H^0(-z)A(-z)]$$

and

$$A^l(Z^2) = \frac{1}{2}[H^l(z)A(z) + H^l(-z)A(-z)]$$

and the reconstruction stage is given by

$$\tilde{A}(z) = [\tilde{H}^0 A^l(z^2) + \tilde{H}^l(z)A^0(z^2)]$$

This can be schematically represented by a diagram similar to the one used to describe the decomposition and reconstruction stages of the fast wavelet algorithm:

$$
\begin{array}{c}
 \quad H^0 \;\rightarrow\; \downarrow_2 \;\rightarrow\; a^0 \downarrow_2 \;\rightarrow\; \uparrow_2 \;\; \tilde{H}^0 \\
\nearrow \searrow \\
a \oplus \; = \tilde{a} \\
\searrow \nearrow \\
 \quad H^l \;\rightarrow\; \downarrow_2 \;\rightarrow\; a^l \downarrow_2 \;\rightarrow\; \uparrow_2 \;\; \tilde{H}^l
\end{array}
$$

Substituting for $A^0(z^2)$ and $A^l(z^2)$, we have

$$\tilde{A}(z) = \frac{1}{2}[\tilde{H}^0(z)H^0(z) + \tilde{H}^l(z)H^l(z)]A(z) + \frac{1}{2}[\tilde{H}^0(z)H^0(-z) + \tilde{H}^l(z)H^l(-z)]A(-z)$$

In the above expression, the second term contains the aliasing effects, as $A(-z)$ corresponds to a shifting of the corresponding Fourier series by π :

$$A(-e^{-i\xi}) = A(e^{-i(\xi+\pi)}) = A(\xi + \pi) = \sum_{n \in \mathbf{Z}} a_n e^{-in(\xi+\pi)},$$

which is caused by aliasing due to sampling at half the Nyquist frequency. To eliminate aliasing, it is, therefore, necessary to have

$$[\tilde{H}^0(z)H^0(-z) + \tilde{H}^l(z)H^l(-z)] = 0.$$

In the earliest schemes filters with real sequences were chosen such that

$$H^l(z) = H^0(-z),$$

$$\tilde{H}^0(z) = H^0(z)$$

and
$$\tilde{H}^l(z) = -H^0(-z).$$

This choice clearly eliminates aliasing, and we have

$$\tilde{A}(z) = \frac{1}{2}[H^0(z)^2 - H^0(-z)^2]A(z).$$

In practice, filters that have finitely many non-zero coefficients in their sequences are preferred. Such filters are called finite impulse response (FIR) filters. However, there are no FIR filters $H^0(z)$ such that $H^0(z)^2 - H^0(-z)^2 = 2$. This means that exact reconstruction is not possible with this choice of filters.

6.2.3   Conjugate quadrature filters (CQF)

An alternative scheme proposed independently by Mintzer, Smith and Barnwell , and Vetterli, not only eliminates aliasing, but also yields exact reconstruction. This scheme is given by the following relations,

$$H^l(z) = z^{-1}H^0(-z^{-1}),$$

$$\tilde{H}^0(z) = H^0(z^{-1}),$$

and

$$\tilde{H}^l(z) = zH^0(-z).$$

This choice again eliminates aliasing and we have

$$\tilde{A}(z) = \frac{1}{2}[H^0(z)H^0(z^{-1}) + H^0(-z)^2 H^0(-z^{-1})]A(z).$$

For $H^0(z)$ with real coefficients, and $z = e^{i\xi}$, we have

$$\tilde{A}(e^{i\xi}) = \frac{1}{2}[|\, H^0(e^{-i\xi})\,|^2 + |\, H^0(-e^{-i\xi})\,|^2]A(e^{i\xi}),$$

i.e.,

$$\tilde{A}(\xi) = \frac{1}{2}[|\, H^0(\xi)\,|^2 + |\, H^0(\xi + \pi)\,|^2]A(\xi)$$

In this case there exist FIR choices for $H^0(z)$ such that

$$\frac{1}{2}[|\, H^0(e^{-i\xi})\,|^2 + |\, H^0(-e^{-i\xi})\,|^2] = \frac{1}{2}[|\, H^0(\xi)\,|^2 + |\, H^0(\xi + \pi)\,|^2] = 1,$$

giving exact reconstruction. Such filters are called conjugate quadrature filters (CQF).

6.2.4 CQFs arising from MRAs

The filter decomposition formulas

$$a_n^l = \sum_{m \in \mathbf{Z}} h_{2n-m}^l a_m \quad \forall n \in \mathbf{Z}$$

and

$$a_n^h = \sum_{m \in \mathbf{Z}} h_{2n-m}^h a_m \quad \forall n \in \mathbf{Z}$$

and the filter reconstruction formula

$$a_n = 2 \sum_{k \in \mathbf{Z}} (h_{2n}^l a_k^l + h_{n-2k}^h a_k^h)$$

are structurally identical to the wavelet decomposition formulas

$$c_n^j = \sum_{m \in \mathbf{Z}} \overline{h_{m-2n}} c_m^{j-1} \quad \forall n \in \mathbf{Z}$$

and

$$d_n^j = \sum_{m \in \mathbf{Z}} \overline{g_{m-2n}} c_m^{j-1} \quad \forall n \in \mathbf{Z},$$

and the wavelet reconstruction formula

$$c_n^{j-1} = \sum_{k \in \mathbf{Z}} [h_{n-2k} c_k^j + g_{n-2k} d_k^j]$$

in Section 5.4.2, where

$$(\tilde{h})_n = \overline{h_{-n}}, (\tilde{g})_n = g_{-n}^-, \text{ and } g_n = (-1)^n h_{-n+1}.$$

If $H(z) = \sum_{n \in \mathbf{Z}} h_n z^n$ and $G(z) = \sum_{n \in \mathbf{Z}} g_n z^n$, then choosing

$$H^0(z) = H(z^{-1}),$$

$$H^l(z) = G(z^{-1}),$$

$$\tilde{H}^0(z) = H(z),$$

and

$$\tilde{H}^l(z) = G(z),$$

and noting that

$$G(z) = -zH(-z^{-1}) \quad (\because \ g_n = (-1)^n h_{-n+1}),$$

we see that this choice corresponds to a CQF scheme. This implies that every orthonormal wavelet basis gives rise to a pair of conjugate quadrature filters, *i.e.*, a subband filtering scheme with exact reconstruction. The converse is, however, not true. We also note that the requirement

$$\frac{1}{2}[|H^0(\xi)|^2 + |H^0(\xi + \pi)|^2] = 1$$

for CQF is nothing but the requirement

$$|m_0(\xi)|^2 + |m_0(\xi + \pi)|^2 = 1$$

on the function

$$m_0(\xi) = \frac{1}{\sqrt{2}} \sum_{n \in \mathbf{Z}} h_n e^{-in\xi}$$

for orthonormal wavelets, because of the choice $\mathbf{H}^0(z) = \mathbf{H}(z^{-1})$. More precisely, the functions m_0 and m_1 occurring in the Fourier transformed versions of the two-scale relations of an MRA,

$$\hat{\phi}(\xi) = m_0(\xi/2)\hat{\phi}(\xi/2)$$

and

$$\hat{\psi}(\xi) = m_1(\xi/2)\hat{\phi}(\xi/2) = -e^{-i\xi}\overline{m_0(\xi/2 + \pi)}\hat{\phi}(\xi/2),$$

correspond to a pair of low-pass and high-pass filters $H_0 = \sqrt{2}m_0$ and $H_1 = \sqrt{2}m_1$ of a CQF scheme.

The question that now arises is whether it is possible to construct orthonormal wavelet bases corresponding to FIR filters . In the next section we will see that compactly supported orthonormal wavelet bases give rise to FIR filters, and we will characterize such filters.

6.3   Compactly Supported Orthonormal Wavelet Bases

6.3.1   The structure of m_0

All examples of orthonormal wavelet bases encountered so far (with the exception of the Haar basis) are constituted of infinitely supported functions because of the orthogonalization operation. To construct orthonormal wavelet bases made up of compactly supported basis functions it is fruitful to start with the sequence

$$h = \{h_n\}_{n \in \mathbf{Z}}$$

of the subband filtering scheme, or equivalently, the function

$$m_0(\xi) = \frac{1}{\sqrt{2}} \sum_{n \in \mathbf{Z}} h_n e^{-in\xi}$$

To ensure that the wavelet basis functions are compactly supported , it is necessary and sufficient to ensure that the basic wavelet ψ is compactly supported. The simplest way to do this is to choose a scaling function ϕ that is compactly supported such that $\{\phi(x - n)\}_{n \in \mathbf{Z}}$ is an orthogonal system of functions. Then, since by definition

$$h_n = \sqrt{2} \int_{-\infty}^{\infty} \phi(x)\overline{\phi(2x - n)} \ \forall n \in \mathbf{Z},$$

it follows that only finitely many of the h_n are nonzero. This in turn implies that ψ is a finite linear combination of compactly supported functions, since

$$\psi(x) = \sqrt{2} \sum_{n \in \mathbf{Z}} (-1)^{n-1} h_{-n+1} \phi(2x - n)$$

and, hence, is itself compactly supported.

If $h = \{h_n\}_{n \in \mathbf{Z}}$ is a finite sequence, the corresponding subband filtering scheme is realizable through only FIR filters . Thus, in the case of compactly supported ϕ, the function

$$m_0(\xi) = \frac{1}{\sqrt{2}} \sum_{n \in \mathbf{Z}} h_n e^{-in\xi}$$

is a trigonometric polynomial. Since

$$\{\phi(x - n)\}_{n \in \mathbf{Z}} = \{\phi_{0,n}\}_{n \in \mathbf{Z}}$$

is an orthonormal set, we have

$$|\, m_0(\xi)\,|^2 + |\, m_0(\xi + \pi)\,|^2 = 1$$

The requirement on the scaling function ϕ that $\hat{\phi}$ be bounded on $\mathbf{R}$ and continuous at $\phi(0)$ with $\hat{\phi}(0) \neq 0$, in order for it to generate an MRA of $L^2(\mathbf{R})$, along with the Fourier transform version of the two-scale relation for ϕ

$$\hat{\phi}(\xi) = m_0(\xi)\hat{\phi}(\xi/2),$$

implies that $m_0(0) = 1$. Using the relation $|\, m_0(\xi)\,|^2 + |\,(\xi + \pi)\,|^2 = 1$ we have

$$m_0(\pi) = 0.$$

This means that

$$m_0(\xi) = \left(\frac{1 + e^i \xi}{2}\right)^M L(\xi), \quad M \geq 1,$$

where $\mathcal{L}$ is a trigonometric polynomial.

For trigonometric polynomials m_0 of the above form that satisfy

$$|\, m_0(\xi)\,|^2 + |\, m_0(\xi + \pi)\,|^2 = 1,$$

the following theorem characterizes $|\, m_0(\xi)\,|^2$ completely.

Theorem 6.5 *For a trigonometric polynomial m_0 given by*

$$m_0(\xi) = (\frac{1 + e^i \xi}{2})^M L(\xi), \quad M \geq 1,$$

$$|\, m_0(\xi)\,|^2 + |\, m_0(\xi + \pi)\,|^2 = 1$$

iff

$$|\, L(\xi)\,|^2 = P(\sin^2 \frac{\xi}{2}),$$

where P is a polynomial such that

$$P(y) = P_M(y) + y^M R(y - 1/2)$$

with

$$P_M(y) = \sum_{m=0}^{M-1} \binom{M + m - 1}{m} y^m$$

and R is an odd polynomial, such that $P(y) \geq 0$ for all $y \in [0, 1]$.

Proof : Let $m_0(\xi) = |\, m_0(\xi)\,|^2$. Then $m_0(\xi)$ is a polynomial in $\cos \xi$, satisfying

$$m_0(\xi) + M_0(\xi + \pi) = 1$$

and

$$m_0(\xi) = \left(\cos^2 \frac{\xi}{2}\right) L(\xi)$$

where $L(\xi) = |\, L(\xi)\,|^2$ is also a polynomial in $\cos \xi$.

It is convenient here to rewrite $L(\xi) = |\, \mathcal{L}(\xi)\,|^2$ as a polynomial P in $\sin^2 \frac{\xi}{2} = (1 - \cos \xi)/2$:

$$m_0(\xi) = (\cos^2 \frac{\xi}{2})^M P(\sin^2 \frac{\xi}{2})$$

Using this expression for $m_0(\xi)$ in $m_0(\xi) + m_0(\xi + \pi) = 1$, and letting $\sin^2 \frac{\xi}{2} = y$, we have

$$(1-y)^M P(y) + y^M P(1-y) = 1,$$

which must be satisfied for all $y \in [0,1]$ and, hence, for all $y \in \mathbf{R}$ (this follows from the fact that the zeros of a nonzero polynomial are isolated).

Applying Bezout's theorem (see the Appendix) to the polynomials $(1-y)^M$ and y^M, there exist unique polynomials Q_1 and Q_2 of degrees less than M, such that

$$(1-y)^M Q_1(y) + y^M Q_2(y) = 1,$$

Substituting y by $1-y$, we have

$$(1-y)^M Q_2(1-y) + y^M Q_1(1-y) = 1.$$

By the uniqueness of the polynomials Q_1 and Q_2, we have

$$Q_2(y) = Q_1(1-y).$$

Thus, $P(y) = Q_1(y)$ is a solution to the equation $(1-y)^M P(y) + y^M P(1-y) = 1$. Solving for $Q_y(y)$, we get

$$Q_1(y) = (1-y)^{-M}[1 - y^M Q_1(1-y)]$$

$$= \sum_{m=0}^{M-1} \binom{M+m-1}{m} y^m + O(y^M),$$

where the latter expression is obtained by computing the first M terms of the Taylor expansion of $(1-y)^{-M}$ about 0. However, since degree of Q_1 is less than M, we indeed have

$$Q_1(y) = \sum_{m=0}^{M-1} \binom{M+m-1}{m} y^m.$$

This is an explicit solution of $(1-y)^M P(y) + y^M P(1-y) = 1$ and is the unique solution of least degree, which will be denoted P_M. If P is any other solution of higher degree, then

$$(1-y)^M[P(y) - P_M(y)] + y^M[P(1-y) - P_M(1-y)] = 0,$$

implying that $P - P_M$ is divisible by y^M; *i.e.*,

$$P(y) - P_M(y) = y^M R(y).$$

Also,

$$R(y) + R(1-y) = 0;$$

i.e.,

$$R(y - \frac{1}{2}) = -R(\frac{1}{2} - y)$$

In order to characterize m_0, it is necessary to compute the square root of the polynomial L. This is achieved by a lemma of Féjer and Riesz.

Lemma 6.6 If A is a positive trigonometric cosine polynomial with real coefficients,

$$A(\xi) = \sum_{m=0}^{M} a_m \cos m\xi,$$

then a trigonometric polynomial $B(\xi) = \sum_{m-=0}^{M} b_m e^{im\xi}$ of order M with real coefficients can be constructed such that $\mid B(\xi) \mid^2 = A(\xi)$.

Proof : Let $A(\xi) = p(\cos \xi)$, where p is a polynomial of degree M with real coefficients. Then, factoring p over $\mathbf{C}$,

$$p(\zeta) = c \prod_{k=1}^{M} (\zeta - \alpha_k)$$

where if α_k is a zero of p, then so is $\bar{\alpha}_k$. Writing

$$A(\xi) = e^{iM\xi} q(e^{-i\xi}),$$

where q is a polynomial of degree $2M$, we have,

$$q(z) = cz^M \prod_{k=1}^{M} (\frac{z + z^{-1}}{2} - \alpha_k) = c \prod_{k=1}^{M} (\frac{1}{2} - \alpha_k z + \frac{1}{2}z^2) \quad \forall \mid z \mid = 1$$

and, hence, also for all $z \in \mathbf{C}$.

Case 1 : If α_k is real, then the zeros of

$$\frac{1}{2} - \alpha_k z + \frac{1}{2}z^2$$

are

$$\alpha_k \pm \sqrt{\alpha_k^2 - 1}.$$

(a) For $\mid \alpha_k \mid \geq 1$, these zeros are real (and equal if $\alpha_k = \pm 1$ and of the form β_k, β_k^{-1}.

(b) For $\mid \alpha_k \mid$, these zeros are complex conjugates of each other, and of absolute value 1 and, hence, of the form $e^{i\lambda_k}, e^{-i\lambda_k}$. Since $\mid \alpha_k \mid < 1$, these zeros correspond to values of ξ for which $A(\xi) = 0$, and since $A(\xi) \geq 0 \; \forall \xi \in$ **R**, these zeros must have even multiplicity.

Case 2 : If α_k is not real, then it is paired with its complex conjugate $\bar{\alpha}_k$, and the polynomial

$$(\frac{1}{2} - \alpha_k z + \frac{1}{2}z^2)(\frac{1}{2} - \bar{\alpha}_k z + \frac{1}{2}z^2)$$

has the four roots $\alpha_k \pm \sqrt{\alpha_k^2 - 1}$ and $\bar{\alpha}_k \pm \sqrt{\bar{\alpha}_k^2 - 1}$, of the form γ_k, γ_k^{-1}, $\bar{\gamma}_k$, $\bar{\gamma}_k^{-1}$. Thus,

$$q(z) = C \left[\prod_{i=1}^{I} (z - \gamma_i)(z - \gamma_i^{-1})(z - \bar{\gamma}_i)(z - \bar{\gamma}_i^{-1}) \right] \cdot \left[\prod_{j=1}^{J} (z - e^{i\alpha_j})^2 (z - e^{-i\alpha_j})^2 \right] \cdot$$

$$\left[\prod_{k=1}^{K} (z - \beta_k)(z - \beta_k^{-1}) \right]$$

For $z = e^{-i\xi}$, $| (e^{-i\xi} - z_0)(e^{-i\xi} - \bar{z}_0^{-1}) | = | z_0 |^{-1} | e^{-i\xi} - z_0 |^2$. Hence,

$$A(\xi) = | A(\xi) | = | q(e^{-i\xi}) |$$

$$= C \left[\prod_{i=1}^{I} | \gamma_i |^{-2} \prod_{j=1}^{J} | \beta_j |^{-1} \right] \cdot | \prod_{i=1}^{I} (\gamma_i - e^{-i\xi})(\bar{\gamma}_i - e^{-i\xi}) |^2 \cdot$$

$$| \prod_{j=1}^{J} (z - e^{i\alpha_j})(z - e^{-i\alpha_j}) |^2 \cdot | \prod_{k=1}^{K} (e^{-i\xi} - \beta_k) |^2 = | B(\xi) |^2$$

where

$$B(\xi) = \left[C \prod_{i=1}^{I} | \gamma_i |^{-2} \prod_{j=1}^{J} | \beta_j |^{-1} \right]^{1/2} \cdot \prod_{i=1}^{I} (\gamma_i - e^{-i\xi}(\bar{\gamma}_i - e^{-i\xi}).$$

$$\prod_{j=1}^{J} (z - e^{i\alpha_j})(z - e^{-i\alpha_j}). \prod_{k=1}^{K} (e^{-i\xi} - \beta_k)$$

This is a trigonometric polynomial with real coefficients.

Thus, we can now construct all possible trigonometric polynomials m_0 of the form

$$m_0(\xi) = (\frac{1 + e^{i\xi}}{2})^M \mathcal{L}(\xi), \; M \geq 1$$

satisfying $| m_0(\xi) |^2 + | m_0(\xi + \pi) |^2 = 1$. However, not every such polynomial gives rise to an orthonormal basis.

6.3.2   Necessary and sufficient conditions for orthonormality

In the following discussion, we will obtain additional conditions under which such m_0 generate orthonormal bases.

Lemma 6.7 If the function m_0 associated with an MRA is a trigonometric polynomial, and if the corresponding scaling function ϕ has compact support, then

$$\hat{\phi}(\xi) = (2\pi)^{-1/2} \prod_{k=1}^{\infty} m_0(2^{-k}\xi)$$

Proof : If ϕ has compact support, then since necessarily $\phi \in L^2(\mathbf{R})$, it is clear that $\phi \in L^2(\mathbf{R})$. Therefore, for all $\xi \in \mathbf{R}$,

$$\hat{\phi}(\xi) = m_0(\xi/2)\hat{\phi}(\xi/2)$$

(Since m_0 is a trigonometric polynomial, it is continuous, and since ϕ is in $L^1(\mathbf{R})$, $\hat{\phi}$ is continuous. Therefore, we may drop the qualifier "a.e." for the above formula). Also, since necessarily $\hat{\phi}(0) \neq 0$, we have $m_0(0) = 1$ and $m_0(\pi) = 0$. As any $n \in \mathbf{Z}$ can be written as $n = 2^p(2m+1)$ for some $p \geq 0$ and $m \in \mathbf{Z}$, we have for all $n \in \mathbf{Z}$,

$$\hat{\phi}(2n\pi) = \hat{\phi}(2\ 2^p(2m+1)\pi)$$

$$= \left[\prod_{k=1}^{p} m_0(2^{p+1-k}(2m+1)\pi)\right] m_0((2m+1)\pi)\hat{\phi}((2m+1)\pi)$$

$$= m_0(\pi)\hat{\phi}((2m+1)\pi) = 0$$

Further, since $\sum_{k\in\mathbf{Z}} \mid \hat{\phi}(\xi+2\pi k) \mid^2 = (2\pi)^{-1}$, ϕ is normalized to

$$\mid \hat{\phi}(0) \mid = (2\pi)^{-1/2}, \ \ i.e., \ \ \left|\int_{-\infty}^{\infty} \phi(x)dx\right| = 1$$

We can choose the phase of $\hat{\phi}$ so that $\int_{-\infty}^{\infty} \phi(x)dx = 1$.

Now we may use the formula $\hat{\phi}(\xi) = m_0(\xi/2)\hat{\phi}(\xi/2)$ repeatedly to obtain

$$\hat{\phi}(\xi) = (2\pi)^{-1/2} \prod_{k=1}^{\infty} m_0(2^{-k}\xi)$$

where the infinite product on the right makes sense, since

$$
\begin{aligned}
| m_0(\xi) | &\leq 1 + | m_0(\xi) - 1 | = 1 + | m_0(\xi) - m_0(0) | \\
&\leq 1 + \sqrt{2} \sum_{n \in Z} | h_n | \ | \sin n\xi/2 | \\
&\leq 1 + \sqrt{2} \sum_{n \in Z} | h_n | \ | n\xi/2 | \leq 1 + C | \xi | \leq e^{C|\xi|}
\end{aligned}
$$

($h_n \neq 0$ for finitely many n), and, therefore,

$$
\prod_{k=1}^{\infty} | m_0(2^{-k}\xi) \leq exp \left(\sum_{k=1}^{\infty} C \, | \, 2^{-k}\xi \, | \right) = e^{C|\xi|}
$$

Thus, the infinite product converges absolutely and uniformly on compact sets.

The above lemma implies that up to a constant phase factor,

$$
\hat{\phi}(\xi) = (2\pi)^{-1/2} \prod_{k=1}^{\infty} m_0(2^{-k}\xi)
$$

is the only possible choice for the scaling function corresponding to a trigonometric polynomial m_0 of the form

$$
m_0(\xi) = \left(\frac{1 + e^{i\xi}}{2} \right)^{M} \mathcal{L}(\xi), \ M \geq 1
$$

and satisfying $| m_0(\xi) |^2 + | m_0(\xi + \pi) |^2 = 1$. It is necessary, however, to check that this sole candidate for a scaling function does indeed satisfy the basic requirements for a scaling function.

(i) The following lemma due to S. Mallat shows that ϕ is square integrable.

Lemma 6.8 If m_0 is a 2π-periodic function such that $| m_0(\xi) |^2 + | m_0(\xi + \pi) |^2 = 1$, and if the sequence of functions

$$
\left\{ (2\pi)^{-1/2} \prod_{j=1}^{k} m_0(2^{-j}\xi) \right\}_{k \geq 1}
$$

converges pointwise a.e., then its limit is in $L^2(\mathbf{R})$, with norm bounded above by 1.

Proof : If

$$
f_k(\xi) = (2\pi)^{-1/2} \prod_{j=1}^{k} m_0(2^{-j}\xi)\chi_{[-\pi,\pi]}(2^{-k}\xi)
$$

then $f_k \to \hat{\phi}$ pointwise a.e., where

$$\hat{\phi}(\xi) = (2\pi)^{-1/2} \prod_{k=1}^{\infty} m_0(2^{-k}\xi)$$

Further,

$$\| f_k \|_2^2 = \int_{-\infty}^{\infty} | f_k(\xi) |^2 \, d\xi = (2\pi)^{-1} \int_{-2^k \pi}^{2^k \pi} \prod_{j=1}^{k} | m_0(2^{-j}\xi) |^2 \, d\xi$$

$$= (2\pi)^{-1} \int_{0}^{2^{k+1} \pi} \prod_{j=1}^{k} | m_0(2^{-j}\xi) |^2 \, d\xi$$

(since m_0 is 2π-periodic)

$$= (2\pi)^{-1} \int_{0}^{2^k \pi} [| m_0(2^{-k}\xi) |^2 + | m_0(2^{-k}\xi + \pi) |^2] \prod_{j=1}^{k-1} | m_0(2^{-j}\xi)^2 d\xi$$

$$= (2\pi)^{-1} \int_{0}^{2^k \pi} \prod_{j=1}^{k-1} | m_0(2^{-j}\xi) |^2 \, d\xi$$

$$= \| f_{k-1} \|_2^2 .$$

Therefore, for all k,

$$\| f_k \|_2^2 = \| f_{k-1} \|_2^2 = \dots = \| f_1 \|_2^2 = 1$$

By Fatou's lemma we then have

$$\int_{-\infty}^{\infty} | \phi(\xi) |^2 \, d\xi \leq \lim_{k \to \infty} \sup \int_{-\infty}^{\infty} | f_k(\xi) |^2 \, d\xi \leq 1$$

(ii) The following lemma due to Deslauriers and Dubuc shows that ϕ has compact support.

Lemma 6.9 If

$$\Lambda(z) = \sum_{k=n}^{N} \lambda_k e^{-ikz}$$

with

$$\sum_{k=n}^{N} \lambda_k = 1$$

then

$$\left| \prod_{k=1}^{\infty} \Lambda(2^{-k}z) \right| \leq C(1+ | z |)^M e^{n|Im\ z|} \quad \text{for } Im\ z \geq 0$$

and

$$\left| \prod_{k=1}^{\infty} \Lambda(2^{-k}z) \right| \le C'(1+\mid z \mid)^{M'} e^{N\mid Im\ z\mid} \quad \text{for } Im\ z \ge 0$$

for some constants C, C', M, M'.

Proof : Let

$$\tilde{\Lambda}(z) = e^{-inz}\Lambda(z) = \sum_{k=0}^{N-n} \lambda_{k+n} e^{-ikz}.$$

Then,

$$\prod_{k=1}^{\infty} \Lambda(2^{-k}z) = e^{inz} \prod_{k=1}^{\infty} \tilde{\Lambda}(2^{-k}z).$$

If $Im\ z \ge 0$, then

$$\mid \tilde{\Lambda}(z) - 1 \mid \le \sum_{k=0}^{N-n} \mid \lambda_{k+n} \mid \ \mid e^{-ikz} - 1 \mid$$

$$\le 2 \sum_{k=0}^{N-n} \mid \lambda_{k+n} \mid \ min(1, n \mid z \mid) \ \le C\ min(1, \mid z \mid).$$

For $Im\ z \ge 0$ if $\mid z \mid \le 1$, then

$$\mid \prod_{k=1}^{\infty} \tilde{\Lambda}(2^{-k}z) \mid \le \prod_{k=1}^{\infty} [1 + C2^{-k}] \le \prod_{k=1}^{\infty} e^{2^{-j}C} = e^{C}$$

If $\mid z \mid \ge 1$, then $\exists k_0 \ge 0 \ni 2^{k_0} \le \mid z \mid < 2^{k_0+1}$ and

$$\mid \prod_{k=1}^{\infty} \tilde{\Lambda}(2^{-k}z) \mid \le \prod_{k=1}^{\infty}(1 + C) \mid \prod_{k=1}^{\infty} \tilde{\Lambda}(2^{-k}2^{-k_0-1}z) \mid \le (1+C)^{k_0+1} e^{C}$$

$$\le e^{C}(1 + C)exp\,[ln(1 + C)ln \mid z \mid /ln2] \le e^{C}(1 + C) \mid z \mid^{ln(1+C)/ln2}$$

Thus, for all $z \ni Im\ z \ge 0$,

$$\mid \prod_{k=1}^{\infty} \tilde{\Lambda}(2^{-k}z) \mid \le e^{C}(1 + C)(1+\mid z \mid)^{ln(1+C)/ln2},$$

i.e.,

$$\mid \prod_{k=1}^{\infty} \Lambda(2^{-k}z) \mid \le C(1+\mid z \mid)^{M} e^{n\mid Im\ z\mid}.$$

Similarly, for all $z \ni Im\, z \leq 0$,

$$| \prod_{k=1}^{\infty} \tilde{\Lambda}(2^{-k}z) | \leq e^{C'}(1+C')(1+|z|)^{ln(1+C')/ln2},$$

i.e.,

$$| \prod_{k=1}^{\infty} \Lambda(2^{-k}z) | \leq C'(1+|z|)^{M'} e^{n|Im\, z|}.$$

Now, applying the Paley-Wiener theorem (see Rudin[35]) for distributions to Λ, we find that it is the Fourier transform of a function with support $\subseteq [n, N]$. Hence, ϕ has compact support.

The above two conditions are not sufficient to ensure that ϕ is a scaling function, as the following example shows. Let

$$m_0(\xi) = \left(\frac{1+e^{-i\xi}}{2}\right)(1+e^{-i\xi}+e^{-2i\xi}) = \frac{1+e^{-3i\xi}}{2} = e^{-3i\xi/2}\cos\frac{3\xi}{2}.$$

Here, m_0 satisfies $|m_0(\xi)|^2 + |m_0(\xi+\pi)|^2 = 1$ and $m_0(0) = 1$.

Using

$$\hat{\phi}(\xi) = (2\pi)^{-1/2}\prod_{k=1}^{\infty}m_0(2^{-k}\xi),$$

we get

$$\hat{\phi}(\xi) = (2\pi)^{-1/2}e^{-3i\xi/2}\frac{\sin 3\xi/2}{3\xi/2}.$$

Now,

$$\sum_{k\in\mathbf{Z}}|\hat{\phi}(\xi+2\pi k)|^2 = (2\pi)^{-1}\left[\frac{1}{3}+\frac{4}{9}\cos\xi+\frac{2}{9}\cos 2\xi\right] = 0 \text{ for } \xi = \frac{2\pi}{3}$$

which implies that $\{\phi_{0,n}\}_{n\in\mathbf{Z}}$ cannot be a Riesz basis. It is, therefore, necessary to ensure that the condition

$$\sum_{k\in\mathbf{Z}}|\hat{\phi}(\xi+2\pi k)|^2 = (2\pi)^{-1},$$

or equivalently,

$$\{\phi_{0,n}\}_{n\in\mathbf{Z}}$$

is an orthonormal system. This is sufficient to guarantee that the subspaces

$$\mathcal{V}_j = \operatorname*{closure}_{L^2(\mathbf{R})}(\{\phi_{j,n}\}_{n\in\mathbf{Z}})$$

generate an MRA, and

$$\psi(x) = \sqrt{2} \sum_{n \in \mathbf{Z}} (-1)^{n-1} h_{-n-1} \phi(2x - n)$$

generates an orthonormal basis of compactly supported wavelets $\{\psi_{j,k}\}_{j,k \in \mathbf{Z}}$ which spans $L^2(\mathbf{R})$.

(iii) The following theorem due to Lawton gives a sufficient condition for $\{\phi_{0,n}\}_{n \in \mathbf{Z}}$ to be an orthonormal system.

Theorem 6.10 *Let*

$$m_0(\xi) = \frac{1}{\sqrt{2}} \sum_{n=0}^{N} h_n e^{-in\xi}$$

be a trigonometric polynomial satisfying $\mid m_0(\xi) \mid^2 + \mid m_0(\xi + \pi) \mid^2 = 1$ *and* $m_0(0) = 1$, *and let*

$$\hat{\phi}(\xi) = (2\pi)^{\frac{-1}{2}} \prod_{k=1}^{\infty} m_0(2^{-k}\xi)$$

If the eigenvalue 1 of the $(2N - 1) \times (2N - 1)$ *matrix*

$$H = \left(\sum_{n=0}^{N} h_n \overline{h_{j-2k+n}} \right)_{-N+1 \le j,k \le N-1}$$

is nondegenerate, then $\{\phi_{0,n}\}_{n \in \mathbf{Z}}$ *is an orthonormal system.*

Proof : Any trigonometric polynomial satisfying

$$\mid m_0(\xi) \mid^2 + \mid m_0(\xi + \pi) \mid^2 = 1 \text{ and } m_0(0) = 1$$

can be transformed to have the form

$$m_0(\xi) = \frac{1}{\sqrt{2}} \sum_{n=0}^{N} h_n e^{-in\xi}$$

by multiplying it by $e^{iM\xi}$ for some $M \in \mathbf{Z}$, corresponding to a shift of ϕ by M. Let

$$\alpha_k = \int_{-\infty}^{\infty} \phi(x) \overline{\phi(x - k)} dx.$$

Since $\text{sup}(\phi) \subseteq [0, N]$, $\alpha_k = 0$ if $\mid k \mid \ge N$. As $\phi(x) = \sqrt{2} \sum_{n \in \mathbf{Z}} h_n \phi(2x - n)$, we have

$$\alpha_k = \int_{-\infty}^{\infty} \phi(x) \overline{\phi(x - k)} dx = \sum_{n,m=0}^{N} h_n \overline{h_m} \int_{-\infty}^{\infty} \phi(2x - n) \overline{\phi(2x - 2k - m)} dx$$

$$= \sum_{n,m=0}^{N} h_n \bar{h}_m \alpha_{2k+m-n}$$

$(h_n = 0 \text{ if } n < 0 \text{ or } n > N)$

$$= \sum_{j=-N+1}^{N-1} (\sum_{n=0}^{N} h_n \overline{h_{j-2k+n}}) \alpha_j \quad \forall -N+1 \leq k \leq N-1$$

If we define the $(2N-1) \times (2N-1)$ matrix

$$H = (\sum_{n=0}^{N} h_n \overline{h_{j-2k+n}})_{-N+1 \leq j,k \leq N-1},$$

then $H\alpha = \alpha$, where $\alpha = (\alpha_{-N+1}, ..., \alpha_0, ..., \alpha_{n-1})$; *i.e.*, $\beta_k = \delta_{k,0}$ is an eigenvector of H with eigenvalue 1. H always has 1 as an eigenvalue, for if $\beta = (0, ..., 0, 1, 0, ..., 0)$, then

$$(H\beta)_j = \sum_{|j|<N} H_{jk}\delta_{j,0} = \sum_n h_n \overline{h_{n-2j}} = \delta_{j,0} = \beta_j \quad \forall \, |\, j \,| < N,$$

i.e., $H\beta = \beta$. Now, if 1 is a nondegenerate eigenvalue of H, then α must be a multiple of β, *i.e.*,

$$\int_{-\infty}^{\infty} \phi(x)\overline{\phi(x-k)}dx = c\delta_{k,0}$$

for some $c \in \mathbf{C}$. Equivalently,

$$\sum_{k \in \mathbf{Z}} |\, \hat{\phi}(\xi + 2\pi k) \,|^2 = (2\pi)^{-1}c.$$

However, since $|\, \hat{\phi}(2\pi k) \,| = 0 \; \forall k \neq 0$ and $\hat{\phi}(0) = (2\pi)^{-1/2}$, it follows that $c = 1$. Thus,

$$\int_{-\infty}^{\infty} \phi(x)\overline{\phi(x-k)}dx = \delta_{k,0},$$

i.e., $\{\phi_{0,n}\}_{n\in\mathbf{Z}}$ is an orthonormal system.

Remarks

(i) By the above theorem, $\{\phi_{0,n}\}_{n\in\mathbf{Z}}$ is not orthonormal if the characteristic equation for H has a multiple zero at 1. This means that the set of vectors $h = (h_0, ..., h_N) \in \mathbf{C}^{N+1}$ for which this happens is a set whose "volume" (measure) is very small compared to that of the set of vectors for which it does not.

In other words, for trigonometric polynomials m_0 that satisfy $\mid m_0(\xi) \mid^2$ $+ \mid m_0(\xi + \pi) \mid^2 = 1$ and $m_0(0) = 1$, the compactly supported function $\phi \in L^2(\mathbf{R})$ defined by

$$\hat{\phi}(\xi) = (2\pi)^{-1/2} \prod_{k=1}^{\infty} m_0(2^{-k}\xi)$$

"almost always" generates an orthonormal set $\{\phi_{0,n}\}_{n \in \mathbf{Z}}$.

(ii) It is not obvious if Lawton's condition is necessary, for it may happen that 1 is a degenerate eigenvalue of H, and yet (for some other reason) the vector α is always equal to the vector β. However, Lawton and Cohen independently proved that the nondegeneracy condition is indeed necessary by showing its equivalence to a necessary and sufficient condition for the orthonormality of $\{\phi_{0,n}\}_{n \in \mathbf{Z}}$ due to Cohen, which involves the structure of the zero set of m_0 (see Cohen[7]).

We summarize our discussion of the necessary and sufficient conditions on the function m_0 associated with an MRA in order for it to generate a compactly supported orthonormal wavelet basis of $L^2(\mathbf{R})$.

- To ensure that the wavelet basis functions are compactly supported, it is necessary and sufficient to ensure that the basic wavelet ψ is compactly supported. The simplest way to do this is to choose a scaling function ϕ that is compactly supported such that $\{\phi(x - n)\}_{n \in \mathbf{Z}}$ is an orthogonal system of functions.

- The 2π-periodic function m_0 associated with an MRA, satisfying $\mid m_0(\xi) \mid^2 + \mid m_0(\xi + \pi) \mid^2 = 1$ and $m_0(0) = 1$, is a trigonometric polynomial iff the scaling function ϕ has compact support.

- If

$$m_0(\xi) = \frac{1}{\sqrt{2}} \sum_{n=0}^{N} h_n e^{-in\xi}$$

is a trigonometric polynomial satisfying $\mid m_0(\xi) \mid^2 + \mid m_0(\xi+\pi) \mid^2 = 1$ and $m_0(0) = 1$, and ϕ, ψ are functions defined by

$$\hat{\phi}(\xi) = (2\pi)^{-1/2} \prod_{k=1}^{\infty} m_0(2^{-k}\xi)$$

and

$$\hat{\psi}(\xi) = -e^{-i\xi/2}\overline{m_0(\xi/2 + \pi)}\hat{\phi}(\xi/2),$$

then ϕ, ψ are compactly supported functions in $L^2(\mathbf{R})$ satisfying

$$\phi(x) = \sqrt{2}\sum_{n=0}^{N} h_n\phi(2x - n)$$

$$\psi(x) = \sqrt{2}\sum_{n=0}^{N}(-1)^{n-1}h_{-n+1}\phi(2x - n).$$

Moreover, $\{\psi_{j,k}(.) = 2^{-j/2}\psi(2^{-j}. - k)\}_{j,k\in\mathbf{Z}}$ is an orthonormal basis of $L^2(\mathbf{R})$ iff the eigenvalue 1 of the $(2N - 1) \times (2N - 1)$ matrix

$$H = (\sum_{n=0}^{N} h_n\overline{h_{j-2k+n}})_{-N+1\leq j,k\leq N-1}$$

is nondegenerate.

The first examples of compactly supported orthonormal wavelets were constructed by Daubechies, and correspond to the case $P \equiv P_N$ in the solution to the Bezout problem . For further examples, see Daubechies ([9] and [8]).

Apart from in the Haar case, there are no closed-form analytic formulas for the compactly supported ϕ and ψ constructed using the methods discussed above. However, if ϕ is continuous, then it can be computed to any degree of accuracy at any given value using the cascade algorithm of Daubechies and Lagarias [11] . We next derive the cascade algorithm.

6.3.3   The cascade algorithm

ϕ is a compactly supported function in $L^1(\mathbf{R})$ with $\int_{-\infty}^{\infty}\phi(x)dx = 1$. Therefore, we have the following theorem.

Theorem 6.11 : *If $f \in C(\mathbf{R})$, then for all $x \in \mathbf{R}$,*

$$\lim_{j\to\infty}\int_{-\infty}^{\infty} f(x + y)\overline{\phi(2^j y)}dy = f(x)$$

If f is uniformly continuous, then this convergence is uniform. If f is Hölder continuous with exponent α, i.e., , $\mid f(x) - f(y) \mid\leq C \mid x - y \mid^{\alpha}$, then the convergence is exponential in j:

$$\mid f(x) - 2^j\int_{-\infty}^{\infty} f(x + y)\overline{\phi(2^j y)}dy \mid\leq C2^{-j\alpha}.$$

Proof : The key observation to proving the theorem is that $2^j \phi(2^j)$ tends to the Dirac δ-function as $j \to \infty$. Indeed,

$$| f(x) - 2^j \int_{-\infty}^{\infty} f(x+y)\overline{\phi(2^j y)}dy |$$

$$| 2^j \int_{-\infty}^{\infty} [f(x) - f(x+y)]\overline{\phi(2^j y)}dy |$$

$$| \int_{-\infty}^{\infty} [f(x) - f(x+2^{-j}z)]\overline{\phi(z)}dz |$$

$$\leq \| \phi \|_1 \cdot \sup_{|u| \leq 2^{-j}R} | f(x) - f(x+u) | \quad \text{(where supp } \phi \subseteq [-R,R] \text{)}$$

If $f \in C(\mathbf{R})$, then the right hand side of the inequality can be made arbitrarily small by choosing j sufficiently large. If f is uniformly continuous, then the choice of j is independent of x, and, hence, the convergence is uniform. If f is Hölder continuous, then the last claim is also clear.

If η_j^0 itself is continuous or Hölder continuous with exponent α, then for any dyadic rational $x = 2^{-J}N$, $N \in \mathbf{Z}$ we have by the above theorem,

$$\phi(x) = \lim_{j \to \infty} 2^j \int_{-\infty}^{\infty} \phi(2^{-J}N + y)\overline{\phi(2^j y)}dy$$

$$= \lim_{j \to \infty} 2^{j/2} \int_{-\infty}^{\infty} \phi(z)\overline{\phi_{-j,2^{j-J}N}}(\bar{z})dz$$

$$= \lim_{j \to \infty} 2^{j/2} \langle \phi, \phi_{-j,2^{j-J}N} \rangle$$

For $j > j_0$ for some j_0,

$$| \phi(2^{-J}\mathbf{N}) - 2^{j/2}\langle \phi, \phi_{-j,2^{j-J}N} \rangle | \leq C2^{-j\alpha},$$

where C, j_0 are dependent on J, N. If 2^{j-J}, which is trivially true for $j \geq J$, then it is easy to compute the inner products

$$\langle \phi, \phi_{-j,2^{j-J}N} \rangle$$

Since $\{\phi_{0,n}\}_{n \in \mathbf{Z}}$ is an orthonormal system, ϕ is the unique function f characterized by

$$\langle f, \phi_{0,n} \rangle = \delta_{0,n}$$

$$\langle f, \psi_{-j,k} \rangle = 0 \text{ for } j > 0, k \in \mathbf{Z}$$

We can use the sequences

$$c^0 = \{c_n^0 = \langle f, \phi_{0,n} \rangle = \delta_{0,n}\}_{n \in \mathbf{Z}}$$

and

$$d^0 = \left\{ d_n^0 = \langle f, \phi_{0,n} \rangle = 0 \right\}_{n \in \mathbf{Z}}$$

as the low-pass and high-pass inputs to the reconstruction algorithm of the subband filtering scheme associated with m_0 to obtain

$$c^{-1} = \left\{ c_n^{-1} = \sum_{k \in \mathbf{Z}} h_{n-2k} c_k^0 = h_n = \langle \phi, \phi_{-1,n} \rangle \right\}_{n \in \mathbf{Z}}$$

Now, using $d^{-1} = \{0\}_{n \in \mathbf{Z}}$ we repeat the above steps with c^{-1} and d^{-1} as the low-pass and high-pass input sequences to obtain

$$c^{-2} = \left\{ c_n^{-2} = \sum_{k \in \mathbf{Z}} h_{n-2k} c_k^{-1} = \sum_{k \in \mathbf{Z}} h_{n-2k} h_k = \langle \phi, \phi_{-2,n} \rangle \right\}_{n \in \mathbf{Z}}$$

At the jth stage, we have

$$c^{-j} = \left\{ c_n^{-j} = \sum_{k \in \mathbf{Z}} h_{n-2k} c_k^{-j+1} \langle \phi, \phi_{-j,n} \rangle \right\}_{n \in \mathbf{Z}}$$

and $d^{-j} = \left\{ d_n^{-j} = 0 \right\}_{n \in \mathbf{Z}}$. The inequality

$$\mid \phi(2^{-J} N) - 2^{j/2} \langle \phi, \phi_{-j, 2^{j-J} N} \rangle \mid \leq C 2^{-j\alpha}$$

for $j > j_0$ for some j_0 implies that the above procedure computes the values of ϕ at dyadic rationals with exponentially rapid convergence. These values can be interpolated to obtain a sequence of functions η_j approximating ϕ.

Define η_j^0 to be the function

$$\eta_j^0 (2^{-j} n) = 2^{j/2} \langle \phi, \phi_{-j,n} \rangle$$

and piecewise constant on

$$\left[2^{-j}(n - 1/2), 2^{-j}(n + 1/2) \right],$$

and η_j^l to be the function

$$\eta_j^l (2^{-j} n) = 2^{j/2} \langle \phi, \phi_{-j,n} \rangle$$

and piecewise linear on

$$\left[2^{-j} n, 2^{-j}(n + 1) \right]$$

The following theorem holds for both the above choices for the approximating function.

Theorem 6.12 *If η_j^0 is Hölder continuous with Hölder exponent α, then there exists $C > 0$ and $j_0 \in \mathbf{N} \ni \forall j \geq j_0$,*

$$\| \phi - \eta_j^\varepsilon \|_\infty \leq C 2^{-\alpha j} \quad \varepsilon = 0,1$$

Proof : Let $x \in \mathbf{R}$. For any j choose n such that $2^{-j}n \leq x \leq 2^{-j}(n+1)$. $\eta_j^\varepsilon(x)$ is a convex linear combination of $2^{j/2}\langle\phi, \phi_{-j,n}\rangle$ and $2^{j/2}\langle\phi, \phi_{-j,n+1}\rangle$ for $\varepsilon = 0,1$. Now if $j > j_0$ for some j_0,

$$| \phi(x) - 2^{j/2}\langle\phi, \phi_{-j,n}\rangle |$$

$$\leq | \phi(x) - \phi(2^{-j}n) | + | \phi(2^{-j}n) - 2^{j/2}\langle\phi, \phi_{-j,n}\rangle |$$

$$\leq C\,| x - 2^{-j}n |^\alpha + C 2^{-j\alpha} \leq 2C 2^{-j\alpha}$$

Similarly,

$$| \phi(x) - 2^{j/2}\langle\phi, \phi_{-j,n+1}\rangle | \leq 2C 2^{-\alpha j},$$

and, hence, for any convex combination of $2^{j/2}\langle\phi, \phi_{-j,n}\rangle$ and $2^{j/2}\langle\phi, \phi_{-j,n+1}\rangle$

This gives the cascade algorithm which rapidly approximates the values of $\phi(x)$ with arbitarily high accuracy :

1. Begin with the sequence $\{\eta_0^\varepsilon(n) = \delta_{0,n}\}_{n\in\mathbf{Z}}$.

2. Compute the sequence

$$\left\{ \eta_j^\varepsilon(2^{-j}n) = \sum_{k\in\mathbf{Z}} h_{n-2k}\eta_{j-1}^\varepsilon(2^{-j}k) \right\}_{n\in\mathbf{Z}}$$

The number of values obtained at each stage is double that of the previous state. Values at the even nodes $2^{-j}(2n)$ are refined from the previous step:

$$\eta_j^\varepsilon(2^{-j}2n) = \sum_{k\in\mathbf{Z}} h_{2(n-k)}\eta_{j-1}^\varepsilon(2^{-j}k)$$

Values at the odd nodes $2^{-j}(2n+1)$ are computed for the first time:

$$\eta_j^\varepsilon(2^{-j}(2n+1)) = \sum_{k\in\mathbf{Z}} h_{2(n-k)+1}\eta_{j-1}^\varepsilon(2^{-j}k)$$

3. Interpolate the sequence

$$\left\{ \eta_j^\varepsilon(2^{-j}n) = \sum_{k\in\mathbf{Z}} h_{n-2k}\eta_{j-1}^\varepsilon(2^{-j}k) \right\}_{n\in\mathbf{Z}}$$

using piecewise constant (if $\varepsilon = 0$) or piecewise linear (if $\varepsilon = 1$) interpolation to obtain $\eta_j^\varepsilon(x)$ $\forall x \in \mathbf{R}$.

6.4   Biorthogonal Wavelets

In Section 6.2 we studied the fast wavelet algorithm in the context of subband filtering schemes and noted that orthonormal wavelet bases associated with MRAs generate pairs of CQFs, which are subband filtering schemes with exact reconstruction, with the functions associated with the MRAs corresponding to the filter functions associated with the CQFs. We then observed that an important and useful subclass of CQFs, namely the FIR CQFs, have polynomials for their filter functions, and, therefore, orthonormal wavelet bases associated with MRAs that give rise to FIR CQFs have polynomial m_0. This in turn implied that the wavelet bases were compactly supported, provided the m_0 satisfied some special conditions.

6.4.1   Linear phase FIR filters

We will continue to use subband filtering as our motivating paradigm to study a new class of wavelet bases, namely, compactly supported biorthogonal wavelet bases. This new class gives rise to FIR CQFs with an additional feature, that of *linear phase* for the filter polynomial, that is very important in many filtering applications. Linear phase is especially important to have in image processing applications such as data compression. In the absence of linear phase, the reconstruction of compressed images induces distortions. We will briefly describe this in the context of wavelet filtering.

Let $\{\mathcal{V}_j\}_{j \in \mathbf{Z}}$ be an MRA of $L^2(\mathbf{R})$. A finite energy analog signal $f \in L^2(\mathbf{R})$ is sampled by approximating it by a function $f_J \in \mathcal{V}_J$ (by either interpolating a discrete sample in $\mathcal{V}_J$ or by projecting f onto $\mathcal{V}_J$), where $|J|$ is large enough to prevent undersampling. f_J can then be decomposed uniquely as follows : $f_J = g_{J-1} + g_{J-2} + \ldots + g_{J-J_0} + f_{J-J_0}$, where $g_j \in \mathcal{W}_j$ and $f_{J-J_0} \in \mathcal{V}_{J-J_0}$. The g_j contain the time-scale information about f_J at the jth scale, and f_{J-J_0} is the projection of f_J onto $\mathcal{V}_{J-J_0}$, where the latter space is chosen such that the time-scale features of the projection of f_J in it are either to be ignored or are not to be tampered with. Compression of the sampled signal is achieved by *thresholding, i.e.,* deleting information (coefficients) of very small magnitudes from the g_j in each of the subspaces $\mathcal{W}_j$, and "adding" together the modified components to obtain the reconstructed signal, thus saving on the amount of information to be stored about the sampled signal f_J. Due to the deletions, exact reconstruction is not possible; therefore, it is necessary to avoid distortion resulting from the *quantization* of the signal. Since the reconstructed signal is a wavelet series, the filtering is linear; hence, distortion can be avoided if the filter

has linear phase or even *generalized* linear phase. A filter has (generalized
) linear phase if its corresponding filter function has (generalized) linear
phase.

Definition 6.13 (i) A function $f \in L^2(\mathbf{R})$ has **linear phase** if $\hat{f}(\xi) = \pm | \hat{f}(\xi) | e^{-i\gamma\xi}$, i.e., where γ is a real-valued constant and the $\pm$ sign is independent of ξ.

(ii) A function $f \in L^2(\mathbf{R})$ has **generalized linear phase** if $\hat{f}(\xi) = F(\xi)e^{-i(\gamma\xi+\theta)}$, a.e., where $F(\xi)$ is a real-valued function and γ, θ are real-valued constants.

As an example, consider the nth order cardinal B-spline $\mathbf{N}_n$. Its Fourier transform is given by

$$\hat{\mathbf{N}}_n(\xi) = (\frac{\sin(\xi/2)}{\xi/2})^n e^{-in\xi/2}.$$

Therefore, $\mathbf{N}_n$ has linear phase, and the phase of $\hat{\mathbf{N}}_n$ is $n/2$.

Definition 6.14 (i) If $\{s_n\}_{n\in\mathbf{Z}} \in l^1(\mathbf{Z})$ and

$$S(\xi) = \sum_{n\in\mathbf{Z}} s_n e^{-in\xi}$$

is the associated Fourier series, then $\{s_n\}_{n\in\mathbf{Z}}$ has linear phase if $S(\xi) = \pm | S(\xi) | e^{-in_0\xi}$, where $n_0 \in 1/2\mathbf{Z} = \{n : 2n \in \mathbf{Z}\}$ is a real-valued constant and the $\pm$ sign is independent of ξ.

(ii) The sequence $\{s_n\}_{n\in\mathbf{Z}}$ has **generalized linear phase** if $S(\xi) = F(\xi)e^{-i(n_0\xi+\theta)}$, where $F(\xi)$ is a real-valued function, $n_0 \in 1/2\mathbf{Z}$ and θ is a real-valued constant.

n_0 is called the phase of the Fourier series associated with $\{s_n\}_{n\in\mathbf{Z}}$.

The following theorem characterizes functions and sequences with generalized linear phase.

Theorem 6.15 *(i) A function $f \in L^2(\mathbf{R})$ has generalized linear phase; that is, $\hat{f}(\xi) = F(\xi)e^{-i(\gamma\xi+\theta)}$, where $F(\xi)$ is a real-valued function and γ, θ are real-valued constants, iff $e^{i\theta}f(x)$ is skew-symmetric w.r.t. γ, i.e.,*

$$e^{i\theta} f(\gamma + x) = \overline{e^{i\theta} f(\gamma - x)}, \quad x \in \mathbf{R}$$

(ii) A sequence $\{s_n\}_{n\in\mathbf{Z}} \in l^1(\mathbf{Z})$ has generalized linear phase, i.e., $S(\xi) = F(\xi)e^{-i(n_0\xi+\theta)}$, where $F(\xi)$ is a real-valued function, $n_0 \in 1/2\mathbf{Z}$ and θ is a

real-valued constant, iff $\left\{e^{i\theta}S_n\right\}_{n\in\mathbf{Z}}$ *is skew-symmetric w.r.t.* n_0, *i.e.*,

$$e^{i\theta}S_n = \overline{e^{i\theta}S_{2n_0-n}}, \quad n \in \mathbf{Z}$$

Proof : (i) Let $f \in L^2(\mathbf{R})$, such that $\hat{f}(\xi) = F(\xi)e^{i(\gamma\xi+\theta)}$. Then,

$$f(x) = \frac{1}{2\pi}\int_{-\infty}^{\infty} F(\xi)e^{-i(\gamma\xi+\theta)}e^{ix\xi}d\xi$$

i.e.,

$$e^{i\theta}f(\gamma-x) = \frac{1}{2\pi}\int_{-\infty}^{\infty} F(\xi)e^{-ix\xi}d\xi$$

As $F(\xi)$ is real-valued, we have

$$\overline{e^{i\theta}f(\gamma-x)} = \overline{\frac{1}{2\pi}\int_{-\infty}^{\infty} F(\xi)e^{-ix\xi}d\xi} = \frac{1}{2\pi}\int_{-\infty}^{\infty} F(\xi)e^{ix\xi}d\xi = e^{i\theta}f(\gamma+x)$$

Conversely, if $e^{i\theta}f(\gamma+x) = \overline{e^{i\theta}f(\gamma-x)}$, $x \in \mathbf{R}$, then

$$e^{i\theta}\hat{f}(\xi)e^{i\gamma\xi} = (e^{i\theta}f(\gamma+x))^{\wedge} = (\overline{e^{i\theta}f(\gamma-x)})^{\wedge} = \overline{e^{i\theta}\hat{f}(\xi)e^{i\gamma\xi}}$$

i.e., , $e^{i\theta}\hat{f}(\xi)e^{i\gamma\xi}$ is real. Letting

$$F(\xi) = e^{i\theta}\hat{f}(\xi)e^{i\gamma\xi},$$

we have

$$\hat{f}(\xi) = F(\xi)e^{i(\gamma\xi+\theta)}.$$

(ii) If $\{s_n\}_Z \in l^1(\mathbf{Z})$ satisfies $S(\xi) = F(\xi)e^{-i(n_0\xi+\theta)}$, then

$$S(\xi)e^{i(n_0\xi+\theta)} = F(\xi) = \overline{F(\xi)} = e^{-i(n_0\xi+\theta)}\overline{S(\xi)}$$

i.e.,

$$S(\xi)e^{i2n_0\xi}e^{i\theta} = \overline{e^{i\theta}S(\xi)},$$

or equivalently,

$$e^{i\theta}s_n = \overline{e^{i\theta}s_{2n_0-n}}, \quad n \in \mathbf{Z}$$

We will restrict ourselves to filters with real coefficients, and, hence, to real-valued functions with generalized linear phase. If $f \in L^2(\mathbf{R})$ is real-valued, then $e^{i\theta}f(\gamma+x) = \overline{e^{i\theta}f(\gamma-x)}$ implies that $\theta = \pi k/2$, $k \in \mathbf{Z}$ and, therefore,

$$f(\gamma+x) = f(\gamma-x), \quad x \in \mathbf{R}$$

(symmetric w.r.t. γ) or

$$f(\gamma+x) = -f(\gamma-x), \quad x \in \mathbf{R}$$

(antisymmetric w.r.t. γ). Thus, we have:

Theorem 6.16 *(i) A real-valued function $f \in L^2(\mathbf{R})$ has generalized linear phase iff it is either symmetric or antisymmetric w.r.t. the phase of $\hat{f}$.*

(ii) A real-valued sequence $\{s_n\}_{n\in\mathbf{Z}} \in l^1(\mathbf{Z})$ has generalized linear phase iff it is symmetric or antisymmetric w.r.t. the phase of its associated Fourier series.

We now examine the connection between the phase property of the two-scale sequence $\{h_n\}_{n\in\mathbf{Z}}$ and the phase property of the corresponding scaling function ϕ.

The following easily verified lemma characterizes sequences with linear phase:

Lemma 6.17 : A real-valued sequence $\{s_n\}_{n\in\mathbf{Z}} \in l^1(\mathbf{Z})$ with Fourier series $S(\xi)$ has linear phase iff there exists $n_0 \in 1/2\mathbf{Z}$ such that $S(\xi)e^{in_0\xi}$ is real-valued, even, and has no sign changes.

However, since we are interested in FIR filters, we focus on finite sequences (*i.e.*, sequences with finitely many nonzero coefficients), in which case a stronger statement can be made.

Lemma 6.18 A real-valued finite sequence $\{s_n\}_{n=0}^N$ has linear phase iff the following conditions are satisfied:

(i) $s_n = s_{N-n} \forall n \in \mathbf{Z}$

and

(ii) The polynomial

$$S(z) = \sum_{n=0}^N s_n z^n$$

has zeros only of even multiplicity on the unit circle.

Proof : By the above lemma, the linearity of the phase of the sequence

$$\{s_n : s_n = 0 \ \ \forall n < 0 \text{ or } n > N\}_{n\in\mathbf{Z}}$$

is equivalent to the function $F(\xi) = S(\xi)e^{in_0\xi}$ being real-valued, even, and having no sign changes, for some $n_0 \in 1/2\mathbf{Z}$

$F(\xi) = F(-\xi)$ implies

$$\sum_{n=0}^{N} s_n e^{in\xi} = S(e^{i\xi}) = e^{i2n_0\xi} S(e^{-i\xi}) = \sum_{n=0}^{N} s_n e^{i(2n_0-n)\xi} = \sum_{n=2n_0-N}^{2n_0} s_{2n_0-n} e^{in\xi}$$

i.e.,

$$2n_0 = N \text{ and } s_n = s_{N-n} \quad \forall n \in \mathbf{Z}$$

$F(\xi) = S(\xi)e^{in_0\xi}$ will have no sign changes iff its real zeros all have even multiplicity, *i.e.,* iff polynomial

$$S(z) = \sum_{n=0}^{N} s_n z^n$$

has zeros only of even multiplicity on the unit circle .

The following theorems characterize the phase properties of scaling functions in terms of those of the associated two-scale sequences.

Theorem 6.19 : *If ϕ is a scaling function with two-scale sequence $\{h_n\}_{n\in\mathbf{Z}} \in l^1(\mathbf{Z})$ and $H(z) = \sum_{n\in\mathbf{Z}} h_n z^n$, then*

(i) ϕ has generalized linear phase iff

$$\overline{H(z)} = z^{-2n_0} H(z)(n_0 \in 1/2\mathbf{Z}), \quad \forall \mid z \mid = 1, \text{ and}$$

(ii) ϕ has linear phase iff

$$H(e^{-i\xi}) = e^{-in_0\xi} \mid H(e^{-i\xi}) \mid, \quad n_0 \in 1/2\mathbf{Z}$$

Proof : If ϕ has generalized linear phase, then $\hat{\phi}(\xi) = F(\xi)e^{-i(\gamma\xi+\theta)}$, a.e., where $F(\xi)$ is a real-valued function and γ, θ are real-valued constants. Thus, $\overline{\hat{\phi}(\xi)} = F(\xi)e^{i(\gamma\xi+\theta)}$ and, hence,

$$\overline{H(e^{-i\xi/2})} = \frac{\overline{\hat{\phi}(\xi)}}{\overline{\hat{\phi}(\xi/2)}} = e^{i\gamma\xi/2} \frac{F(\xi)}{F(\xi/2)}$$

$$= e^{i\gamma\xi} \frac{\hat{\phi}(\xi)}{\hat{\phi}(\xi/2)} = e^{i\gamma\xi} H(e^{-i\xi/2}) \text{ a.e.}$$

This means that $2\gamma \in \mathbf{Z}$, and letting $\gamma = n_0$, we have

$$\overline{H(z)} = z^{-2n_0} H(z)(n_0 \in \frac{1}{2}\mathbf{Z}), \quad \forall \mid z \mid = 1.$$

If ϕ has linear phase, then $\theta = 0$ and $F(\xi)$ does not change sign. Therefore,

$$H(e^{-i\xi/2}) = \frac{\hat{\phi}(\xi)}{\hat{\phi}(\xi/2)} = e^{-i\gamma\xi/2} \frac{F(\xi)}{F(\xi/2)}$$

$$= e^{-i\gamma\xi/2} \mid \frac{F(\xi)}{F(\xi/2)} \mid = e^{-in_0\xi/2} \mid H(e^{-i\xi/2}) \mid,$$

and so

$$H(e^{-i\xi}) = e^{-in_0\xi} \mid H(e^{-i\xi}), \quad n_0 \in \frac{1}{2}\mathbf{Z}.$$

Conversely, if $\overline{H(z)} = z^{-2n_0} H(z) \ (n_0 \in \frac{1}{2}\mathbf{Z}), \ \forall \mid z \mid = 1$, then

$$\overline{\hat{\phi}(\xi)} = \prod_{k=1}^{\infty} \overline{H(e^{-i2^{-k}\xi})} = \prod_{k=1}^{\infty} (\overline{H(e^{-i2^{-k}\xi})}e^{-i2n_0 2^{-k}\xi})$$

$$= e^{-i2n_0\xi} \prod_{k=1}^{\infty} \overline{H(e^{-i2^{-k}\xi})} = e^{-i2n_0\xi}\overline{\hat{\phi}(\xi)}.$$

Therefore, $F(\xi) = \hat{\phi}(\xi)e^{in_0\xi}$ is real-valued, and as $\hat{\phi}(\xi) = F(\xi)e^{-in_0\xi}$, ϕ has generalized linear phase. If $H(e^{-i\xi}) = e^{in_0\xi} \mid H(e^{-i\xi}) \mid$, $n_0 \in \frac{1}{2}\mathbf{Z}$, then

$$\hat{\phi}(\xi) = \prod_{k=1}^{\infty} H(e^{-i2^{-k}\xi}) = \prod_{k=1}^{\infty} (\mid H(e^{-i2^{-k}\xi}) \mid e^{-in_0 2^{-k}\xi})$$

$$= e^{-in_0\xi} \left| \prod_{k=1}^{\infty} H(e^{-i2^{-k}\xi}) \right| = \mid \hat{\phi}(\xi) \mid e^{-in_0\xi}$$

and, hence, ϕ has linear phase.

The following corollary is immediate.

Corollary 6.20 A scaling function ϕ has generalized linear phase iff there exists $n_0 \in \frac{1}{2}\mathbf{Z}$ such that $\phi(n_0 + x) = \overline{\phi(n_0 - x)}$ a.e.

Returning to the case that interests us most, if the two-scale sequence $\{h_n\}_{n \in \mathbf{Z}}$ is real-valued and finite, then by the above theorem and the lemma before it we have:

Theorem 6.21 *If ϕ is a real-valued scaling function with real-valued finite two-scale sequence $\{h_n\}_{n=0}^{N}$, then*

(i) ϕ has generalized linear phase iff $h_{N-n} = h_n \ \forall n \in \mathbf{Z}$; and

(ii) ϕ has linear phase iff $h_{N-n} = h_n$ $\forall n \in \mathbf{Z}$, and all the zeros of the associated polynomial

$$H(z) = \sum_{n=0}^{N} h_n z^n,$$

that lie on the unit circle have even multiplicities. Thus, the scaling function ϕ has linear phase iff the function m_0 has linear phase.

Using the fact that $\hat{\phi}(\xi) = m_1(\xi/2)\hat{\phi}(\xi/2)$ and arguments similar to the ones thus far, it can be concluded that ψ has (generalized) linear phase if the function m_1 has (generalized) linear phase.

6.4.2 Compactly supported orthonormal wavelets are asymmetric

Unfortunately, except for the Haar basis , all compactly supported orthonormal wavelet bases are *necessarily* nonsymmetric. The theorem below demonstrates this fact. We first require two lemmas. As noted earlier, an MRA does not uniquely determine the functions ϕ and ψ. Indeed,

Lemma 6.21 If $\{e_n(x) = e(x-n)\}_{n\in\mathbf{Z}}$ and $\{\tilde{e}_n(x) = \tilde{e}(x-n)\}_{n\in\mathbf{Z}}$ are orthonormal bases of the same subspace M of $L^2(\mathbf{R})$, then $\hat{\tilde{e}}(\xi) = \nu(\xi)\hat{e}(\xi)$, $|\nu(\xi)| = 1$ for some 2π-periodic function $\nu(\xi)$.

Proof : Since $\{e_n(x) = e(x-n)\}_{n\in\mathbf{Z}}$ is an o.n. basis of M and $\tilde{e} \in M$,

$$\tilde{e} = \sum_{n\in\mathbf{Z}} \nu_n e_n$$

Also,

$$\sum_{n\in\mathbf{Z}} |\nu_v|^2 = \|\tilde{e}\|^2 = 1$$

Therefore, $\hat{\tilde{e}}(\xi) = \nu(\xi)\hat{e}(\xi)$, with

$$\nu(\xi) = \sum_{n\in\mathbf{Z}} \nu_n e^{-in\xi}$$

The orthonormality of the e_n is equivalent to

$$2\pi \sum_{n\in\mathbf{Z}} |\hat{e}(\xi + 2\pi n)|^2 = 1 \quad \text{a.e.}$$

and that of the $\tilde{e}_n$, to

$$2\pi \sum_{n \in \mathbf{Z}} \mid \hat{\tilde{e}}(\xi + 2\pi n) \mid^2 = 1 \quad \text{a.e.}$$

That is,

$$\sum_{n \in \mathbf{Z}} \mid \hat{e}(\xi+2\pi n) \mid^2 (1- \mid \nu(\xi+2\pi n) \mid^2) = (1- \mid \nu(\xi) \mid^2) \sum_{n \in \mathbf{Z}} \mid \hat{e}(\xi+2\pi n) \mid^2 = 0$$

Thus, $\mid \nu(\xi) \mid = 1$

However, if the two bases in the above lemma are compactly supported, then $\nu(\xi)$ would be a trigonometric polynomial, leading to further restrictions on the choice of the bases, as the following lemma indicates.

Lemma 6.22 If $\{\nu_n\}_{n \in \mathbf{Z}}$ is a sequence with finitely many nonzero terms, and if $\mid \nu(\xi) \mid = 1$, then $\nu_n = \nu \delta_{n,n_0} \; \forall n \in \mathbf{Z}$, for some $\nu \in \mathbf{C}$ and $n_0 \in \mathbf{Z}$, i.e., $\nu(\xi) = \nu e^{-in_0\xi}$, $\mid \nu \mid = 1$.

Proof : Since $\mid \nu(\xi) \mid = 1$,

$$\sum_{n \in \mathbf{Z}} \nu_n \overline{\nu_{n+m}} = \delta_{0,m}$$

Let n_1 and n_2 be the indices of the first and the last nonzero terms ν_{n_1} and ν_{n_2}, respectively, of the sequence $\{\nu_n\}_{n \in \mathbf{Z}}$. Since

$$\sum_{n \in \mathbf{Z}} \nu_n \overline{\nu_{n+n_2-n_1}} = \delta_{0,n_2-n_1'}$$

we have $\delta_{0,n_2-n_1} = \nu_n \bar{\nu}_{n_2} \neq 0$. Therefore, $n_1 = n_2$.

The above lemmas imply that the compactly supported scaling function ϕ and the associated wavelet ψ are each uniquely determined up to a shift for a given MRA.

In particular, if ϕ and $\tilde{\phi}$ are any two compactly supported scaling functions generating orthonormal wavelet bases for the same MRA, then the corresponding constant $\nu = 1$, since

$$\hat{\phi}(0) = (2\pi)^{-1} = \hat{\tilde{\phi}}(0) = \nu \hat{\phi}(0)$$

Therefore, $\nu(\xi) = e^{-in_0\xi}$, and $\tilde{\phi}(x) = \phi(x - n_0)$.

Theorem 6.23 *If a scaling function ϕ and the associated wavelet ψ of an MRA are real and compactly supported, and if ψ is symmetric or antisymmetric about some point, then ψ is the Haar function.*

Proof : ϕ can be shifted such that

$$h_n = \sqrt{2} \int_{-\infty}^{\infty} \phi(x)\phi(2x - n)dx = 0 \quad \forall n < 0, h_0 \neq 0$$

Since ϕ is real, so are the h_n. If N is the largest index for which h_n is nonzero, then N is necessarily odd, for indeed if $N = 2n_0$, then since

$$| m_0(\xi) |^2 + | m_0(\xi + \pi) |^2 = 1$$

we have

$$\sum_{n \in \mathbf{Z}} h_n h_{n+2m} = \delta_{0,m}$$

wherein letting $m_0 = n_0$, we get $h_N = 0$, a contradiction.

Since $h_n = 0 \ \forall n < 0$ and $n > N$, by a result of Lemarié and Malouyres [25] , we have supp $\phi = [0, N]$ and sup $\psi = [-\frac{N-1}{2}, \frac{N-1}{2} + 1]$. The axis of symmetry/anti-symmetry is, therefore, $x = \frac{1}{2}$, and either $\psi(1 - x) = \psi(x)$ or $\psi(1 - x) = -\psi(x) \ \forall x \in \mathbf{R}$. Hence,

$$\psi_{j,k}(-x) = \pm 2^{-j/2}\psi(2^{-j}x + k + 1) = \pm\psi_{j,-(k+1)}(x)$$

This implies that the spaces $\mathcal{W}_j$ are invariant under the action of the map $x \mapsto -x$, and since

$$\mathcal{V}_j = \underset{L^2(\mathbf{R})}{\text{closure}}(\oplus_{k>j}\mathcal{W}_k)$$

the $\mathcal{V}_j$ are invariant under $x \mapsto -x$ also. Therefore, if $\tilde{\phi}(x) = \phi(N - x)$, then

$$\left\{\tilde{\phi}(-k)\right\}_{k \in \mathbf{Z}}$$

is also an orthonormal basis for $\mathcal{V}_0$. Moreover, $\hat{\tilde{\phi}}(0) = \hat{\phi}(0) = 1$, and supp$\tilde{\phi}$ = suppϕ. By the foregoing lemmas and the observations following them, we have $\hat{\tilde{\phi}} = \hat{\phi}$. i.e., $\phi(x) = \phi(N - x)$.

Therefore,

$$h_n = \sqrt{2} \int_{-\infty}^{\infty} \phi(x)\phi(2x - n)dx$$

$$= \sqrt{2} \int_{-\infty}^{\infty} \phi(N - x)\phi(N - 2x + n)dx$$

$$= \sqrt{2} \int_{-\infty}^{\infty} \phi(y)\phi(2y - N + n)dy = h_{N-n}$$

Now,

$$\delta_{0,m} = \sum_{n \in \mathbf{Z}} h_n h_{n+2m}$$

$$= \sum_{k \in \mathbf{Z}} h_{2k} h_{2k+2m} + \sum_{k \in \mathbf{Z}} h_{2k+1} h_{2k+2m+1}$$

$$= \sum_{k \in \mathbf{Z}} h_{2k} h_{2k+2m} + \sum_{k \in \mathbf{Z}} h_{2n_0-2k} h_{2n_0-2k-2m} \quad (N = 2n_0 + 1)$$

$$= 2 \sum_{k \in \mathbf{Z}} h_{2k} h_{2k+2m}$$

By Lemma 6.22, we get $h_{2k} = \nu \delta_{k,k_0}$ for some $k_0 \in \mathbf{Z}$, $| \nu | = 1/\sqrt{2}$. Since by hypothesis $h_0 \neq 0$, $h_{2k} = \nu \delta_{0,k}$. Also, since $h_n = h_{N-n}$, $h_{2k+1} = \nu \delta_{k,n_0}$. As $m_0(0) = 1$ implies

$$\sum_{n \in \mathbf{Z}} h_n = \sqrt{2},$$

we have $\nu = 1/\sqrt{2}$. Thus, $h_0 = 1/\sqrt{2}$ and $h_n = 0 \ \forall n \neq 0, N$. *i.e.*, $m_0(\xi) = (1 + e^{-in\xi})/2$, and using $\hat{\phi}(\xi) = m_0(\xi/2)\hat{\phi}(\xi/2)$, we get

$$\hat{\phi}(\xi) = \frac{(1 - e^{-iN\xi})}{\sqrt{2\pi} iN\xi},$$

or equivalently $\frac{\phi(x)=1}{N\chi_{[0,N]}}$. If $N > 1$, then the $\phi(.-n)$ are not orthonormal, which contradicts the hypothesis of the theorem. Hence, $N = 1$, and ϕ is indeed the scaling function for the Haar basis.

Since the Haar function is discontinuous and has poor frequency localization, it is not an interesting scaling function for most practical purposes.

If ϕ is allowed to be a complex valued function, then it is possible to have both symmetry as well as compact support.

6.4.3   Dual FIR filters with exact reconstruction

The above discussed incompatibility of symmetry/antisymmetry with compactness of support for real orthonormal wavelets is entirely analogous to the situation of incompatibility of symmetry and exact reconstruction property, familiar in FIR subband filtering, if the same FIR filters are used for both decomposition and reconstruction. However, it is also known that if one were to allow different FIR filters for the decomposition and reconstruction stages of a subband filtering scheme, then one can have linear phase (i.e., symmetry) as well as exact reconstruction. In the context of the subband filtering scheme associated with a multiresolution analysis, this amounts to the following configuration

$$\bar{H} \;\rightarrow\; \downarrow_2 \;\rightarrow\; c^j \downarrow_2 \;\rightarrow\; \uparrow_2 \;\rightarrow \tilde{H}$$

$$c^{j-1} \; \oplus \;\; = c^{j-1}$$

$$\bar{G} \;\rightarrow\; \downarrow_2 \;\rightarrow\; d^j \downarrow_2 \;\rightarrow\; \uparrow_2 \;\rightarrow \tilde{G}$$

with two pairs of *dual filters* H, G and $\tilde{H}$, $\tilde{G}$, replacing the earlier (non-symmetric) scheme with two filters H, G, represented by

$$\bar{H} \;\rightarrow\; \downarrow_2 \;\rightarrow\; c^j \downarrow_2 \;\rightarrow\; \uparrow_2 \;\rightarrow H$$

$$c^{j-1} \; \oplus \;\; = c^{j-1}$$

$$\bar{G} \;\rightarrow\; \downarrow_2 \;\rightarrow\; d^j \downarrow_2 \;\rightarrow\; \uparrow_2 \;\rightarrow G$$

The above filter pairs are dual in the sense that they are required to satisfy certain reciprocal properties with respect to each other, in order for the scheme to have exact reconstruction property. However it is not clear at this stage if this modified configuration corresponds to a multiresolution analysis or is associated with a wavelet basis.

It turns out that if the filter pairs H, G and $\tilde{H}$, $\tilde{G}$ satisfy certain conditions, then the modified filtering scheme with these different pairs of decomposition and reconstruction filters corresponds to two dual wavelet bases associated with two MRAs.

We will now examine the constraints imposed on these four filters by the requirement of an exact reconstruction scheme, and see if we can construct compactly supported wavelet bases corresponding to such a scheme. We first examine the effect of introducing separate reconstruction filters on the decomposition and reconstruction formulas of Section 5.4.3.

As we have four filters instead of two in the new scheme, the decomposition formulas

$$c_k^j = \sum_{n \in \mathbf{Z}} \overline{h_{n-2k}} c_n^{j-1}$$

and

$$d_k^j = \sum_{n \in \mathbf{Z}} \overline{g_{n-2k}} c_n^{j-1}$$

and the reconstruction formula

$$c_n^{j-1} = \sum_{k \in \mathbf{Z}} \left[h_{n-2k} c_k^j g_{n-2k} d_k^j \right]$$

are replaced by the decomposition formulas

$$c_k^j = \sum_{n \in \mathbf{Z}} h_{n-2k} c_n^{j-1}$$

and

$$d_k^j = \sum_{n \in \mathbf{Z}} g_{n-2k} c_n^{j-1}$$

and the reconstruction formula

$$c_n^j = \sum_{k \in \mathbf{Z}} \left[\tilde{h}_{n-2k} c_k^j + \tilde{g}_{n-2k} d_k^j \right]$$

Using Z-transforms we can rewrite the decomposition formulas as

$$C^{j-1}(z^2) = \frac{1}{2} \left[\bar{H}(z) C^j(z) + \bar{H}(-z) C^j(-z) \right]$$

$(C^j(z) = \sum_{n \in \mathbf{Z}} c_n^j z^n$) and

$$D^{j-1}(z^2) = \frac{1}{2} \left[\bar{G}(z) C^j(z) + \bar{G}(-z) C^j(-z) \right]$$

$(D^j(z) = \sum_{n \in \mathbf{Z}} d_n^j z^n$),

while the reconstruction formula can be rewritten as

$$C^j(z) = \left[\bar{H}(z) C^{j-1}(z^2) + \bar{G}(z) D^{j-1}(z^2) \right].$$

Combining these two formulas we have

$$C^j(z) = \frac{1}{2} \left[\tilde{H}(z) \bar{H}(z) + \tilde{G}(z) \bar{G}(z) \right] C^j(z) + \frac{1}{2} \left[\tilde{H}(z) \bar{H}(-z) + \tilde{G}(z) \bar{G}(-z) \right] C^j(-z)$$

Therefore, we must have

$$\tilde{H}(z) \bar{H}(z) + \tilde{G}(z) \bar{G}(z) = 2$$

and

$$\tilde{H}(z) \bar{H}(-z) + \tilde{G}(z) \bar{G}(-z) = 0$$

where the filter functions H, G, $\tilde{H}$ and $\tilde{G}$ are polynomials (in the extended sense that finitely many negative powers of z are also allowed) with real coefficients, since the filters are all real and FIR.

Since $\tilde{H}(z) \bar{H}(-z) + \tilde{G}(z) \bar{G}(-z) = 0$, $\bar{H}$ and $\bar{G}$ have no zeros in common. Therefore, there exists a polynomial P (also in the extended sense) such that

$$\tilde{G}(z) = \bar{H}(-z) P(z)$$

and
$$\tilde{H}(z) = -\bar{G}(-z)P(z)$$
and substituting in
$$\tilde{H}(z)\bar{H}(z) + \tilde{G}(z)\bar{G}(z) = 2$$
$$[H(-z)G(z) - H(z)G(-z)]\,\bar{P}(z) = 2$$
This implies that $2/\bar{P}(z)$ is a polynomial, and Therefore, $P(z) = pz^q$, $p \in \mathbf{C}$, $q \in \mathbf{Z}$. Thus,
$$\tilde{G}(z) = pz^q\bar{H}(-z)$$
and
$$G(z) = -p^{-1}(-1)^q z^q \bar{\tilde{H}}(-z)$$
Now, since any choice of p and q will satisfy the required conditions, we choose $p = q = 1$. Therefore,
$$\tilde{G}(z) = z\bar{H}(-z)$$
and
$$G(z) = z\bar{\tilde{H}}(-z)$$
Substituting these back in $\tilde{H}(z)\bar{H}(z) + \tilde{G}(z)\bar{G}(z) = 2$, we obtain
$$H(z)\bar{\tilde{H}}(z) + H(-z)\bar{\tilde{H}}(-z) = 2.$$
Expressing these formulas in terms of filter coefficients, we get
$$g_n = (-1)^{n+1}\tilde{h}_{-n+1}$$
$$\tilde{g}_n = (-1)^{n+1}h_{-n+1}$$
and
$$\sum_{n\in\mathbf{Z}} h_n\tilde{h}_{n+2m} = \delta_{0,m}$$
Clearly, these equations are generalized versions of the equation $g_n = (-1)^{n+1}h_{-n+1}$ relating the coefficients of the two-scale relations for the functions ϕ and ψ, and the equation
$$\sum_{n\in\mathbf{Z}} h_n\overline{h_{n+2m}} = \delta_{0,m'}$$
which is the "coefficient" version of the equation
$$|\,m_0(\xi)\,|^2 + |\,m_0(\xi+\pi)\,|^2 = 1,$$
in the case of an orthonormal wavelet basis associated with an MRA. Thus, the filter pairs H, G and $\tilde{H}$, $\tilde{G}$ are dual in the sense that they are symmetrically related to each other.

6.4.4 Dual scaling functions and wavelets

Since the functions m_0 and m_1, associated with an MRA, were related in the orthonormal case to the filter functions H and G by $m_0(\xi) = \frac{1}{\sqrt{2}}H(e^{-i\xi})$ and $m_1(\xi) = \frac{1}{\sqrt{2}}G(e^{-i\xi})$, we define

$$m_0(\xi) = \frac{1}{\sqrt{2}}\sum_n h_n e^{-in\xi}, \quad m_1(\xi) = \frac{1}{\sqrt{2}}\sum_n g_n e^{-in\xi}$$

and

$$\tilde{m}_0(\xi) = \frac{1}{\sqrt{2}}\sum_n \tilde{h}_n e^{-in\xi}, \quad \tilde{m}_1(\xi) = \frac{1}{\sqrt{2}}\sum_n \tilde{g}_n e^{-in\xi}.$$

The equations relating the coefficients of the filter polynomials can be rewritten in terms of the trigonometric polynomials m_0, m_1, $\tilde{m}_0$ and $\tilde{m}_1$ as follows

$$m_1(\xi) = e^{-i\xi}\overline{\tilde{m}_0(\xi + \pi)}$$

and

$$\tilde{m}_1(\xi) = e^{-i\xi}\overline{m_0(\xi + \pi)},$$

and $H(z)\tilde{\bar{H}}(z) + H(-z)\tilde{\bar{H}}(-z) = 2$ becomes

$$\overline{m_0(\xi)}\tilde{m}_0(\xi) + \overline{m_0(\xi + \pi)}\tilde{m}_0(\xi + \pi) = 1$$

From the dual pairs m_0, m_1 and $\tilde{m}_0$, $\tilde{m}_1$ we will construct dual pairs ϕ, ψ and $\tilde{\phi}$, $\tilde{\psi}$ of scaling function and wavelet as in the orthonormal case by defining

$$\hat{\phi}(\xi) = m_0(\xi/2)\hat{\phi}(\xi/2), \quad \hat{\psi}(\xi) = m_1(\xi/2)\hat{\phi}(\xi/2)$$

$$\hat{\tilde{\phi}}(\xi) = \tilde{m}_0(\xi/2)\hat{\tilde{\phi}}(\xi/2), \quad \hat{\tilde{\psi}}(\xi) = \tilde{m}_1(\xi/2)\hat{\tilde{\phi}}(\xi/2)$$

As seen before, in order for ψ, $\tilde{\psi}$ to give rise to Riesz bases it is necessary that $\hat{\psi}(0) = \hat{\tilde{\psi}}(0) = 0$, *i.e.*, , $m_1(0) = \tilde{m}_1(0) = 0$ or, equivalently,

$$m_0(\pi) = \tilde{m}_0(\pi) = 0$$

By $\overline{m_0(\xi)}\tilde{m}_0(\xi) + \overline{m_0(\xi + \pi)}\tilde{m}_0(\xi + \pi) = 1$, we, therefore, have

$$\overline{m_0(0)}\tilde{m}_0(0) = 1,$$

or, equivalently,

$$(\sum_n h_n)(\sum_n \tilde{h}_n) = 2.$$

If we normalize the filter polynomials H and $\tilde{H}$ by setting $H(0) = \tilde{H}(0) = \sqrt{2}$, then

$$m_0(0) = \tilde{m}_0(0) = 1$$

Arguments similar to the ones in the orthonormal case give

$$\hat{\phi}(\xi) = (2\pi)^{-1/2} \prod_{j=1}^{\infty} m_0(2^{-j}\xi),$$

and

$$\hat{\tilde{\phi}}(\xi) = (2\pi)^{-1/2} \prod_{j=1}^{\infty} \tilde{m}_0(2^{-j}\xi).$$

These infinite products converge uniformly on compact sets, and ϕ, $\tilde{\phi}$ have compact supports $[N_1, N_2]$, $\left[\tilde{N}_1, \tilde{N}_2\right]$, where N_1, N_2 and $\tilde{N}_1$, $\tilde{N}_2$ are, respectively, the indices of the first and the last nonzero coefficients of the trigonometric polynomials $m_0(\xi)$ and $\tilde{m}_0(\xi)$. Since ψ and $\tilde{\psi}$ are finite linear combinations of ϕ and $\tilde{\phi}$, they too are compactly supported.

6.4.5 Biorthogonal Riesz bases of wavelets and associated MRAs

All of this, however, is not sufficient to guarantee that

$$\left\{\psi_{j,k} : \psi_{j,k}(x) = 2^{-j/2}\psi(2^{-j}x - k)\right\}_{j,k\in\mathbf{Z}}$$

and

$$\left\{\tilde{\psi}_{j,k} : \tilde{\psi}_{j,k}(x) = 2^{-j/2}\tilde{\psi}(2^{-j}x - k)\right\}_{j,k\in\mathbf{Z}}$$

are wavelet Riesz bases, and if they are, that they are indeed dual to each other in the sense that $\langle\psi_{j,k}, \tilde{\psi}_{j',k'}\rangle = \delta_{j,j'}\delta_{k,k'}$. If both of these conditions are satisfied, then

$$\left(\left\{\tilde{\psi}_{j,k}\right\}_{j,k\in\mathbf{Z}}, \{\psi_{j,k}\}_{j,k\in\mathbf{Z}}\right)$$

is called a pair of *biorthogonal* wavelet bases.

If

$$\left(\left\{\tilde{\psi}_{j,k}\right\}_{j,k\in\mathbf{Z}'}, \{\psi_{j,k}\}_{j,k\in\mathbf{Z}}\right)$$

is a biorthogonal pair of Riesz bases, then they are associated with a pair of MRAs of $L^2(\mathbf{R})$,

$$\{\mathcal{V}_j : \mathcal{V}_j \subset \mathcal{V}_{j-1}\}_{j\in\mathbf{Z}} \text{ and } \left\{\tilde{\mathcal{V}}_j : \tilde{\mathcal{V}}_j \subset \tilde{\mathcal{V}}_{j-1}\right\}_{j\in\mathbf{Z}},$$

with

$$\mathcal{V}_j = \underset{L^2(\mathbf{R})}{\text{closure}}(span\,\{\phi_{j,k}\}_{k\in\mathbf{Z}}), \quad \mathcal{W}_j = \underset{L^2(\mathbf{R})}{\text{closure}}(span\,\{\psi_{j,k}\}_{k\in\mathbf{Z}}),$$

and

$$\tilde{\mathcal{V}}_j = \underset{L^2(\mathbf{R})}{\text{closure}}(span\left\{\tilde{\phi}_{j,k}\right\}_{k\in\mathbf{Z}}), \quad \tilde{\mathcal{W}}_j = \underset{L^2(\mathbf{R})}{\text{closure}}(span\left\{\tilde{\psi}_{j,k}\right\}_{k\in\mathbf{Z}}).$$

Also,

$$\mathcal{V}_{j-1} = \mathcal{V}_j + \mathcal{W}_j \; \forall j \in \mathbf{Z} \text{ and } \tilde{\mathcal{V}}_{j-1} = \tilde{\mathcal{V}}_j + \tilde{\mathcal{W}}_j \; \forall j \in \mathbf{Z},$$

where the direct sum decompositions $\mathcal{V}_{j-1} = \mathcal{V}_j + \mathcal{W}_j$ mean that

$$\forall f_{j-1} \in \mathcal{V}_{j-1}, \; \exists! f_j \in \mathcal{V}_j \; \& \; g_j \in \mathcal{W}_j \ni f_{j-1} = f_j + g_j.$$

However, we do not have the orthogonal decompositions

$$\mathcal{V}_{j-1} = \mathcal{V}_j \oplus \mathcal{W}_j \; \forall j \in \mathbf{Z} \text{ and } \tilde{\mathcal{V}}_{j-1} = \tilde{\mathcal{V}}_j \oplus \tilde{\mathcal{W}}_j \; \forall j \in \mathbf{Z},$$

as the dual Riesz bases

$$\left\{\psi_{j,k}\right\}_{j,k\in\mathbf{Z}'}, \left\{\tilde{\psi}_{j,k}\right\}_{j,k\in\mathbf{Z}}$$

are not orthonormal. Instead, since

$$\langle \psi_{j,k} \tilde{\psi}_{j,k} \rangle = \delta_{j,j'}\delta_{k,k'},$$

we have

$$\mathcal{V}_j \perp \tilde{\mathcal{W}}_j \; \forall j \in \mathbf{Z} \text{ and } \tilde{\mathcal{V}}_j \perp \mathcal{W}_j \; \forall j \in \mathbf{Z}, \text{ and, hence, } \mathcal{W}_j \perp \tilde{\mathcal{W}}_j, \; \forall j, j' \in \mathbf{Z}$$

i.e., while we have lost orthogonality within each MRA, we have orthogonality "across" the MRAs.

In what follows, we derive conditions on the polynomials $m_0(\xi)$ and $\tilde{m}_0(\xi)$ that ensure that

$$(\left\{\tilde{\psi}_{j,k}\right\}_{j,k\in\mathbf{Z}}, \{\psi_{j,k}\}_{j,k\in\mathbf{Z}})$$

is a pair of biorthogonal wavelet bases.

Note : All lemmas, theorems and definitions involving m_0, ϕ, ψ stated below are also valid when replaced with their dual counterparts.

6.4.6 Conditions for biorthogonality

If $m_0(\xi)$ is a trigonometric polynomial satisfying $m_0(0) = 1$ and $m_0(\pi) = 0$, then define the associated transition operator T_0 on 2π periodic functions, by

$$(T_0 f)(\xi) = |\, m_0(\tfrac{\xi}{2})\,|^2 \, f(\tfrac{\xi}{2}) + |\, m_0(\tfrac{\xi}{2} + \pi)\,|^2 \, f(\tfrac{\xi}{2} + \pi)$$

Lemma 6.24 If

$$m_0(\xi) = \sum_{n=0}^{N} c_n e^{in\xi},$$

and T_0 the corresponding transition operator, then the space

$$E^N = \left\{ \sum_{n=-N}^{N} c_n e^{in\xi} \mid (c_{-N}, ..., c_N) \in \mathbf{C}^{2N+1} \right\},$$

and its subspace

$$F^N = \left\{ \sum_{n=-N}^{N} c_n e^{in\xi} \mid \sum_{n=-N}^{N} c_n = 0 \right\}$$

are invariant under the action of T_0.

Proof : If

$$m_0(\xi) = \sum_{n=0}^{N} c_n e^{in\xi},$$

then

$$\mid m_0 \mid^2 = \sum_{n=-N}^{N} \mu_n e^{in\xi} \sum E^N.$$

If $f \in E^N$, then

$$f(\xi) = \sum_{n=-N}^{N} f_n e^{in\xi},$$

and, therefore,

$$\mid m_0(\tfrac{\xi}{2}) \mid^2 f(\tfrac{\xi}{2}) = \sum_{n=-2N}^{2N} \gamma_n e^{in\xi/2}$$

and

$$\mid m_0(\tfrac{\xi}{2} + \pi) \mid^2 f(\tfrac{\xi}{2} + \pi) = \sum_{n=-2N}^{2N} \gamma_n e^{in(\xi/2+\pi)} = \sum_{n=-2N}^{2N} (-1)^n \gamma_n e^{in\xi/2}.$$

Since

$$(T_0 f)(\xi) = \mid m_0(\tfrac{\xi}{2}) \mid^2 f(\tfrac{\xi}{2}) + \mid m_0(\tfrac{\xi}{2} + \pi) \mid^2 f(\tfrac{\xi}{2} + \pi)$$

we have $T_0 f \in E^N$.

Now, $f \in F^N$ iff $f \in E^N$ and $f(0) = 0$. Therefore, if $f \in F^N$, we have

$$(T_0 f)(0) = \mid m_0(0) \mid^2 f(0) + \mid m_0(\pi) \mid^2 f(\pi) = 0,$$

and, hence, $T_0 f \in F^N$

If

$$| m_0(\xi) |^2 = \sum_{n=-N}^{N} \mu_n e^{in\xi},$$

then T_0 can be represented by the following matrix over E^N :

$$(2\mu_{i-2j})_{-N \le i, j \le N} = 2 \begin{pmatrix} \mu_N & 0 & 0 & \cdots & 0 \\ \mu_{N-2} & \mu_{N-1} & \mu_N & \cdots & \vdots \\ M & \vdots & \vdots & & 0 \\ \mu_{-N} & \mu_{-N-1} & \vdots & & \mu_N \\ 0 & 0 & \mu_{-N} & \cdots & \mu_N \\ M & \vdots & \vdots & \vdots & \vdots \\ 0 & 0 & \cdots & \cdots & \mu_{-N} \end{pmatrix}$$

For all $n \ge 1$, define

$$\hat{\phi}_n(\xi) = (2\pi)^{-1/2} \prod_{j=1}^{n} m_0(2^{-j}\xi) \cdot \chi_{[-2^n \pi, 2^n \pi]}(\xi)$$

(and similarly, $\hat{\tilde{\phi}}_n$).

Lemma 6.25 If $f(\xi)$ is any 2π-periodic function, then

$$\int_{-\pi}^{\pi} T_0^n f(\xi) d\xi = \int_{-2^n \pi}^{2^n \pi} f(2^{-n}\xi) \prod_{j=1}^{n} | m_0(2^{-j}\xi) |^2 \, d\xi$$

$$= \int_{-\infty}^{\infty} f(2^{-n}\xi) | \hat{\phi}_n(\xi) |^2 \, d\xi \quad \forall n \ge 0.$$

Proof : (By induction). The above equation is trivially true for $n = 0$. If it is true for some $n = m \ge 0$, then

$$\int_{-\pi}^{\pi} T_0^{m+1} f(\xi) d\xi = \int_{-\pi}^{\pi} T_0^m (T_0 f)(\xi) d\xi$$

$$= \int_{-2^m \pi}^{2^m \pi} (T_0 f)(2^{-m}\xi) \prod_{j=1}^{m} | m_0(2^{-j}\xi) |^2 \, d\xi$$

$$= \int_{-2^m \pi}^{2^m \pi} [| m_0(2^{-m-1}) |^2 \, f(2^{-m-1}\xi) + | m_0(2^{-m-1}\xi + \pi) |^2 \, f(2^{-m-1}\xi + \pi)]$$

$$\prod_{j=1}^{m} \mid m_0(2^{-j}\xi) \mid^2 d\xi$$

$$= \int_{-2^m\pi}^{2^m\pi} f(2^{-(m+1)}\xi) \prod_{j=1}^{m+1} \mid m_0(2^{-j}\xi) \mid^2 d\xi.$$

Hence, the equation is true $\forall n \geq 0$.

The following theorem is central to the construction of biorthogonal wavelet bases.

Theorem 6.26 *Suppose λ is the largest eigenvalue of T_0 restricted to F^N. If $\mid \lambda \mid < 1$, then*

(1) ϕ_n converges to ϕ in $L^2(\mathbf{R})$,

(2) $\sup_{\xi \in \mathbf{R}} \sum_{n \in \mathbf{Z}} \mid \phi(\xi + 2n\pi) \mid^{2-\sigma} < \infty$ for some $\sigma > 0$, and $\sup_{\xi \in \mathbf{R}} (1+ \mid \xi \mid)^{\sigma} \mid \phi(\xi) \mid < \infty$.

Proof : (1) ϕ_n converges to ϕ in $L^2(\mathbf{R})$.

(i) $\phi \in L^2(\mathbf{R})$:

If $p(\xi) = 1 - \cos\xi$, then $p \in F^N$. By the above lemma,

$$\int_{-\infty}^{\infty} p(2^{-n}\xi) \mid \hat{\phi}_n(\xi) \mid^2 d\xi = \int_{-\pi}^{\pi} T_0^n p(\xi) d\xi$$

$$\leq \sqrt{2\pi} \parallel T_0^n p \parallel_{L^2(\mathbf{R})} \leq C(\frac{1+ \mid \lambda \mid}{2})^n = C2^{-n\varepsilon}$$

where $\varepsilon = \frac{1}{\log 2}[\log 2 - \log(1+ \mid \lambda \mid)] > 0$, as $\mid \lambda \mid < 1$.

Since $p(\xi) \geq 0$ and $p(\xi) \geq 1 \; \forall \pi/2 \leq \mid \xi \mid \leq \pi$,

$$\int_{2^{n-1}\pi \leq \mid \xi \mid \leq 2^n\pi} \mid \hat{\phi}_n(\xi) \mid^2 d\xi \leq C2^{-n\varepsilon}$$

Also, since

$$\mid \hat{\phi}(\xi) \mid = \mid \hat{\phi}_n(\xi)\hat{\phi}(2^{-n}\xi) \mid \leq \mid \hat{\phi}_n(\xi) \mid \max_{\mid \xi \mid \leq \pi}(\mid \hat{\phi}(\xi) \mid) \; \forall \mid \xi \mid \leq 2^n\pi,$$

we have

$$\int_{2^{n-1}\pi \leq \mid \xi \mid \leq 2^n\pi} \mid \hat{\phi}_n(\xi) \mid^2 d\xi \leq C2^{-n\varepsilon} \;\;, \text{ and hence } \phi \in L^2(\mathbf{R})$$

(ii) $\phi_n \longrightarrow_{L^2(\mathbf{R})} \phi$:

$m_0(0) = \hat{\phi}(0) = 1$. Therefore, $\exists \alpha \in [0, \pi] \ni |\xi| \le \alpha \Rightarrow |\hat{\phi}(\xi)| \ge C > 0$.

Let $\phi_n = \phi_n^1 + \phi_n^2$, where

$$\hat{\phi}_n^1(\xi) = \hat{\phi}_n(\xi)\chi_{[-2^n\alpha, 2^n\alpha]}(\xi)$$

and

$$\hat{\phi}_n^2(\xi) = \hat{\phi}_n(\xi)\left[\chi_{[-2^n\pi, 2^n\pi]}(\xi) - \chi_{[-2^n\alpha, 2^n\alpha]}(\xi)\right]$$

Since $\hat{\phi}_n \longrightarrow_{pointwise} \hat{\phi}$, we have $\hat{\phi}_n^1 \longrightarrow_{pointwise} \hat{\phi}$.

As

$$|\hat{\phi}(\xi)| = |\hat{\phi}_n(\xi)\hat{\phi}(2^{-n}\xi)|$$

and

$$|\hat{\phi}_n^1(\xi)| \le |\hat{\phi}_n(\xi)|$$

using

$$|\hat{\phi}(\xi)| \ge C > 0 \ \forall |\xi| \le \alpha$$

we get

$$|\hat{\phi}_n^1(\xi)| \le \frac{|\hat{\phi}(\xi)|}{C}.$$

Applying Lebesgue's dominated convergence theorem, we see that $\phi_n^1 \longrightarrow_{L^2(\mathbf{R})} \phi$.

Also,

$$\int_{-\infty}^{\infty} |\hat{\phi}_n^2(\xi)|^2 \, d\xi = \int_{2^n\alpha \le |\xi| \le 2^n\pi} |\hat{\phi}_n(\xi)|^2 \, d\xi,$$

and as p is increasing on $[0, p]$,

$$\le \frac{1}{p(\alpha)} \int_{2^n\alpha \le |\xi| \le 2^n\pi} |\hat{\phi}_n(\xi)|^2 \, p(2^{-n}\xi)d\xi$$

$$\le \frac{1}{p(\alpha)} \int_{-\infty}^{\infty} |\hat{\phi}_n(\xi)|^2 \, p(2^{-n}\xi)d\xi$$

$$\le C2^{-n\varepsilon}.$$

Thus, $\phi_n^2 \longrightarrow_{L^2(\mathbf{R})} 0$, and, therefore, $\phi_n \longrightarrow_{L^2(\mathbf{R})} \phi$.

(2) $\sup_{\xi \in \mathbf{R}} \sum_{n \in \mathbf{Z}} |\hat{\phi}(\xi + 2n\pi)|^{2-\sigma} < \infty$ for some $\sigma > 0$, and $\sup_{\xi \in \mathbf{R}}(1 + |\xi|)^{\sigma} |\hat{\phi}(\xi)| < \infty$.

(i) $\sup_{\xi \in \mathbf{R}} \sum_{n \in \mathbf{Z}} |\hat{\phi}(\xi + 2n\pi)|^{2-\sigma} < \infty$ for some $\sigma > 0$.

Since $m_0(\pi) = m_0(-\pi) = 0$, we have $\hat{\phi}(2n\pi) = 0$ $\forall n \neq 0$. Now, for any fixed $\xi \in \mathbf{R}$, $\exists q \in \mathbf{Z} \ni 2q\pi \leq \xi \leq 2(q+1)\pi$. Thus,

$$|\hat{\phi}(\xi + 2n\pi)|^{2-\sigma} = |\hat{\phi}(\xi + 2n\pi)|^{2-\sigma} - |\hat{\phi}(2(q+n)\pi)|^{2-\sigma}$$

$$= \int_{2(q+n)\pi}^{\xi+2n\pi} \frac{d}{d\eta}(|\hat{\phi}(\eta)|^{2-\sigma})d\eta$$

$$\leq \int_{2(q+n)\pi}^{\xi+2n\pi} |\frac{d}{d\eta}(|\hat{\phi}(\eta)|^{2-\sigma})| \, d\eta$$

$$\leq \int_{2(q+n)\pi}^{2(q+n+1)\pi} |\frac{d}{d\eta}(|\hat{\phi}(\eta)|^{2-\sigma})| \, d\eta$$

Hence,

$$\sum_{n\in\mathbf{Z}} |\hat{\phi}(\xi + 2n\pi)|^{2-\sigma} \leq \int_{\mathbf{R}} |\frac{d}{d\xi}(|\hat{\phi}|^{2-\sigma})| \, d\xi$$

$$= \int_{\mathbf{R}} |\frac{d}{d\xi}(|\hat{\phi}|^{2})^{1-\sigma/2}| \, d\xi$$

$$= \int_{\mathbf{R}} |(1 - \frac{\sigma}{2})\frac{d}{d\xi}(|\hat{\phi}|^{2})| \, |\hat{\phi}|^{-\sigma/2} \, d\xi$$

$$\leq \int_{\mathbf{R}} |2 - \sigma| |\frac{d\hat{\phi}}{d\xi}| |\hat{\phi}|^{1-\sigma/2} \, d\xi$$

$$\leq C(\int_{\mathbf{R}} |\frac{d\hat{\phi}}{d\xi} d\xi)^{(\frac{1}{2})}(\int_{\mathbf{R}} |\hat{\phi}|^{2-\sigma} \, d\xi)^{1/2}$$

Now,

$$(\int_{\mathbf{R}} |\frac{d\hat{\phi}}{d\xi}|^{2} \, d\xi)^{1/2} = \|(\hat{\phi})'\|_{2} = 2\pi \|x\phi\|_{2} < \infty$$

since $\phi \in L^{2}(\mathbf{R})$ and has compact support. By Hölder's inequality, we get for all $n > 0$

$$\int_{2^{n-1}\pi \leq |\xi| \leq 2^{n}\pi} |\hat{\phi}(\xi)|^{2-\sigma} \, d\xi$$

$$\leq \left[\int_{2^{n-1}\pi \leq |\xi| \leq 2^{n}\pi} |\hat{\phi}(\xi)|^{2} \, d\xi \leq C2^{-n\varepsilon}\right]^{1-\sigma/2} (2^{n}\pi)^{\sigma/2}.$$

Using

$$\int_{2^{n-1}\pi \leq |\xi| \leq 2^{n}\pi} |\hat{\phi}(\xi)|^{2} \, d\xi \leq C2^{-n\varepsilon},$$

we have

$$\int_{2^{n-1}\pi \leq |\xi| \leq 2^{n}\pi} |\hat{\phi}(\xi)|^{2-\sigma} \, d\xi \leq C2^{-n[\varepsilon(1-\sigma/2)+\sigma/2]}.$$

Choosing $\sigma < \frac{2\varepsilon}{1+\varepsilon}$, we have $\varepsilon(1 - \frac{\sigma}{2}) + \frac{\sigma}{2} > 0$, and, hence,

$$\int_{\pi \le |\xi|} |\hat{\phi}(\xi)|^{2-\sigma} \, d\xi < \infty$$

Also,

$$\int_{|\xi| \le \pi} |\hat{\phi}(\xi)|^{2-\sigma} \, d\xi < \infty$$

since $\hat{\phi}$ is an entire function. Thus,

$$\sup_{\xi \in \mathbf{R}} \sum_{n \in \mathbf{Z}} |\hat{\phi}(\xi + 2n\pi)|^{2-\sigma} < \infty$$

for some $\sigma > 0$.

(ii) $\sup_{\xi \in \mathbf{R}}(1 + |\xi|)^{\sigma} |\hat{\phi}(\xi)| < \infty$

For all $n > 0$, if $2^{n-1}\pi \le |\xi| \le 2^n \pi$,

$$|\hat{\phi}(\xi)|^2 \le \int_{2^{n-1}\pi \le |\xi| \le 2^n \pi} \left| \frac{d}{d\xi}(|\hat{\phi}|^2) \right| d\xi$$

$$\le C \left(\int_{\mathbf{R}} \left| \frac{d\hat{\phi}}{d\xi} \right|^2 d\xi \right)^{1/2} \left(\int_{2^{n-1}\pi \le |\xi| \le 2^n \pi} |\hat{\phi}(\xi)|^2 | \, d\xi \right)^{1/2}$$

$$\le C 2^{-n\varepsilon/2} \le C'(1 + |\xi|)^{-\varepsilon/2},$$

where C' is independent of n. Since $\hat{\phi}$ is an entire function, a similar estimate holds for $|\xi| \le \pi$, and, hence, choosing $\sigma \le \varepsilon/4$, we get

$$\sup_{\xi \in \mathbf{R}}(1 + |\xi|)^{\sigma} |\hat{\phi}(\xi)| < \infty$$

Remark 6.27 The above theorem is sharp in the sense that the converse is also true (see Cohen[6]). Therefore, if the theorem also holds for the operator $\tilde{T}_0$ associated with the polynomial $\tilde{m}_0$, then it is necessary and sufficient for m_0 and $\tilde{m}_0$ to generate a pair of biorthogonal Riesz bases, as the remaining results below collectively imply.

Lemma 6.28 If $\phi_n \longrightarrow_{L^2(\mathbf{R})} \phi$ and $\tilde{\phi}_n \longrightarrow_{L^2(\mathbf{R})} \tilde{\phi}$, then

$$\langle \phi(x - p), \tilde{\phi}(x - q) \rangle = \delta_{p,q}$$

and

$$\langle \psi_k^j, \tilde{\psi}_{k'}^{j'} \rangle = \delta_{j,j'} \delta_{k,k'}$$

Proof : $\forall n > 0$,

$$\langle \phi_n(x-p), \tilde{\phi}_n(x-q) \rangle = \frac{1}{2\pi} \int_{-2^n \pi}^{2^n \pi} (\prod_{k=1}^{n} m_0(2^{-k}\xi)\overline{\tilde{m}_0 2^{-k}\xi}) e^{i(q-p)\xi} d\xi)$$

$$= \frac{2^n}{2\pi} \int_{-\pi}^{\pi} (\prod_{k=0}^{n-1} m_0(2^k\xi)\overline{\tilde{m}_0(2^k\xi)} e^{i2^n(q-p)\xi} d\xi$$

$$= \frac{2^n}{2\pi} \int_{-\pi/2}^{\pi/2} (\prod_{k=2}^{n-1} m_0(2^k\xi)\overline{\tilde{m}_0 2^k\xi}))$$

$$= \left[m_0(\xi)\overline{\tilde{m}_0(\xi)} + m_0(\xi+\pi)\overline{\tilde{m}_0(\xi+\pi)} \right] e^{i2^n(q-p)\xi} d\xi$$

$$= \langle \phi_{n-1}(x-p), \tilde{\phi}_{n-1}(x-q) \rangle = ... = \frac{1}{2\pi} \int_{-\pi}^{\pi} e^{i(q-p)\xi} d\xi = \delta_{p,q}$$

Therefore, $\langle \phi(x-p), \tilde{\phi}(x-q) \rangle = \delta_{p,q}$, or, equivalently,

$$\sum_{n \in \mathbf{Z}} (\overline{\hat{\phi}}\hat{\tilde{\phi}})(\xi + 2n\pi) = (2\pi)^{-1}.$$

Now,

$$\sum_{n \in \mathbf{Z}} (\overline{\hat{\psi}}\hat{\tilde{\psi}})(\xi + 2n\pi) = \sum_{n \in \mathbf{Z}} (\bar{m}_1 \tilde{m}_1 \overline{\hat{\phi}}\hat{\tilde{\phi}})(\xi/2 + n\pi)$$

$$= (\bar{m}_1 \tilde{m}_1)(\xi/2) \sum_{n \in \mathbf{Z}} (_1\overline{\hat{\phi}}\hat{\tilde{\phi}})(\xi/2+2n\pi) + \bar{m}_1 \tilde{m}_1(\xi/2+\pi) \sum_{n \in \mathbf{Z}} (_1\overline{\hat{\phi}}\hat{\tilde{\phi}})(\xi/2+\pi+2n\pi)$$

$$= [\bar{m}_1 \tilde{m}_1(\xi/2) + \bar{m}_0 \tilde{m}_0(\xi/2)] \sum_{n \in \mathbf{Z}} (_1\overline{\hat{\phi}}\hat{\tilde{\phi}})(\xi/2 + 2n\pi) = (2\pi)^{-1},$$

and

$$\sum_{n \in \mathbf{Z}} (\overline{\hat{\psi}}\hat{\tilde{\phi}})(\xi + 2n\pi) = \sum_{n \in \mathbf{Z}} (\bar{m}_1 \tilde{m}_0 \overline{\hat{\phi}}\hat{\tilde{\phi}})(\xi/2 + n\pi)$$

$$= [(\bar{m}_1 \tilde{m}_0)(\xi/2) + (\bar{m}_1 \tilde{m}_0)(\xi/2 + \pi)] \sum_{n \in \mathbf{Z}} (_1\overline{\hat{\phi}}\hat{\tilde{\phi}})(\xi/2 + 2n\pi) = 0$$

since

$$(\bar{m}_1 \tilde{m}_0)(\xi/2) \qquad\qquad\qquad + (\bar{m}_1 \tilde{m}_0)(\xi/2 + \pi)$$
$$= \left[e^{i\xi/2}\tilde{m}_0(\tfrac{\xi}{2} + \pi)\tilde{m}_0(\tfrac{\xi}{2}) + e^{i(\xi/2+\pi)}\tilde{m}_0(\tfrac{\xi}{2})\tilde{m}_0(\tfrac{\xi}{2} + \pi) \right]$$
$$= 0$$

Thus,

$$\langle \psi(x-p), \tilde{\psi}(x-q) \rangle = \delta_{p,q}$$

and
$$\langle \psi(x-p), \tilde{\phi}(x-q) \rangle = 0$$

Let
$$V_j = \underset{L^2(\mathbf{R})}{\mathrm{closure}}(span \{\phi_{j,k}\}_{k \in \mathbf{Z}})$$

$$W_j = \underset{L^2(\mathbf{R})}{\mathrm{closure}}(span \{\psi_{j,k}\}_{k \in \mathbf{Z}})$$

$$\tilde{V}_j = \underset{L^2(\mathbf{R})}{\mathrm{closure}}(span \{\tilde{\phi}_{j,k}\}_{k \in \mathbf{Z}})$$

and
$$\tilde{W}_j = \underset{L^2(\mathbf{R})}{\mathrm{closure}}(span \{\tilde{\psi}_{j,k}\}_{k \in \mathbf{Z}})$$

Then,
$$\langle \psi(x-p), \tilde{\phi}(x-q) \rangle = 0 \Rightarrow \tilde{V}_j \perp W_j \quad \forall j \in \mathbf{Z}$$

Since $\forall j' < j \ \tilde{W}_j \subset V_j$, we have $\forall j' < j \ \tilde{W}_j \perp W_j$, and by symmetry, $\forall j < j' \ \tilde{W} \perp W_j$. Hence, along with $\langle \psi(x-p), \tilde{\psi}(x-q) \rangle = \delta_{p,q}$, this implies $\langle \psi_{j,k}, \tilde{\psi}_{j',k'} \rangle = \delta_{j,j'}, \delta_{k,k'}$.

Lemma 6.29 If $\sup_{\xi \in \mathbf{R}} \sum_{n \in \mathbf{Z}} |\hat{\phi}(\xi + 2n\pi)|^{2-\sigma} < \infty$ for some $\sigma > 0$, and $\sup_{\xi \in \mathbf{R}}(1 + |\xi|)^{\sigma} |\hat{\phi}(\xi)| < \infty$, then there exists a constant C such that

$$\sum_{j,k \in \mathbf{Z}} |\langle f, \psi_{j,k} \rangle|^2 \le C \|f\|^2 \quad \forall f \in L^2(\mathbf{R})$$

Proof : Since $\hat{\psi}(\xi) = m_1(\xi/2)\hat{\phi}(\xi/2)$, and m_1 is bounded, we have

$$\sum_{n \in \mathbf{Z}} |\hat{\psi}(\xi + 2n\pi)|^{2-\sigma} = \sum_{n \in \mathbf{Z}} |m_1(\xi/2 + n\pi)|^{2-\sigma} |\hat{\phi}(\xi/2 + n\pi)|^{2-\sigma}$$

$$= \sum_{n \in \mathbf{Z}} |m_1(\xi/2 + 2n\pi)|^{2-\sigma} |\hat{\phi}(\xi/2 + 2n\pi)|^{2-\sigma}$$

$$+ \sum_{n \in \mathbf{Z}} |m_1(\xi/2 + (2n+1)\pi)|^{2-\sigma} |\hat{\phi}(\xi/2 + (2n+1)\pi)|^{2-\sigma}$$

$$= |m_1(\frac{\xi}{2})|^{2-\sigma} \sum_{n \in \mathbf{Z}} |\hat{\phi}(\xi/2 + 2n\pi)|^{2-\sigma} +$$

$$+ |m_1(\frac{\xi}{2} + \pi)|^{2-\sigma} \sum_{n \in \mathbf{Z}} |\hat{\phi}(\xi/2 + \pi + 2n\pi)|^{2-\sigma} < \infty.$$

Since $\hat{\psi}$ is an entire function, and $\hat{\psi}(0)$, we have $|\hat{\psi}(\xi)| \le C |\xi|$ for $|\xi|$ sufficiently small. Therefore,

$$\sum_{j \le 0} |\hat{\psi}(2^j \xi)|^{\sigma}$$

is uniformly bounded for $|\xi| \leq 2\pi$. Now for all $j > 0$,

$$\sup_{2^j\pi \leq |\xi| \leq 2^{j+1}\pi} |\hat{\psi}(\xi)|^2 \leq \int_{2^j\pi \leq |\eta| \leq 2^{j+1}\pi} |\frac{d}{d\eta}| \hat{\psi}(\eta)|^2| \, d\eta$$

$$\leq 2 \int_{2^j\pi \leq |\eta| \leq 2^{j+1}\pi} |\hat{\psi}(\eta)| \, |\frac{d}{d\eta}\hat{\psi}(\eta)| \, d\eta$$

$$\leq C \left[\int_{2^j\pi \leq |\eta| \leq 2^{j+1}\pi} |\hat{\psi}(\eta)|^2 \, d\eta\right]^{\frac{1}{2}} \cdot \left[\int |x\psi(x)|^2 \, dx\right]^{1/2} \leq C2^{-ji\varepsilon} \ (\varepsilon > 0),$$

i.e., $\sup_{\pi \leq |\xi| \leq 2\pi} |\hat{\psi}(2^j\xi)|^\sigma \leq C2^{-j\alpha} \ \alpha > 0.$

Therefore,

$$\sup_{\pi \leq |\xi| \leq 2\pi} |\hat{\psi}(2^j\xi)|^\sigma < \infty$$

is uniformly bounded for $|\xi| \leq 2\pi$, and, hence,

$$\sup_{|\xi| \leq 0} \sum_{j \in \mathbf{Z}} |\hat{\psi}(2^j\xi)|^\sigma < \infty$$

and, therefore,

$$\sup \sum_{j \in \mathbf{Z}} |\hat{\psi}(2^j\xi)|^\sigma < \infty$$

since $\hat{\psi}(0) = 0$.

$$\sum_{k \in \mathbf{Z}} |\langle f, \psi_k^j \rangle|^2 = \frac{1}{4\pi^2} \sum_{k \in \mathbf{Z}} 2^j |\int_{-\infty}^{\infty} \hat{f}(\xi)\overline{\hat{\psi}(2^j\xi)} e^{-i2^j k\xi} \, d\xi|^2$$

$$= \frac{1}{4\pi^2} \sum_{k \in \mathbf{Z}} 2^{-j} |\int_{-\infty}^{\infty} \hat{f}(2^{-j}\xi)\overline{\hat{\psi}(\xi)} e^{-ik\xi} \, d\xi|^2$$

$$= \frac{2^{-j}}{2\pi} \int_0^{2\pi} |\sum_{k \in \mathbf{Z}} \frac{1}{2\pi} \left[\int_{-\infty}^{\infty} \hat{f}(2^{-j}\xi)\overline{\hat{\psi}(\xi)} e^{-ik\xi} \, d\xi\right] e^{ik\xi} |^2 \, d\xi$$

(by Parseval's formula)

$$= \frac{2^{-j}}{2\pi} \int_0^{2\pi} \left|\sum_{k \in \mathbf{Z}} \left[\frac{1}{2\pi} \int_0^{2\pi} (\sum_{m \in \mathbf{Z}} (\hat{f}(2^{-j}(\xi + 2m\pi))\overline{\hat{\psi}(\xi + 2m\pi)}) e^{-ik\xi} \, d\xi\right] e^{ik\xi}\right|^2 \, d\xi$$

$$= \frac{2^{-j}}{2\pi} \int_0^{2\pi} |\sum_{m \in \mathbf{Z}} \hat{f}(2^{-j}(\xi + 2m\pi))\overline{\hat{\psi}(\xi + 2m\pi)}|^2 \, d\xi$$

(by Poisson's summation formula)

$$\leq \frac{2^{-j}}{2\pi} \int_0^{2\pi} (\sum_{m\in\mathbf{Z}} |\hat{f}(2^{-j}(\xi+2m\pi))| \, |\hat{\psi}(\xi+2m\pi)|^{\sigma/2} |\hat{\psi}(\xi+2m\pi)|^{1-\sigma/2})^2 d\xi$$

$$\leq \frac{2^{-j}}{2\pi} \int_0^{2\pi} (\sum_{m\in\mathbf{Z}} |\hat{f}(2^{-j}(\xi+2m\pi))| \, |\hat{\psi}(\xi+2m\pi)|^{\sigma})(\sum_{m\in\mathbf{Z}} |\hat{\psi}(\xi+2m\pi)|^{2-\sigma}) d\xi$$

$$\leq C \frac{2^{-j}}{2\pi} \int_{-\infty}^{\infty} |\hat{f}(2^{-j}\xi)|^2 |\hat{\psi}(\xi)|^{\sigma} d\xi$$

(using $\sum_{n\in\mathbf{Z}} |\hat{\psi}(\xi + 2n\pi)|^{2-\sigma} \leq C$)

$$\leq C \frac{1}{2\pi} \int_{-\infty}^{\infty} |\hat{f}(\xi)|^2 |\hat{\psi}(2^j\xi)|^{\sigma} \, d\sigma$$

Summing over all the scales $j \in \mathbf{Z}$, we have

$$\sum_{j,k\in\mathbf{Z}} |\langle f, \psi_{j,k}\rangle|^2 \leq C \frac{1}{2\pi} \int_{-\infty}^{\infty} |\hat{f}(\xi)|^2 \sum_{j\in\mathbf{Z}} |\hat{\psi}(2^j\xi)|^{\sigma} \, d\xi \leq C \, \| f \|^2$$

(using $\sum_{j\in\mathbf{Z}} |\hat{\psi}(2^j\xi)|^{\sigma} \leq C$)

For each $j \in \mathbf{Z}$, define formally the *approximation* operators $\mathcal{P}_j$ and $\tilde{\mathcal{P}}_j$, and the *detail* operators $\mathcal{P}_j$ and $\tilde{\mathcal{P}}_j$ as follows

$$\mathcal{P}_j = \sum_{k\in\mathbf{Z}} \langle f, \tilde{\phi}_{j,k}\rangle \phi_{j,k}$$

$$\tilde{\mathcal{P}}_j = \sum_{k\in\mathbf{Z}} \langle f, \phi_{j,k}\rangle \tilde{\phi}_{j,k}$$

$$\mathcal{P}_j = \sum_{k\in\mathbf{Z}} \langle f, \tilde{\psi}_{j,k}\rangle \psi_{j,k}$$

$$\tilde{\mathcal{P}}_j = \sum_{k\in\mathbf{Z}} \langle f, \psi_{j,k}\rangle \tilde{\psi}_{j,k}$$

Lemma 6.30 The operators $\mathcal{P}_j$, $\tilde{\mathcal{P}}_j$, $\mathcal{P}_j$ and $\tilde{\mathcal{P}}_j$ are bounded on $L^2(\mathbf{R})$, with their norm being independent of j.

Proof : Since $\mathcal{P}_j(f(y))(x) = \mathcal{P}_0(f(2^j y))(2^{-j}x)$, it suffices to show that $\mathcal{P}_0$ is a bounded operator on $L^2(\mathbf{R})$, and further we will have $\| \mathcal{P}_j \| = \| \mathcal{P}_0 \|$ $\forall j \in \mathbf{Z}$.

As $\phi, \tilde{\phi} \in L^2(\mathbf{R})$ and have compact support,

$$\tau(\xi) = \sum_{n \in \mathbf{Z}} |\hat{\phi}(\xi + 2n\pi)|^2 = \sum_{n \in \mathbf{Z}} \langle \phi(x), \phi(x - n) \rangle e^{in\xi}$$

and

$$\tilde{\tau}(\xi) = \sum_{n \in \mathbf{Z}} |\hat{\tilde{\phi}}(\xi + 2n\pi)|^2 = \sum_{n \in \mathbf{Z}} \langle \tilde{\phi}(x), \tilde{\phi}(x - n) \rangle e^{in\xi}$$

are trigonometric polynomials, and, hence, are uniformly bounded over $\mathbf{R}$.

For every sequence $\{s_n\}_{n \in \mathbf{Z}} \in l^2(\mathbf{Z})$,

$$\int_{-\infty}^{\infty} |\sum_{n \in \mathbf{Z}} s_n \phi(x - n)|^2 \, dx = \frac{1}{2\pi} \int_{-\infty}^{\infty} |\sum_{n \in \mathbf{Z}} s_n e^{-in\xi}|^2 |\hat{\phi}(\xi)|^2 \, d\xi$$

$$= \frac{1}{2\pi} \int_{-\infty}^{\infty} |\sum_{n \in \mathbf{Z}} s_n e^{-in\xi}|^2 \sum_{n \in \mathbf{Z}} |\hat{\phi}(\xi + 2n\pi)|^2 \, d\xi$$

$$\leq \max(\tau(\xi)) \sum_{n \in \mathbf{Z}} |s_n|^2$$

For every $f \in L^2(\mathbf{R})$,

$$\sum_{n \in \mathbf{Z}} |\langle f, \phi(x - n) \rangle|^2 = \frac{1}{4\pi} \sum_{n \in \mathbf{Z}} \left| \int_{-\infty}^{\infty} f(\xi) \overline{\hat{\tilde{\phi}}(xi)} e^{in\xi} d\xi \right|^2$$

$$= \frac{1}{2\pi} \int_{-\pi}^{\pi} \sum_{n \in \mathbf{Z}} \left| \hat{f}(\xi + 2n\pi) \overline{\hat{\tilde{\phi}}(\xi + 2n\pi)} \right|^2 \, d\xi$$

$$\leq \frac{1}{2\pi} \int_{-\pi}^{\pi} (\sum_{n \in \mathbf{Z}} |\hat{f}(\xi + 2n\pi)|^2)(\sum_{n \in \mathbf{Z}} |\hat{\tilde{\phi}}(\xi + 2n\pi)|^2) d\xi$$

$$\leq \max(\tilde{\tau}(\xi)) \| f \|^2$$

i.e., $\left\{ \langle f, \tilde{\phi}(x - n) \rangle \right\}_{n \in \mathbf{Z}} \in l^2(\mathbf{Z})$.

Combining the above two bounds, we get for all $f \in L^2(\mathbf{R})$

$$\| \mathcal{P}_0 f \|_{L^2}^2 = \| \sum_{n \in \mathbf{Z}} \langle f, \tilde{\phi}(x - n) \rangle \phi(x - n) \|_{L^2}^2 \leq \max(\tilde{\tau}(\xi)) \max(\tau(\xi)) \| f \|_{L^2}^2$$

i.e.,

$$\| P_0 \| \leq [\max(\tilde{\tau}(\xi)) \max(\tau(\xi))]^{1/2}$$

and, hence, is bounded.

A similar argument holds for the operators $\tilde{\mathcal{P}}_j$, $\mathcal{P}_j$ and $\tilde{\mathcal{P}}_j$

Lemma 6.31 For all $j \in \mathbf{Z}$

$$\mathcal{P}_{j-1} = \mathcal{P}_j + \mathcal{P}_j \text{ and } \tilde{\mathcal{P}}_{j-1} = \tilde{\mathcal{P}}_j + \tilde{\mathcal{P}}_j$$

Proof : Let $\{h_n\}_{n \in \mathbf{Z}'}$, $\{g_n\}_{n \in \mathbf{Z}'}$, $\{\tilde{h}_n\}_{n \in \mathbf{Z}'}$, $\{\tilde{g}_n\}_{n \in \mathbf{Z}'}$ be the coefficients of the functions m_0, m_1, $\tilde{m}_0$, $\tilde{m}_1$, respectively. Then, for all $f \in L^2(\mathbf{R})$ we have

$$\mathcal{P}_0(f) + \mathcal{P}_0(f) = \sum_{k \in \mathbf{Z}} \langle f, \tilde{\phi}(x-k) \rangle \phi(x-k) + \sum_{k \in \mathbf{Z}} \langle f, \tilde{\psi}(x-k) \rangle \psi(x-k)$$

$$= 4 \sum_{m,n \in \mathbf{Z}} (\sum_{k \in \mathbf{Z}} \bar{\tilde{h}}_{n-2k} h_{m-2k} + \sum_{k \in \mathbf{Z}} \bar{\tilde{g}}_{n-2k} g_{m-2k} \langle f, \tilde{\phi}(2x-n) \rangle \phi(2x-m)$$

$$= \qquad 4 \sum_{m,n \in \mathbf{Z}} (\sum_{k \in \mathbf{Z}} \bar{\tilde{h}}_{n-2k} h_{m-2k} + (-1)^{n+m}$$
$$\sum_{k \in \mathbf{Z}} \bar{\tilde{h}}_{1-m-2k} h_{1-n-2k}) \langle f, \tilde{\phi}(2x-n) \rangle \phi(2x-m)$$

$$= 4 \sum_{m,n \in \mathbf{Z}} s_{n,m} \langle f, \tilde{\phi}(2x-n) \rangle \phi(2x-m)$$

If $m - n = 2l$, then

$$s_{n,m} = \sum_{k \in \mathbf{Z}} \bar{\tilde{h}}_k h_{k+m-n} = \sum_{k \in \mathbf{Z}} \bar{\tilde{h}}_k h_{k+2l} = \frac{1}{2} \delta_{0,l} = \frac{1}{2} \delta_{n,m}$$

since $\overline{m_0(\xi)} \tilde{m}_0(\xi) + \overline{m_0(\xi+\pi)} \tilde{m}_0(\xi+\pi) = 1$, and, similarly, if $m-n = 2l+1$, then

$$s_{n,m} = \sum_{k \in \mathbf{Z}} \bar{\tilde{h}}_{n-2k} h_{m-2k} - \sum_{k \in \mathbf{Z}} \bar{\tilde{h}}_{1-m+2k} h_{1-n+2k} = 0 = \frac{1}{2} \delta_{n,m}$$

Thus, $\forall m, n \in \mathbf{Z}$,

$$s_{n,m} = \frac{1}{2} \delta_{n,m}$$

and, therefore,

$$\mathcal{P}_0(f) + \mathcal{P}_0(f) = 2 \sum_{k \in \mathbf{Z}} \langle f, \tilde{\phi}(2x-k) \rangle \phi(2x-k)$$

$$\sum_{k \in \mathbf{Z}} \langle f, \sqrt{2} \tilde{\phi}(2x-k) \rangle \sqrt{2} \phi(2x-k) = \mathcal{P}_{-1}(f)$$

Now, since

$$\mathcal{P}_j(f(y))(x) = \mathcal{P}_0(f(2^j y))(2^{-j} x)$$

and
$$P_j(f(y))(x) = P_0(f(2^j y))(2^{-j}x)$$

rescaling the above identity, we have

$$P_j(f) + P_j(f) = P_{j-1}(f)$$

A similar argument holds for the operators $\tilde{P}_j$, and $\tilde{P}_j$.

Lemma 6.32 For any $f \in L^2(\mathbf{R})$,

$$\lim_{j\to\infty} \| P_j(f) \|_{L^2} = \lim_{j\to\infty} \| \tilde{P}_j(f) \|_{L^2} = 0$$

and

$$\lim_{j\to\infty} \| P_j(f) - f \|_{L^2} = \lim_{j\to\infty} \| \tilde{P}_j(f) - f \|_{L^2} = 0$$

Proof : If $f \in L^2(\mathbf{R})$, then given any $\varepsilon > 0$ there exists a simple function

$$g = \sum_{m=1}^{n} \gamma_m \chi_{I_m}$$

such that $\| f - g \| \le \varepsilon$. By lemma 6.30, we have

$$\| P_j(f) \| = \| \tilde{P}_j(f - g + g) \| \le \| P_j(g) \| + C\varepsilon$$

and

$$\| P_j(f) - f \| \le \| \tilde{P}_j(f - g) - (f - g) + P_j(g) - g \|$$
$$\| P_j(g) - g \| + (C + 1)\varepsilon$$

Therefore, it suffices to prove the lemma for the case $f = \chi_{[a,b]}$.

Arguing as in lemma 6.30, we have

$$\| P_j(f) \|^2 \le C \sum_{k\in\mathbf{Z}} | \langle f, \tilde{\phi}_{j,k} \rangle |^2$$

$$= C \sum_{k\in\mathbf{Z}} | \int_a^b \overline{\tilde{\phi}_{j,k}} dx |^2$$

$$\le C \sum_{k\in\mathbf{Z}} \int_a^b 2^{-j} | \tilde{\phi}(2^{-j}x - k) |^2 \, dx$$

$$= C \sum_{k\in\mathbf{Z}} \int_{2^{-j}a-k}^{2^{-j}b-k} 2^{-j} | \tilde{\phi}(y) |^2 \, dy \text{ as } j \to \infty$$

As ϕ, $\tilde{\phi} \in L^2(\mathbf{R})$ and have compact support, the sums

$$\sum_{n \in \mathbf{Z}} \phi(x - n) \text{ and } \sum_{n \in \mathbf{Z}} \tilde{\phi}(x - n)$$

converge pointwise a.e. and are functions in $L^2([0,1])$. Since $\hat{\phi}(2n\pi) = \hat{\tilde{\phi}}(2n\pi) = \delta_{0,n'}$, we have

$$\sum_{n \in \mathbf{Z}} \phi(x - n) = 1 \text{ and } \sum_{n \in \mathbf{Z}} \tilde{\phi}(x - n) = 1 \text{ a.e.}$$

and

$$\int_{-\infty}^{\infty} \phi(x)dx = \int_{-\infty}^{\infty} \tilde{\phi}(x)dx = 1$$

Let $\operatorname{supp}\phi$, $\operatorname{supp}\tilde{\phi} \subset [-r, r]$. For $f = \chi_{[a,b]}$ and $j < 0$ with $|j|$ consider

$$\mathcal{P}_j(f) = \sum_{k \in \mathbf{Z}} \langle f, \tilde{\phi}_{j,k} \rangle \phi_{j,k}$$

If $x \notin [a - 2^{j+1}r, b + 2^{j+1}r]$, then $\phi_{j,k}(x) = 0$ or $\langle f, \tilde{\phi}_{j,k} \rangle = 0$, and, hence, $\mathcal{P}_j(f)(x) = 0$. If $x \in [a + 2^{j+1}r, b - 2^{j+1}r]$, then $\phi_{j,k}(x) = 0$ or

$$\langle f, \tilde{\phi}_{j,k} \rangle = \int_{-\infty}^{\infty} \overline{\tilde{\phi}_{j,k}} dx = 2^{j/2}$$

and, hence,

$$\mathcal{P}_j(f)(x) = \sum_{k \in \mathbf{Z}} 2^{j/2} \phi_{j,k}(x) = \sum_{k \in \mathbf{Z}} \phi(2^{-j}x - k) = 1$$

i.e., outside

$$\Delta_j = [a - 2^{j+1}r, a + 2^{j+1}r] \cup [b - 2^{j+1}r, b + 2^{j+1}r], \quad \mathcal{P}_j(f) = f$$

When $x \in \Delta_j$, $f(x) = \chi_{\Delta_{j+1}}(x)f(x)$ since $\Delta_j \subset \Delta_{j+1}$. Therefore, $\mathcal{P}_j(f)(x) = \mathcal{P}_j(f_j)(x)$. Hence,

$$\int_{\Delta_j} |\mathcal{P}_j(f)(x) - f(x)|^2 \, dx = \int_{\Delta_j} |\mathcal{P}_j(f_j)(x) - f_j(x)|^2 \, dx$$

$$\leq \int_{-\infty}^{\infty} |\mathcal{P}_j(f_j)(x) - f_j(x)|^2 \, dx \leq C \| f_j \|^2 \text{ as } j \to -\infty$$

A similar argument holds for the operator $\tilde{\mathcal{P}}_j$

The following theorem is an immediate consequence of the above three lemmas.

Theorem 6.33 *For any $f \in L^2(\mathbf{R})$,*

$$f = \lim_{j \to \infty} \sum_{j=-J}^{J} \sum_{k \in \mathbf{Z}} \langle f, \psi_{j,k} \rangle \tilde{\psi}_{j,k} \ \lim_{j \to \infty} \sum_{j=-J}^{J} \sum_{k \in \mathbf{Z}} \langle f, \tilde{\psi}_{j,k} \rangle \psi_{j,k}$$

where the convergence is with respect to the L^2-norm.

Proof : Since for all $j \in \mathbf{Z}$ $\mathcal{P}_{j-1} = \mathcal{P}_j + P_j$ and $\tilde{\mathcal{P}}_{j-1} = \tilde{\mathcal{P}}_j + \tilde{P}_j$, we have for $J > 0$

$$\mathcal{P}_{-J-1} = \mathcal{P}_J + \sum_{j=-J}^{J} P_j \text{ and } \tilde{\mathcal{P}}_{-J-1} = \tilde{\mathcal{P}}_J + \sum_{j=-J}^{J} \tilde{P}_j$$

Using the above lemma, we have

$$\lim_{J \to \infty} \| f - \sum_{j=-J}^{J} P_j(f) \|_{L^2} = \lim_{J \to \infty} \| f - \mathcal{P}_{-J-1}(f) + \mathcal{P}_J(f) \|_{L^2}$$

$$\leq \lim_{J \to \infty} \| f - \mathcal{P}_{-J-1}(f) \|_{L^2} + \leq \lim_{J \to \infty} | \mathcal{P}_J(f) \|_{L^2} = 0$$

i.e.,

$$\leq \lim_{J \to \infty} \| f - \sum_{j=-J}^{J} \sum_{k \in \mathbf{Z}} \langle f, \tilde{\psi}_{j,k} \rangle \psi_{j,k} \|_{L^2} = 0$$

and, similarly,

$$\leq \lim_{J \to \infty} \| f - \sum_{j=-J}^{J} \sum_{k \in \mathbf{Z}} \langle f, \psi_{j,k} \rangle \tilde{\psi}_{j,k} \|_{L^2} = 0$$

Theorem 6.34 $\{\psi_{j,k}\}_{j,k \in \mathbf{Z}}$, $\left\{\tilde{\psi}_{j,k}\right\}_{j,k \in \mathbf{Z}}$ *are a pair of biorthogonal Riesz bases iff*

(i) $\forall j, j', k, k' \in \mathbf{Z}$, $\langle \psi_{j,k}, \tilde{\psi}_{j',k'} \rangle = \delta_{j,j'} \delta_{k,k'}$

(ii) $\exists C, \tilde{C} > 0 \ni \forall j \in L^2(\mathbf{R})$,

$$\sum_{j,k \in \mathbf{Z}} | \langle f, \psi_{j,k} \rangle |^2 \leq C \| f \|^2 \text{ and } \sum_{j,k \in \mathbf{Z}} | \langle f, \tilde{\psi}_{j,k} \rangle |^2 \leq \tilde{C} \| f \|^2$$

Proof: If $\forall j, j', k, k' \in \mathbf{Z}$, $\langle psi_{j,k}, \tilde{\psi}_{j',k'} \rangle = \delta_{j,j'} \delta_{k,k'}$, then for all $f \in \text{closure}_{L^2(\mathbf{R})}(span \{\psi_{j,k} : (j,k) \in \mathbf{Z}^2 \backslash (j_0, k_0)\})$, $\langle f, \tilde{\psi}_{j_0,k_0} \rangle = 0$. This would imply that $\psi_{j_0,k_0} \ni \text{closure}_{L^2(\mathbf{R})} (span \{\psi_{j,k} : (j,k) \in \mathbf{Z}^2 \backslash (j_0, k_0)\})$, and, hence, the $\psi_{j,k}$ are linearly independent. Conversely, since

$$\psi_{j_0,k_0} = \lim_{j \to \infty} \sum_{j=-J}^{J} \sum_{k \in \mathbf{Z}} \langle \psi_{j_0,k_0}, \tilde{\psi}_{j,k} \rangle \psi_{j,k'}$$

$$\psi_{j_0,k_0}(1 - \langle \psi_{j_0,k_0}, \tilde{\psi}_{j_0,k_0} \rangle) \in \underset{L^2(\mathbf{R})}{\text{closure}}(span\,\{\psi_{j,k} : (j,k) \in \mathbf{Z}^2 \backslash (j_0,k_0)\})$$

But since the $\psi_{j,k}$ are linearly independent, we must have $\langle \psi_{j_0,k_0}, \tilde{\psi}_{j_0,k_0} \rangle = 1$. Similarly, by the linear independence of the $\psi_{j,k}$ we have $\langle \psi_{j_0,k_0}, \tilde{\psi}_{j,k} \rangle = 0 \forall (j,k) \neq (j_0,k_0)$. Thus, we have

$$\forall j, j', k, k' \in \mathbf{Z}, \quad \langle \psi_{j,k}, \tilde{\psi}_{j',k'} \rangle = \delta_{j,j'} \delta_{k,k'}$$

Since for all $f \in L^2(\mathbf{R})$

$$f = \lim_{J \to \infty} \sum_{j=-J}^{J} \sum_{k \in \mathbf{Z}} \langle f, \tilde{\psi}_{j,k} \rangle \psi_{j,k'}$$

$$\| f \|^2 = \lim_{J \to \infty} \sum_{j=-J}^{J} \sum_{k \in \mathbf{Z}} \langle f, \tilde{\psi}_{j,k} \rangle \langle \psi_{j,k}, f \rangle$$

$$\leq \left(\sum_{j,k \in \mathbf{Z}} | \langle f, \psi_{j,k} \rangle |^2 \right)^{1/2} \left(\sum_{j,k \in \mathbf{Z}} | \langle f, \tilde{\psi}_{j,k} \rangle |^2 \right)^{1/2}$$

(Cauchy-Schwarz inequality)

$$C^{1/2} \| f \| \left(\sum_{j,k \in \mathbf{Z}} \left| \langle f, \tilde{\psi}_{j,k} \rangle \right|^2 \right)^{1/2}$$

(using $\sum_{j,k \in \mathbf{Z}} | \langle f, \psi_{j,k} \rangle |^2 \leq C \| f \|^2$)

i.e.,

$$C^{-1} \| f \| \leq \left(\sum_{j,k \in \mathbf{Z}} \left| \langle f, \tilde{\psi}_{j,k} \rangle \right|^2 \right)^{1/2}$$

and, similarly,

$$\tilde{C}^{-1} \| f \| \leq \left(\sum_{j,k \in \mathbf{Z}} |\langle f, \psi_{j,k} \rangle|^2 \right)^{1/2}$$

This, along with $\forall j, j', k, k' \in \mathbf{Z}, \langle \psi_{j,k}, \tilde{\psi}_{j',k'} \rangle = \delta_{j,j'} \delta_{k,k'}$ implies that $\{\psi_{j,k}\}_{j,k \in \mathbf{Z}}$, $\{\tilde{\psi}_{j,k}\}_{j,k \in \mathbf{Z}}$ are a pair of biorthogonal Riesz bases. Thus, any $f \in L^2(\mathbf{R})$ can be written uniquely as a biorthogonal wavelet series in either of the dual wavelet bases

$$f = \sum_{j,k \in \mathbf{Z}} \langle f, \tilde{\psi}_{j,k} \rangle \psi_{j,k} = \sum_{j,k \in \mathbf{Z}} \langle f, \psi_{j,k} \rangle \tilde{\psi}_{j,k}$$

with the series converging unconditionally.

We have, thus, obtained necessary and sufficient conditions on the polynomials m_0 and m_1, and hence on the corresponding filter coefficients, in order for a pair of dual FIR filters of a subband filtering scheme to correspond to a pair of biorthogonal wavelet bases and an associated pair of MRAs. The main object behind this exercise was the design of linear phase FIR wavelet filters for a subband filtering scheme with exact reconstruction.

6.4.7　Symmetry for m_0 and $\tilde{m}_0$

Unlike the case of compactly supported orthonormal wavelet bases, both m_0 and $\tilde{m}_0$ can be symmetric in the case of biorthogonal wavelet bases. Indeed,

(i) If the filter corresponding to m_0 is symmetric and has an *odd* number of taps, that is to say,

$$m_0(-\xi) = e^{2ik\xi} m_0(\xi)$$

then m_0 can be expressed as

$$m_0(\xi) = e^{-ik\xi} p_0(\cos \xi)$$

for some polynomial p_0 (in the ordinary sense). Then $\tilde{m}_0$ can also be represented similarly

$$\tilde{m}_0(\xi) = e^{-ik\xi} \tilde{p}_0(\cos \xi)$$

where $\tilde{p}_0$ is any polynomial satisfying

$$p_0(z)\overline{\tilde{p}_0(z)} + p_0(-z)\overline{\tilde{p}_0(z)} = 1$$

(so that $\overline{m_0(\xi)}\tilde{m}_0(\xi) + \overline{m_0(\xi + \pi)}\tilde{m}_0(\xi+\pi) = 1$, i.e., $H(z)\overline{\tilde{H}}(z) + H(-z)\overline{\tilde{H}}(z) = 2$)

Such a $\tilde{p}_0$ always exists, by Bezout's theorem (see Appendix), provided $p_0(z)$ and $p_0(-z)$ have no zeros in common.

(ii) If the filter corresponding to m_0 is symmetric and has an *even* number of taps, that is to say,

$$m_0(-\xi) = e^{2ik\xi + i\xi} m_0(\xi)$$

then m_0 can be expressed as

$$m_0(\xi) = e^{-ik\xi - i\xi/2} \cos(\frac{\xi}{2}) p_0(\cos \xi)$$

for some polynomial p_0. As before, $\tilde{m}_0$ can also be represented as

$$\tilde{m}_0(\xi) = e^{-ik\xi - i\xi/2} \cos(\frac{\xi}{2}) \tilde{p}_0(\cos \xi)$$

where $\tilde{p}_0$ is any polynomial satisfying

$$\cos^2 \frac{\xi}{2} p_0(\cos \xi) \overline{\tilde{p}_0(\cos \xi)} + \sin^2 \frac{\xi}{2} p_0(-\cos \xi) \overline{\tilde{p}_0(\cos \xi)} = 1$$

since $m_0(\xi) \overline{\tilde{m}_0(\xi)} + m_0(\xi + \pi) \overline{\tilde{m}_0(\xi + \pi)} = 1$ must be satisfied. *i.e.*,

$$(\frac{1 + \cos \xi}{2}) p_0(\cos \xi) \overline{\tilde{p}_0(\cos \xi)} + (\frac{1 - \cos \xi}{2}) p_0(-\cos \xi) \overline{\tilde{p}_0(-\cos \xi)} = 1$$

or, equivalently,

$$q_0(z) \overline{\tilde{p}_0(z)} + q_0(-z) \overline{\tilde{p}_0(-z)} = 1 \quad \text{where } q_0(z) = (\frac{1 + z}{2}) p_0(z)$$

Again, such a $\tilde{p}_0$ always exists, by Bezout's theorem, provided $p_0(z)$ and $p_0(-z)$ have no zeros in common.

From the above considerations, it is clear that compactly supported biorthogonal bases are easier to construct than compactly supported orthonormal bases, since finding a symmetric $\tilde{m}_0$ for a specified symmetric m_0 via Bezout's theorem involves solving only linear equations, as against the spectral factorization in the orthonormal case.

The Battle-Lemarié construction of orthonormal spline wavelet bases discussed in Section 6.1.2. gave rise to infinitely supported spline wavelets because of the orthonormalization process.

We now briefly discuss the construction of compactly supported symmetric biorthogonal spline wavelet bases using the above observations.

6.4.8 Biorthogonal spline wavelets with compact support

We choose $\phi^N = (*)^{N+1} \chi_{[0,1]}$, the cardinal B-spline of order $N + 1$, as a scaling function. The corresponding m_0 is then given by the polynomial

$$m_0^N(\xi) = \left(\frac{1 + e^{-i\xi}}{2} \right)^{N+1}$$

If $N = 2M - 1$, then

$$m_0^N(\xi) = e^{-iM\xi} \left(\frac{1 + \cos\xi}{2}\right)^M$$

with

$$p_0^N(\cos\xi) = (\frac{1 + \cos\xi}{2})^M$$

Therefore, we must have $\tilde{m}_0^N(\xi) = e^{-iM\xi}\tilde{p}_0^N(\cos\xi)$, such that $p_0(-z)\overline{\tilde{p}_0(-z)} + p_0(z)\ \overline{\tilde{p}_0(z)} = 1$, *i.e.*,

$$(\frac{1 + z}{2})^M \overline{\tilde{p}_0(z)} + (\frac{1 - z}{2})^M \overline{\tilde{p}_0(-z)} = 1$$

If $N = 2M$, then

$$m_0^N(\xi) = e^{-iM\xi - i\xi/2} \cos\frac{\xi}{2}(\frac{1 + \cos\xi}{2})^M$$

with

$$p_0^N(\cos\xi) = (\frac{1 + \cos\xi}{2})^M$$

Therefore, we must have $\tilde{m}_0^N(\xi) = e^{-iM\xi} \cos\frac{\xi}{2}\tilde{p}_0^N(\cos\xi)$ such that $(\frac{1+z}{2})$ $p_0(z)\overline{\tilde{p}_0(z)} + (\frac{1-z}{2})p_0(-z)\overline{\tilde{p}_0(-z)} = 1$, *i.e.*,

$$(\frac{1 + lz}{2})^{M+1}\overline{\tilde{p}_0(z)} + (\frac{1 - z}{2})^{M+1}\overline{\tilde{p}_0(-z)} = 1$$

In either case, we have

$$(\frac{1 + z}{2})^L \overline{\tilde{p}_0(z)} + (\frac{1 - z}{2})^L \overline{\tilde{p}_0(-z)} = 1$$

If we were to write $\tilde{p}_0^N(\cos\xi) = q(\sin^2\frac{\xi}{2})$, then we would have

$$(\frac{1 + z}{2})^L q(\frac{1 - z}{2}) + (\frac{1 - z}{2})^L q(\frac{1 + z}{2}) = 1$$

Letting $h = (\frac{1-z}{2})$, we have

$$(1 - y)^L q(y) + (y)^L q(1 - y) = 1$$

By Theorem 6.5, the solution to this Bezout equation is given by

$$q(y) = \sum_{l=0}^{L-1} \frac{L + l - 1}{l} y^i$$

for the case $R \equiv 0$.

Letting $y = sin^2\frac{\xi}{2}$, we get $(\cos\xi 2)^{2L}q(\sin^2\frac{\xi}{2}) + (\sin\xi 2)^{2L}q(\cos^2\frac{\xi}{2}) = 1$, i.e.,

$$\left[\frac{1+e^{i\xi}}{2}\right]^{2L}\left[e^{-iL\xi}q(\sin^2\xi/2)\right] + \left[\frac{1-e^{i\xi}}{2}\right]^{2L}\left[(-1)e^{-iL\xi}q(\cos^2\xi/2)\right] = 1$$

This formula gives solutions for all $N \leq 2L-1$, or, alternatively, for a fixed N,

$$\tilde{m}_0^{N,L}(\xi) = e^{-iL\xi}\left[\frac{1+e^{i\xi}}{2}\right]^{2L-N-1} q(\sin^2\frac{\xi}{2})$$

is a dual polynomial for m_0^N, for $2L \geq N+1$.

For $N > 0$ ($N = 0$ is the Haar Case), the value of L has to be sufficiently large (depending on N) to ensure a square integrable dual scaling function $\tilde{\phi}^{N,L}$, or more precisely, $\tilde{m}_0^{N,L}$ must satisfy the conditions of theorem 6.26. For instance, when $N = 1$, choosing $L = 1$ yields $\tilde{m}_0^{-1,1}(\xi) = e^{-i\xi}$ and, therefore, $\tilde{m}_0^{-1,1}(\xi) = e^{-i\xi} \notin L^2(\mathbf{R})$. Thus, for a given value of N, the optimal choice for L is the smallest value such that $\tilde{m}_0^{N,L}$ satisfies the conditions of Theorem 6.26, which are both necessary and sufficient.

Finally, we note that the FIR dual wavelet filters obtained from compactly supported biorthogonal wavelet bases do not suffer from the following serious shortcommings of FIR CQFs obtained from compactly supported orthonormal wavelet bases:

(a) Linear phase is absent (except in the Haar case).

(b) Spectral factorization (Fejer-Riesz lemma) must be used for the extraction of m_0 from $|m_0|^2$, which is not generalizable for higher dimensional cases.

(c) Being solutions of the quadratic equation $|m_0(\xi)|^2 + |m_0(\xi+\pi)|^2 = 1$, the m_0 in general have algebraic numbers for coefficients, which are difficult to approximate.

(d) Spline constructions are not possible (except in the Haar case).

Hence, they are preferred for signal processing applications.

Chapter 7

Application of Wavelets to Image Processing

In this chapter we address two applications of wavelets to image processing that are due to Mallat [27], Mallat and Zhong [28], and Froment and Mallat [15]. Both applications are compact image encoding techniques.

Image encoding techniques are broadly classified under two categories, namely, first-generation image encoding techniques and second-generation image encoding techniques. The most important first-generation techniques use an orthonormal basis for representing an image, while second-generation techniques exploit image characteristics and the psychophysics of human visual perception.

The first application that we shall consider is a first-generation image encoding technique that is based on the fast wavelet algorithm discussed in Chapter 5. The second is a second-generation image encoding technique based on multiscale edge representation of an image, combined with a texture encoding technique based on the fast pyramidal wavelet algorithm.

We briefly discuss image encoding techniques and introduce the Laplacian pyramidal algorithm of Burt and Adelson, which will be compared with the wavelet pyramidal algorithm of Mallat.

An *image* is a compactly supported, real-valued function in $L^2(\mathbf{R}^2)$. In practice, it is discretely sampled and represented by a matrix $c =$

$(c(i,j))_{R \times C}$ with R rows and C columns, with integer entries ranging from 0 to $K - 1$, for some positive integer K. The discrete spatial positions (i,j) are called *pixels*. The entries $c(i,j)$ are called the *pixel values* or *gray-scale levels* and represent an amplitude discretization of the image intensity (gray-level value) at the corresponding pixels.

In general, an image can contain complex features at several different scales, and, hence, representing it at an adequate resolution often requires a very large matrix. Faithful representation, storage, and transmission of images are thus costly exercises in terms of computation, memory, and time.

A feature commonly found in images is the high correlation between the values of neighboring pixels, often resulting in high redundancy of information content in the images. The representation of images verbatim as discrete pixels is thus, frequently, inefficient. The principal task of an efficient coding scheme for images is, therefore, the elimination of this redundancy, leading to significant data compression. This is,indeed, one of the aims of several image decomposition and reconstruction schemes designed to achieve image compression. Typically, these schemes fall into two broad categories, namely, the causal and the noncausal predictive schemes.

In causal predictive coding schemes, pixels are encoded row-wise sequentially (raster format). Before encoding each pixel value, it is "predicted" from already encoded pixel values. This predicted value represents redundant information, and, therefore, is subtracted from the actual value of the pixel, and only the difference (prediction error) is encoded. These encoding schemes are referred to as *causal* since they employ only previously encoded pixels for prediction. A pixel value is decoded by computing its predicted value from previously decoded pixel values and adding it to the encoded prediction error.

Noncausal (neighborhood-based) predictive schemes are more natural and result in more accurate prediction, leading to higher compression. It is, however, harder to implement such schemes as they do not permit sequential coding but involve application of image transforms or the solution of large systems of simultaneous equations, resulting in high computational overheads.

The following scheme due to Burt and Adelson [1] is a hybrid of the causal and noncausal schemes in that it enjoys the ease of computation of a causal scheme, while retaining the advantages and naturalness of a noncausal scheme. Moreover, it is inherently multiresolutional and, therefore, easily adapted to wavelet-based image analysis techniques.

7.1 The Burt-Adelson Pyramidal Decomposition Scheme

The basic idea of this scheme is as follows. Let $c^0 = (c^0(i,j))_{R \times C}$ be the actual image represented in matrix form. If a low-pass filter is applied to c^0, then, since the resulting image does not contain the high spatial frequency components of c^0, it may be appropriately downsampled without loss of information (detail) to obtain a "reduced" version c^1 of c^0 with lower resolution and sampling rate than that of c^0. By interpolating values between sample points, the reduced image c^1 can be "expanded" into an image $\tilde{c}^0$, which has the same size (number of sample points) as that of c^0, and is a "blurred" version of it.

The predicted value of the pixel (i,j) is given by $\tilde{c}^0(i,j)$. The prediction error is then obtained by subtracting the predicted value from the actual value of the pixel:

$$d^0 = c^0 - \tilde{c}^0$$

Instead of encoding c^0, the images d^0 and c^1 are encoded, as this results in data compression, since (1) d^0 is largely decorrelated, and, therefore, may be represented pixel-wise with fewer bits than c^0, and (2) c^1 is obtained from c^0 by low-pass filtering, and, hence, may be encoded at a reduced sample rate. The sample rate reduction is proportional to the band limit reduction. This process is iterated to achieve further data compression. At the next step c^1 is low-pass filtered to obtain c^2, which is expanded to obtain the predicted (blurred) image $\tilde{c}^1$ of c^1. The error image d^1 is again obtained by subtracting the blurred image from c^1; *i.e.*, $d^1 = c^1 - \tilde{c}^1$. This process yields a sequence of error images d^0, d^1, ..., d^n, each smaller than the previous by a constant scale factor. Stacking these images on one another results in a pyramidal configuration. The various layers of this pyramidal data structure correspond to distinct ranges of spatial frequency of the original image c^0. Each of the layers can be obtained by convolving the original image with the difference of two distinct weighting functions, and as this operation is similar to the application of Laplacian operators in image enhancement techniques, the Burt-Adelson scheme is also called the Laplacian pyramid scheme. We next describe in detail the various steps involved in this image coding scheme.

First, the original image $c^0 = (c^0(i,j))_{R \times C}$ is low-pass filtered to obtain the image c^1, which is reduced both in resolution and sampling rate compared with c^0. This process is repeated to obtain a sequence of images c^0, c^1, ..., c^L. The low-pass filtering is equivalent to convolving the image with a localized and symmetric weighting function and down-sampling the resulting image. Since a typical and important family of weighting functions used resemble the Gaussian distribution, the sequence c^0, c^1, ..., c^L is called a Gaussian pyramid. The original image c^0 forms the bottom or zeroth level of the Gaussian pyramid. The low-passed version, c^1, of c^0 forms level 1. The value of each pixel in c^1 is computed by taking a weighted average of the pixels in c^1, with the weighting function centered at the corresponding pixel in c^0. Burt and Adelson employ a weighting function of size 5-by-5, but the size of the weighting function is not critical. The level-to-level filtering is effected by the function REDUCE: $c^i = \text{REDUCE}\ (c^{l-1})$ with REDUCE defined by:

For all levels $0 < l < L$ and all nodes (i, j) such that $0 \le i < C_l$ and $0 \le j < R_l$,

$$c^l(i, j) = \sum_{m,n} w(m - 2i, n - 2j)c^{l-1}(m, n),$$

where L is the number of levels in the pyramid and R_l and C_l are the dimensions of the image at the lth level. The dimensions of images are reduced by half from level to level. The dimensions R and C of the original image c^0 are chosen such that there exist integers M_R and M_C satisfying $R = 2^L M_R + 1$ and $C = 2^L M_C + 1$. Thus, the dimensions R_l and C_l of c^l satisfy $R_l = 2^{L-l} M_R + 1$ and $C_l = 2^{L-l} M_C + 1$.

The weighting function w is subjected to certain conditions:

(a) $w(m, n) = \tilde{w}(m)\tilde{w}(n)$ (separability)

(b) $\sum_n \tilde{w}(n) = 1$ (normality)

(c) $\tilde{w}(n) = \tilde{w}(-n) \forall n$ (symmetry)

(d) $\sum_n \tilde{w}(2n) = \sum_n \tilde{w}(2n + 1)$ (equal contribution property)

Condition (d) arises from the requirement that all nodes (pixels) at a given level must contribute the same total weight to nodes at the next higher level. That is,

$$\sum_{i,j} w(m - 2i, n - 2j)$$

is independent of m and n. By property (a), however, this is equivalent to

$$\sum_k \tilde{w}(n - 2k)$$

being independent of n. Considering the cases of even and odd n separately and equating the two, we get condition (d).

Successive iterations of low-pass filtering to generate the various levels of the Gaussian pyramid is equivalent to convolving the original image with a sequence of *equivalent weighting functions* h_l, $0 < l < L$:

$$c^l = h_l * c^0$$

i.e.,

$$c^l(i,j) = \sum_{m,n} h_l(m - 2^l i, n - 2^l j)c^0(m,n)$$

In Figure 7.1, the Gaussian pyramidal decomposition is demonstrated on the "Lena" image for the first six levels, with the original image at level zero of size 257×257 pixels, and the low-pass filtered image at level 5, of size 9×9 pixels:

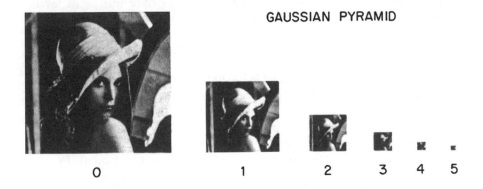

GAUSSIAN PYRAMID

Figure 7.1: Gaussian Pyramid. (c) IEEE 1983. (Adapted from Burt, P. J. and Adelson, E. H., The Laplacian pyramid as a compact image code, *IEEE Trans. Commun.* 31(4): 535-536, April, 1983.)

The Gaussian pyramid is thus a multiresolution low-pass filter. The pyramid generation algorithm reduces the filter band limit by an octave, while reducing the sample interval by the same factor. The algorithm is very rapid, taking fewer computations to obtain a set of filtered images than the FFT algorithm to obtain a single filtered image.

Each level of the Gaussian pyramid corresponds to the predicted image of the level below it. In order to obtain the error image corresponding to the predicted image, the latter has to be subtracted from the image corresponding to the level below it. This subtraction cannot be directly performed, since these two images do not have the same size (number of pixels). The predicted image has to be, therefore, expanded to match the size of the actual image while maintaining the resolution. This is achieved by the function EXPAND, $\tilde{c}^{l-1} =$ EXPAND (c^l), with EXPAND defined by:

For all levels $0 < l < L$ and all nodes (i, j) such that $0 \leq i < C_l$ and $0 \leq j < R_l$,

$$\tilde{c}^{l-1}(i, j) = \sum_{m,n} w(i - 2m, j - 2n) c^l(m, n)$$

The effect of EXPAND is to magnify an $(M + 1) \times (N + 1)$ array into a $(2M + 1) \times (2N + 1)$ array and interpolate the new pixel values between existing pixel values by taking weighted averages of the existing pixel values. Thus, $\tilde{c}^{l-1}$ is a blurred version of c^{l-1} and has the same size as c^{l-1} but with the resolution of c^l. We may now obtain the error images at levels $0 \leq l \leq L$ by subtracting the blurred image from the actual image

$$d^1 = c^l - \tilde{c}^1 = c^1 - \text{EXPAND}(c^{l+1})$$

Since c^L is the highest level in the Gaussian pyramid, there is no image predicted for it. Therefore, the error image d^L is identified with the predicted image c^L.

The Laplacian pyramid is then a sequence of error images $d^0, d^1, ..., d^L$, obtained from the Gaussian pyramid $c^0, c^1, ..., c^L$, by the procedure

$$d^l = c^l - \text{EXPAND}(c^{l+1}) \ \forall 0 \leq l < L; \ d^L = c^L$$

Figure 7.2 illustrates the first four levels of the Gaussian and Laplacian pyramidal decomposition for the Lena image. The first row of Figure 7.2 shows levels 0 to 3 of the Gaussian pyramid in Figure 7.1 with each higher level expanded by the technique of interpolation described above. The second row of Figure 7.2 shows levels 0 to 3 of the Laplacian pyramid, where each level is obtained from the levels 0 to 4 of the Gaussian pyramid by subtracting from the corresponding Gaussian image the (expanded) Gaussian image at the next higher level. The Laplacian pyramid is thus a multiresolution band-pass filter.

The process of recovering the original image c^0 from the Laplacian pyramid is achieved by a simple transformation of the above formula. That is, since $c^L = d^L$ and $c^l = d^l + \text{EXPAND } (c^{l+1}) \; \forall 0 \le l \le L$, we can recover c^0 by iteratively applying the above formula.

Using operator notation, we may write the Laplacian pyramidal scheme more concisely.

Define the operator
$$W : l^2(\mathbf{Z}^2) \to l^2(\mathbf{Z}^2)$$
by
$$(Ws)_{i,j} = \sum_{m,n} w(m - 2i; n - 2j)s_{m,n} \; \forall \{s_{i,j}\}_{i,j \in \mathbf{Z}} \in l^2(\mathbf{Z}^2)$$

Then,
$$c^l = Wc^{l-1} \forall 1 \le l \le L;$$
$$\tilde{c}^l = W^* c^{l+1} = W^* W c^l \; \forall 0 \le l < L;$$
$$d^l = c^l - W^* c^{l+1} = (I - W^* W)c^l \; \forall 0 \le l < L;$$

where W^* is the adjoint of the operator W, and is given by
$$(W^* s)_{i,j} = \sum_{m,n} w(i - 2m, j - 2n)s_{m,n} \; \forall \{s_{i,j}\}_{i,j \in \mathbf{Z}} \in l^2(\mathbf{Z}^2)$$

The decomposition formulas
$$c^1 = Wc^{l-1} \; \forall 1 \le l \le L$$
and
$$d^l = (I - W^* W)c^l \; \forall 0 \le l < L$$
and the reconstruction formula
$$c^{l-1} = d^{l-l} + W^* c^l$$

employ the same filter coefficients, and all the operations are linear. This makes the algorithm very easy to implement.

If the number of nodes (pixels) in the original image is N, then the total number of nodes in the Laplacian pyramid is given by
$$\sum_{i=0}^{L} 4^{-i} N \le \frac{4}{3} N$$

Thus, while the Laplacian pyramid has more entries than the original image, because of decorrelation of the node values the decomposed image can be greatly compressed, using variable length coding of pixel values to minimize entropy and quantization of pixel values.

The Laplacian pyramidal algorithm bears a strong resemblance to Mallat's fast wavelet algorithm arising out of an MRA, discussed in Section 7.4.3. In both cases the initial sequence is decomposed into a pyramid of sequences, each successive level representing a progressively coarser resolution of the original sequence. Also, the differences between successive approximations are computed. In the following, we will study the similarities and differences between the two schemes. We will see that the algorithm of Mallat, while resembling the Laplacian pyramid algorithm, has certain advantages over it. For the sake of simplicity, we will employ the one-dimensional version of the Laplacian pyramid algorithm, wherein the arrays c^l, d^l, w are all one-dimensional sequences. The forms of the decomposition and reconstruction formulas, however, remain the same.

7.1.1   The smoothing function H_∞

Before we compare the two algorithms, we take a closer look at the coefficient sequence $\{w(n)\}_{n\in\mathbf{Z}}$ of the operator $W : l^2(\mathbf{Z}) \to l^2(\mathbf{Z})$. Associated with this sequence are the equivalent weighting functions h_l, $1 \leq l \leq L$, which when convolved with the original image sequence c^0, generate the Gaussian pyramid levels c^1, ..., c^L; that is, $c^l = h_l * c^0$ $\forall 1 \leq l \leq L$. If we represent the equivalent weighting function $h_l = \{w(n)\}_{n\in\mathbf{Z}}$ by the piecewise constant function

$$\omega_1(x) = \sum_{n\in\mathbf{Z}} w(n)\chi_{[-1/2,1/2]}(x - n),$$

then $h_2 = h_1 * h_1$ is represented by the piecewise constant function ω_2, obtained by replacing the characteristic functions $\chi_{[-1/2,1/2]}(x - n)$ in ω_1 by the translates $\omega_1(x - n)$ of ω_1. *i.e.*,

$$\omega_2(x) = \sum_{n\in\mathbf{Z}} w(n)\omega_1(x - n) = \sum_{n\in\mathbf{Z}} w(n) \sum_{m\in\mathbf{Z}} w(m)\chi_{[-1/2,1/2]}(x - m - n)$$

Similarly, since $h_3 = h_2 * h_1$, it is represented by the piecewise constant function ω_3, obtained from ω_2 by replacing the characteristic functions $\chi_{[-1/2,1/2]}(x - m - n)$ by the translates $\omega_1(x - m - n)$ of ω_1. *i.e.*,

$$\omega_3(x) = \sum_{n\in\mathbf{Z}} w(n) \sum_{m\in\mathbf{Z}} w(n)\omega_1(x - m - n)$$

$$= \sum_{n \in \mathbf{Z}} w(n) \sum_{m \in \mathbf{Z}} w(m) \sum_{k \in \mathbf{Z}} w(k) \chi_{[-1/2,1/2]}(x - k - m - n)$$

$$= \sum_{n \in \mathbf{Z}} w(n) \omega_2(x - n)$$

This process can be repeated indefinitely to obtain a sequence of piecewise constant functions $\omega_1, \omega_2, \ldots, \omega_l, \ldots$ representing the equivalent weighting functions $h_1, h_2, \ldots, h_l, \ldots$ and satisfying the relation

$$\omega_{l+1}(x) = \sum_{n \in \mathbf{Z}} w(n) \omega_l(x - n)$$

To prove this relation, it is sufficient to note that convolution is a commutative operation, *i.e.*, $a * b = b * a \ \forall a, b \in l^2(\mathbf{Z})$. Therefore, $h_{l+1} = h_l * h_1 = h_1 * h_l$. This implies that we may obtain ω_{l+1} from ω_l by replacing the characteristic functions $\chi_{[-1/2,1/2]}(x - n)$ in ω_1 by the translates $\omega_l(x - n)$ of ω_l.

If the sequence $\{w(n)\}_{n \in \mathbf{Z}}$ has finitely many nonzero terms, then the supports of the functions ω_l are finite, and the support of ω_l is twice the size of the support of ω_{l-1}. To compare the profiles of these functions, it is, therefore, necessary to scale each function ω_l down by a factor of 2 before constructing the next function from it. Also, to normalize the area under these functions to one, each of the scaled down functions is multiplied by a factor of 2. Thus, we obtain a new sequence of functions $h_1, h_2, ..., h_l ...$ where

$$h_1(x) = 2\omega_1(2x)$$

and

$$h_{l+1}(x) = 2 \sum_{n \in \mathbf{Z}} w(n) h_l(2x - n) \ \forall l \geq 1$$

If we let $h_0(x) = \chi_{[-1/2,1/2]}(x)$, then we have a sequence of functions $h_0, h_1, \ldots, h_l, \ldots \in L^2(\mathbf{R})$ representing the sequence $e, h_1, h_2, \ldots, h_l, \ldots$ where $e = \{\delta_{0n}\}_{n \in \mathbf{Z}}$ and satisfying the two-scale relations

$$h_l(x) = 2 \sum_{n \in \mathbf{Z}} w(n) h_{l-1}(2x - n) \ \forall l \geq 1$$

Using the Fourier transform, this can be rewritten as

$$\hat{h}_l(\xi) = W(\xi/2) \hat{h}_{l-1}(\xi/2) \forall l \geq 1$$

where

$$W(\xi) = \sum_{n \in \mathbf{Z}} w(n) e^{-in\xi}$$

Using the above formula repeatedly, and noting that

$$\hat{h}_0(\xi) = (2\pi)^{-1/2} \frac{\sin(\xi/2)}{\xi/2}$$

we get

$$\hat{h}_l(\xi) = (2\pi)^{-1/2} [\prod_{j=1}^{l} W(2^{-j}\xi)] \frac{\sin(2^{-l-1}\xi)}{2^{-l-1}\xi}$$

The pointwise limit $\hat{h}_\infty(\xi) = \lim_{l\to\infty} \hat{h}_l(\xi)$ exists and is equal to

$$(2\pi)^{-1/2} \prod_{j=1}^{\infty} W(2^{-j}\xi)$$

only if the infinite product is meaningful. For purposes of filtering, only finitely many of the terms of the coefficient sequence $\{w(n)\}_{n\in\mathbb{Z}}$ are nonzero. Hence, by Lemma 6.9 and the Paley-Wiener theorem for distributions,

$$\hat{h}_\infty(\xi) = (2\pi)^{-1/2} \prod_{j=1}^{\infty} W(2^{-j}\xi)$$

is the Fourier transform of a function with compact support. h_∞ is the low-pass filter function of the pyramidal algorithm, and W is the transfer function of the low-pass discrete filter $\{w(n)\}_{n\in\mathbb{Z}}$. It is desirable to have h_∞ as smooth as possible, along with good localization about the origin in space. The regularity of h_∞ has been studied in terms of the transfer function W in Daubechies [8]. The role of h_∞ corresponds to that of the scaling function ϕ of the MRA associated with a one-dimensional fast wavelet algorithm, and W corresponds to the function m_0.

7.2 Mallat's Wavelet-Based Pyramidal Decomposition Scheme

In this section, we discuss Mallat's fast wavelet algorithm described at the end of Chapter 5 in the context of image decomposition and reconstruction. We shall see that the two-dimensional fast wavelet algorithm is a pyramidal algorithm like the Laplacian pyramidal algorithm and is computationally as efficient. It also enjoys the additional features of optimality of storage space and orientation sensitivity, which are lacking in the Laplacian scheme. We first address the one-dimensional fast wavelet algorithm.

7.2.1 The one-dimensional fast wavelet algorithm

In the case of Mallat's fast wavelet algorithm, the decomposition formulas are given by

$$c_k^j = \sum_{n \in \mathbf{Z}} h_{n-2k} c_n^{j-1}$$

and

$$d_k^j = \sum_{n \in \mathbf{Z}} g_{n-2k} c_n^{j-1}$$

or

$$c^j = h \ \# \ c^{j-1}$$

and

$$d^j = g \ \# \ c^{j-1}$$

and the reconstruction formulas are given by

$$c_n^{j-1} = \sum_{k \in \mathbf{Z}} [h_{n-2k} c_k^j + g_{n-2k} d_k^j]$$

or

$$c^{j-1} = h^* \ \# \ c^j + g^* \ \# \ d^j$$

Here,

$$\{c_n^0\}_{n \in \mathbf{Z}}$$

is the input sequence, the sequences

$$\{c_n^j\}_{n \in \mathbf{Z}} \text{ and } \{d_n^j\}_{n \in \mathbf{Z}}$$

are the successive approximation and difference sequences, respectively, and the sequences

$$\{h_n\}_{n \in \mathbf{Z}} \text{ and } \{g_n\}_{n \in \mathbf{Z}}$$

are given by

$$h_n = \langle \phi, \phi_{-1,n} \rangle, \ \ g_n = \langle \psi, \phi_{-1,n} \rangle \ \forall n \in \mathbf{Z}$$

where ϕ and ψ are the (real-valued) scaling function and orthonormal wavelet of the associated multiresolution analysis $\{\mathbf{V}\}_{j \in \mathbf{Z}}$ of $L^2(\mathbf{R})$, respectively (see Section 7.4).

Given a sequence

$$c^0 = \{c_n^0\}_{n \in \mathbf{Z}} \in l^2(\mathbf{Z})$$

(which is the one-dimensional equivalent of a discretely sampled image), it is identified with a function $f \in \mathbf{V}_0$, given by

$$f = \sum_{n \in \mathbf{Z}} c_n^0 \phi_{0n} = \sum_{n \in \mathbf{Z}} c_n^0 \phi(x - n)$$

Thus, $c_n^0 = \langle f, \phi_{0,n} \rangle \ \forall n \in \mathbf{Z}$.

The successive (coarser) approximations of f are given by the projections $\mathbf{P}_j f$ onto the subspaces $\mathbf{V}_j$, $j \geq 1$, of $L^2(\mathbf{R})$, where

$$\mathbf{P}_j f = \sum_{n \in \mathbf{Z}} c_n^j \phi_{j,n}, \quad c_n^j = \langle f, \phi_{j,n} \rangle$$

The difference in information between two successive coarse approximations is given by the projections $\mathbf{Q}_j f$ onto the subspaces $\mathbf{W}_j$, $j \geq 1$, of $L^2(\mathbf{R})$, where

$$\mathbf{Q}_j f = \sum_{n \in \mathbf{Z}} d_n^j \psi_{j,n}, \quad d_n^j = \langle f, \psi_{j,n} \rangle$$

Further, since $\mathbf{V}_{j-1} = \mathbf{V}_j \oplus \mathbf{W}_j$, we have

$$\mathbf{P}_{j-1} f = \mathbf{P}_j f + \mathbf{Q}_j f.$$

$$\mathbf{P}_j f = \sum_{n \in \mathbf{Z}} c_n^j \phi_{j,n}$$

implies that

$$c_n^j = \langle \mathbf{P}_j f, \phi_{j,n} \rangle$$

Since $\mathbf{V}_{j-1} \supset \mathbf{V}_j$,

$$\langle \mathbf{P}_j f, \phi_{j,n} \rangle = \langle \mathbf{P}_{j-1} f, \phi_{j,n} \rangle = \sum_{k \in \mathbf{Z}} c_k^{j-1} \langle \phi_{j-1}, \phi_{j,n} \rangle = \sum_{k \in \mathbf{Z}} h_{n-2k} c_k^{j-1}$$

i.e.,

$$c_n^j = \sum_{k \in \mathbf{Z}} h_{n-2k} c_k^{j-1}$$

Similarly, we have

$$d_k^j = \sum_{n \in \mathbf{Z}} g_{n-2k} c_n^{j-1}$$

These are the decomposition formulas given above. Using these and the identity $\mathbf{P}_{j-1} f = \mathbf{P}_j f + \mathbf{Q}_j f$ we have the reconstruction formula

$$c_n^{j-1} = \sum_{k \in \mathbf{Z}} [h_{n-2k} c_k^j + g_{n-2k} d_k^j]$$

Thus, corresponding to the Laplacian pyramidal decomposition d^0, d^1, ..., d^{L-1}, c^L of the sequence c^0, we have the wavelet decomposition d^1, d^2, ..., d^L, c^L of the sequence c^0.

The similarities between the two decomposition schemes are obvious. Apart from the slight structural differences in the decomposition and the reconstruction formulas of the two schemes, there are two key differences that give Mallat's algorithm a significant advantage over the Laplacian pyramidal scheme. These are (1) economy of representation, and (2) orientation sensitivity.

Economy of representation

If N is the size (number of nonzero entries) of the sequence c^0, then in the Laplacian pyramidal scheme the size of the error image d^j is $2^{-j}N$ $0 \leq j \leq L - 1$ and the size of c^L is $2^{-L}N$. This gives a total size of $N + 2^{-1}N + ... + 2^{-L+1}N + 2^{-L}N$ for the decomposition, which is greater than the size of the original sequence and could be nearly (at most) twice its size. In the two-dimensional (image) case, this is bounded above by $\frac{4}{3}N$.

In the wavelet scheme also, the size of the difference sequence d^j is $2^{-j}N$ $1 \leq j \leq L$ and the size of c^L is $2^{-L}N$. However, in this case the total size of the decomposition is given by

$$2^{-1}N + ... + 2^{-L}N + 2^{-L}N = N$$

Thus, the size of the decomposition is the same as the original sequence. This is true in the two-dimensional case also.

The reason for this difference in representation sizes in the two schemes is due to the redundancy in representation across scales in the error images of the Laplacian pyramidal scheme, which is absent in the wavelet scheme. That is, the sequences (vectors) d^l and c^{l+1} are not "orthogonal" to each other $\forall 0 \leq< L$, *i.e.*, they overlap in their information content. Hence the sum of their sizes is greater than the size of their "sum" in the sense of the reconstruction formula of the pyramidal scheme. This redundancy is absent in the wavelet scheme, wherein the difference sequence d^j represents the difference in the information contents of the successive approximation sequences c^{j-1} and c^j, in the sense that there is no overlap or redundancy in representation between the sequences d^j and c^j. This is because the decomposition is with respect to an orthonormal basis

$$\{\psi_{j,k}\}_{j,k\in\mathbf{Z}}$$

and $\mathbf{V}_{j-1} = \mathbf{V}_j \oplus \mathbf{W}_j \; \forall j \in \mathbf{Z}$. Thus, the original sequence c^0 is decomposable into two uncorrelated sequences, namely the difference sequence d^1, and the approximation sequence c^1, both of which are half the size of c^0 when c^0 is a finite sequence. That is, the size of the decomposed representation is equal to the size of the original sequence. As will be seen later, in the two-dimensional case the image is decomposed into four subimages, each of $\frac{1}{4}$ size of that of the original image, whereas, in the case of the Laplacian pyramidal scheme, the corresponding decomposition increases the size of the representation by a factor of $\frac{3}{2}$ in the one-dimensional case and by a factor of $\frac{5}{4}$ in the two-dimensional (image) case.

Orientation sensitivity

To understand orientation sensitivity of Mallat's algorithm for image decomposition and reconstruction, we must discuss it in its proper setting, namely in two dimensions. We, therefore, digress briefly to describe (1) a method due to P. G. Lemarié of generating a multiresolution analysis for $L^2(\mathbf{R}2)$ from that of $L^2(\mathbf{R})$ by taking a "tensor product of the MRA of $L^2(\mathbf{R})$ with itself", and (2)) the corresponding fast wavelet algorithm.

7.2.2   An MRA of $\mathbf{L}^2(\mathbf{R}^2)$

If
$$\{\mathbf{V}_j\}_{j\in\mathbf{Z}}$$

is an MRA of $L^2(\mathbf{R})$, and ϕ and ψ are the (real-valued) scaling function and orthonormal wavelet associated with the MRA, then the tensor product of the subspace $\mathbf{V}_0$ with itself is given by

$$\begin{aligned}\mathbf{V}_0 \otimes \mathbf{V}_0 \;=\;& \{f \in L^2(R^2) : f(x,y)\\ =\;& \sum_{r,s\in\mathbf{Z}} c_{r,s}\phi(x-r)\phi(y-s), \{c_r,s\}_{r,s\in\mathbf{Z}} \in l^2(\mathbf{Z}^2)\}\end{aligned}$$

$$\underset{\mathbf{L}^2(\mathbf{R}^2)}{\text{closure}}(span\langle\{\phi_{0,r}(x)\phi_{0,s}(y)\}_{r,s\in\mathbf{Z}}\rangle)$$

Using the inner product on $L^2(R^2)$, given by

$$\langle f, g\rangle = \int_{R^2}\int f\bar{g}\,dx\,dy,$$

and Fubini's theorem, it can be shown that

$$\{\phi_{0,r}(x)\phi_0, s(y)\}_{r,s\in\mathbf{Z}}$$

is an orthonormal basis of $\mathbf{V}_0 \otimes \mathbf{V}_0$, if

$$\{\phi_{0,n}\}_{n\in\mathbf{Z}}$$

is an orthonormal basis of $\mathbf{V}_0$. Defining $\mathbf{V}_j \otimes \mathbf{V}_j$ by

$$f(x,y) \in \mathbf{V}_j \otimes \mathbf{V}_j \Leftrightarrow f(2^j x, 2^j y) \in \mathbf{V}_0 \otimes \mathbf{V}_0,$$

it is easy to verify that

$$\{V_j \otimes V_j\}_{j \in \mathbf{Z}}$$

is an MRA of $L^2(\mathbf{R}^2)$, in the sense that it satisfies properties (a)- (d) of Definition 7.17, appropriately modified for $L^2(\mathbf{R}^2)$.

Since $\mathbf{V}_{-1} = \mathbf{V}_0 \oplus \mathbf{W}_0$, by using the properties of tensor products it can be shown that

$$\mathbf{V}_{-1} \otimes \mathbf{V}_{-1} = (\mathbf{V}_0 \otimes \mathbf{V}_0) \oplus (\mathbf{V}_0 \otimes \mathbf{W}_0) \oplus (\mathbf{W}_0 \otimes \mathbf{V}_0) \oplus (\mathbf{W}_0 \otimes \mathbf{W}_0)$$

(prove!)

Again, it is easily verified that the spaces $\mathbf{V}_0 \otimes \mathbf{V}_0$, $\mathbf{V}_0 \otimes \mathbf{W}_0$, $\mathbf{W}_0 \otimes \mathbf{V}_0$, and $\mathbf{W}_0 \otimes \mathbf{W}_0$ have as bases the sets of functions

$$\{\phi_{0,r}(x)\phi_{0,s}(y)\}_{r,s \in \mathbf{Z}'}$$

$$\{\phi_{0,r}(x)\psi_{0,s}(y)\}_{r,s \in \mathbf{Z}'}$$

$$\{\psi_{0,r}(x)\phi_{0,s}(y)\}_{r,s \in \mathbf{Z}}$$

and

$$\{\psi_{0,r}(x)\psi_{0,s}(y)\}_{r,s \in \mathbf{Z}'},$$

respectively. If we let $\psi^0 = \phi$, and $\psi^1 = \psi$, then for

$$\sigma \in \{0,1\} \times \{0,1\} = \{(\sigma_1, \sigma_2) : \sigma_i = 0 \text{ or } 1; \ i = 1,2\}$$

we define $\psi^\sigma(x,y) = \psi^{\sigma_1}(x)\psi^{\sigma_2}(y)$. Then, the above bases can be rewritten as

$$\{\Psi_{0;r,s}^{(0,0)}\}_{r,s \in \mathbf{Z}'}, \{\Psi_{0;r,s}^{(0,1)}\}_{r,s \in \mathbf{Z}'}, \{\Psi_{0;r,s}^{(1,0)}\}_{r,s \in \mathbf{Z}'} \text{ and } \{\Psi_{0;r,s}^{(1,1)}\}_{r,s \in \mathbf{Z}'},$$

respectively. Also letting $\mathbf{V}_0 = \mathbf{W}_0^0$ and $\mathbf{W}_0 = \mathbf{W}_0^1$, we define $\tilde{\mathbf{W}}_0^\sigma = \mathbf{W}_0^{\sigma_1} \otimes \mathbf{W}_0^{\sigma_2}$, for $\sigma \in \{0,1\}^2$. Similarly, for any $j \in \mathbf{Z}$, we have the spaces $\tilde{\mathbf{W}}_j^\sigma = \mathbf{W}_j^{\sigma_1} \otimes \mathbf{W}_j^{\sigma_2}$ spanned by the orthonormal bases

$$\{\psi_{j;r,s}^\sigma\}_{r,s \in \mathbf{Z}'}$$

where

$$\psi_{j;r,s}^\sigma(x,y) = \psi_{j,r}^{\sigma_1}(x)\psi_{j,s}^{\sigma_2}(y)$$

Thus, we have a direct sum decomposition of $L^2(\mathbf{R}^2)$

$$L^2(\mathbf{R}^2) = \bigoplus_{j \in \mathbf{Z}} [\bigoplus_{\substack{\sigma \in \{0,1\}^2 \\ \sigma \neq (0,0)}} \tilde{\mathbf{W}}_j^\sigma],$$

and $\{\psi^{\sigma}_{j;r,s} : j,r,s \in \mathbf{Z}; \; \sigma \in \{0,1\}^2\backslash(0,0)\}$ is an orthonormal basis for $L^2(\mathbf{R}^2)$. This technique can be generalized to obtain an MRA of $L^2(\mathbf{R}^n)$.

Indeed, if $\Sigma = \{0,1\}^n = \{(\sigma_1,...,\sigma_n)\colon \sigma_i = 0 \text{ or } 1 \text{ for } 1 \le i \le n \text{ and } \theta = (0,0,...,0) \in \Sigma$, then

$$L^2(\mathbf{R}^n) = \bigoplus_{j\in\mathbf{Z}}[\oplus_{\sigma\in\Sigma\backslash\{\theta\}}\tilde{\mathbf{W}}^{\sigma}_j]$$

with

$$\{\psi^{\sigma}_{j;k}\vec{r} : \; j \in \mathbf{Z}, \vec{k} \in \mathbf{Z}^n; \; \sigma \in \Sigma\backslash\{\theta\}\}$$

as an orthonormal basis of $L^2(\mathbf{R}^n)$, where

$$\psi^{\sigma}_{j;\vec{k}}(x,y) = \psi^{\sigma_1}_{j;k_1}(x)\psi^{\sigma_2}_{j;k_2}(x)...\psi^{\sigma_n}_{j;k_n}(x), \; \forall j \in \mathbf{Z}, \; \forall \vec{k} = (k_1,...,k_n) \in \mathbf{Z}^n.$$

7.2.3   The two-dimensional wavelet algorithm

The two-dimensional fast wavelet transform of a discretely sampled image is computed using the same scheme as used in the one-dimensional case. First, each row of the image array undergoes decomposition, resulting in an image whose horizontal resolution is reduced by a factor of two and whose scale is doubled. The high-pass (wavelet filter) component of the decomposition characterizes the high-frequency information with *horizontal* orientation. Next, the high-pass and low-pass subimages obtained by the row decomposition are each separately filtered columnwise to obtain four subimages corresponding to low-low-pass, low-high-pass, high-low-pass, and high-high-pass row-column filtering. These correspond to the coefficient arrays of the image with respect to the subbases

$$\{\Psi^{(0,0)}_{1;r,s}\}_{r,s\in\mathbf{Z}'}, \{\Psi^{(0,1)}_{1;r,s}\}_{r,s\in\mathbf{Z}'}, \{\Psi^{(1,0)}_{1;r,s}\}_{r,s\in\mathbf{Z}'} \text{ and } \{\Psi^{(1,1)}_{1;r,s}\}_{r,s\in\mathbf{Z}'},$$

respectively. The latter three subbases capture the horizontal, vertical and diagonal (corners) features, respectively, of the image. To highlight this spatial orientation selectivity feature of these subbases, we rename the corresponding "mother" wavelets a little more suggestively.

We let

$$\Psi^{\mathbf{H}}(x,y) = \psi^{(0,1)}(x,y) = \phi(x)\psi(y).$$
$$\Psi^{\mathbf{V}}(x,y) = \psi^{(1,0)}(x,y) = \psi(x)\phi(y).$$
$$\Psi^{\mathbf{D}}(x,y) = \psi^{(1,1)}(x,y) = \psi(x)\psi(y).$$

Also, we let

$$\Phi(x,y) = \psi^{(0,0)}(x,y) = \phi(x)\phi(y).$$

Thus $\mathbf{\Phi}$ is the two-dimensional scaling function, and $\mathbf{\Psi}^H$, $\mathbf{\Psi}^V$ and $\mathbf{\Psi}^D$ are the three orthonormal wavelets corresponding to the three orientation selectivities. The corresponding subspaces generated by these functions at various scales are renamed matchingly

$$\tilde{\mathbf{W}}_j^H = \tilde{\mathbf{W}}_j^{(0,1)}, \tilde{\mathbf{W}}_j^V = \tilde{\mathbf{W}}_j^{(1,0)}, \tilde{\mathbf{W}}_j^D = \tilde{\mathbf{W}}_j^{(1,1)}, \text{ and} \tilde{\mathbf{W}}_j = \tilde{\mathbf{W}}_j^{(0,0)}.$$

After decomposition of the image into a low-pass subimage and three high-pass subimages as described above, the low-pass subimage is again subjected to the row-column filtering operation to obtain a further (coarser) decomposition, and this process is repeated either until the low-pass image has no more interesting features or a desired number of times.

If f is the representation of the image $\{c_{n,m}^0\}_{m,n\in\mathbf{Z}} \in l^2(\mathbf{Z}^2)$ in $\tilde{\mathbf{V}}_0$, *i.e.*,

$$f = \sum_{m,n\in\mathbf{Z}} c_{n,m}^0 \Phi_{0;n,m},$$

the orthogonal projections of f onto the subspaces $\tilde{\mathbf{V}}_j$, $\tilde{\mathbf{W}}_j^H$, $\tilde{\mathbf{W}}_j^V$ and $\tilde{\mathbf{W}}_j^D$ are given by

$$\mathbf{P}_j f = \sum_{m,n\in\mathbf{Z}} c_{n,m}^j \Phi_{j;n,m}, \quad c_{n,m}^j = \langle f, \Phi_{j;n,m}\rangle,$$

$$\mathbf{Q}_j^H f = \sum_{m,n\in\mathbf{Z}} d_{n,m}^{H;j} \Psi_{j;n,m}^H, \quad d_{n,m}^{H;j} = \langle f, \Psi_{j;n,m}^H\rangle,$$

$$\mathbf{Q}_j^V f = \sum_{m,n\in\mathbf{Z}} d_{n,m}^{V;j} \Psi_{j;n,m}^V, \quad d_{n,m}^{V;j} = \langle f, \Psi_{j;n,m}^V\rangle,$$

and

$$\mathbf{Q}_j^D f = \sum_{m,n\in\mathbf{Z}} d_{n,m}^{D;j} \Psi_{j;n,m}^D, \quad d_{n,m}^{D;j} = \langle f, \Psi_{j;n,m}^D\rangle,$$

respectively. These projections satisfy the relation

$$\mathbf{P}_{j-1}f = \mathbf{P}_j f + \mathbf{Q}_j^H f + \mathbf{Q}_j^V f + \mathbf{Q}_j^D f$$

The reconstruction algorithm is also similar to that of the one-dimensional case. Each subimage at a given scale is subjected to a synthesis by means of the one-dimensional filters, first operating on the columns and then on the rows of the subimage, and the resulting subimages are added up to obtain the low-pass subimage at the next finer scale. This process is repeated until the original resolution level is reached.

The two-dimensional fast wavelet algorithm can be easily formulated from its one-dimensional counterpart. Corresponding to the operators $h\#$

and $g\#$ of the one-dimensional fast wavelet algorithm, we can define the row and column operator pairs $h_r\#$, $g_r\#$ and $h_c\#$, $g_c\#$, respectively, where each pair operates on the second and first indices of the image array, respectively. That is, for any image

$$s^j = \{s^j_{n,m}\}_{m,n\in\mathbf{Z}}$$

defining

$$_rS^j = h_r\#s^{j-1}$$

by

$$(_rs^j)_{n,k} = \sum_{m\in\mathbf{Z}} h_{m-2k}s^{j-1}_{n,m}, \quad t^j = g_r\#s^{j-1}$$

by

$$(_rt^j)_{n,k} = \sum_{m\in\mathbf{Z}} g_{m-2k}s^{j-1}_{n,m}$$

and

$$_cs^j = h_c\#s^{j-1}$$

by

$$(_cs^j)_{k,m} = \sum_{n\in\mathbf{Z}} h_{n-2k}s^{j-1}_{n,m}$$

$$_ct^j = g_c\#s^{j-1}$$

by

$$(_ct^j)_{k,m} = \sum_{n\in\mathbf{Z}} g_{n-2k}s^{j-1}_{n,m},$$

we can express the subimages at scale j obtained from a low-pass subimage

$$c^{j-1} = \{c^{j-1}_{n,m}\}_{m,n\in\mathbf{Z}}$$

at scale $j - 1$ by the following decomposition formulas:

$$c^j = h_c\#h_r\#c^{j-1}$$

$$d^{H;j} = g_c\#h_r\#c^{j-1}$$

$$d^{V;j} = h_c\#g_r\#c^{j-1}$$

$$d^{D;j} = g_c\#g_r\#c^{j-1}.$$

The presence of oriented features are reflected in the magnitudes and profiles of the coefficients of these subimages. Thus, orientation selectivity is an additional feature which is a natural fallout of the choice of the MRA of $L^2(\mathbf{R}^2)$ and does not affect the complexity of the algorithm or the storage size. Since the process of decomposition downsamples the subimages by a factor of 2 in each dimension, each of the resulting four subimages requires a fourth of the number of pixels of the previous image for storage. Thus, the total size of the representation remains the same at each decomposition stage. The corresponding reconstruction formulas at each scale are given by

$$c^{j-1} = h_c^* \# h_r^* \# c^j + h_r^* \# g_c^* \# d^{H;j} + g_r^* \# h_c^* \# d^{V;j} + g_r^* \# g_c^* \# d^{D;j}$$

where the reconstruction operators

$$h_c^* \# h_r^* \#, \ h_r^* \# g_c^* \#, \ g_r^* \# h_c^* \#, \ g_r^* \# g_c^* \#$$

are the adjoints of the decomposition operators

$$h_r \# h_c, \ g_c \# h_r, \ h_c \# g_r, \ g_c \# g_r \#,$$

respectively.

The following schematic diagrams illustrate the decomposition and reconstruction stages of the two-dimensional fast wavelet algorithm

Image decomposition stage: Rows and columns of the low-pass subimage c^j at scale 2^j are convolved with the one-dimensionalimensional CQF filters $\bar{H}$ and $\bar{G}$ to obtain high-pass subimages $d^{D;j+1}$, $d^{V;j+1}$, $d^{H;j+1}$ and the low pass subimage c^{j+1} at scale 2^{j+1}.

$_2 \downarrow_1$: Delete every other column.

$_1 \downarrow_2$: Delete every other row.

Image reconstruction stage: Rows and columns of the high-pass subimages $d^{D;j+1}$, $d^{V;j+1}$, $d^{H;j+1}$ and the low-pass subimage c^{j+1} at scale 2^{j+1}, c^j are convolved with the one-dimensionalimensional CQF filters H and G to obtain the low-pass subimage c^j at scale 2^j.

$_2 \uparrow_1$: Insert a column of zeros between every adjacent pair of columns.

$_1 \uparrow_2$: Insert a row of zeros between every adjacent pair of rows.

$\oplus, \times 2$: Add, and multiply by 2.

Thus, we see that Mallat's fast wavelet algorithm in two dimensions is not only spatially more economical (in fact, optimally so) than the Laplacian pyramidal algorithm, but also is sensitive to oriented spatial features, which the Laplacian algorithm is not. It is also computationally as efficient as the Laplacian pyramidal algorithm.

The two-dimensional fast wavelet algorithm is also a pyramidal algorithm, the pyramid generated containing the three high-pass subimages at resolution level 1 at the lowest tier and the low-pass subimage of the coarsest resolution level at the highest tier. Each intermediate tier contains three high-pass subimages corresponding to an intermediate resolution level. This is illustrated by the following diagram.

Figure 7.3(a) schematically illustrates the pyramidal structure of the two-dimensional fast wavelet algorithm, while Figure 7.3(b) illustrates the pyramidal decomposition of the Lena image using the two-dimensional wavelet algorithm with respect to an orthonormal wavelet basis. The figure shows the decomposition of the image at the resolution levels $j=1,2,3$ into 3 spatially oriented detail images at each level and a low-pass image at level $j=3$. Figure 7.3(c) illustrates the reconstructed image.

This algorithm is ideally suited for various applications such as image compression, filtering, and multiscale edge detection. To enable fast computation, it is necessary to keep the filter lengths small. Therefore, FIR filters corresponding to compactly supported scaling functions and orthonormal wavelets are used (see Chapter 6). These filters cannot be arbitrarily short, as this affects the smoothness of the image representation. Also, it is preferable to have linear phase for these filters, as this leads to an algorithm which does not need to correct for phase. Nonlinear phase filters lead to distortions in the encoding scheme, which may affect the image quality. However, we have seen in the previous chapter that it is not possible to have wavelet filters of both linear phase (corresponding to symmetric scaling function and orthonormal wavelet) and finite length (corresponding to compactly supported scaling function and orthonormal wavelet) without sacrificing orthonormality. Thus, in order to satisfy both these requirements on the filters, it is necessary to relax the orthonormality of the wavelets to biorthogonality to avoid redundancy of representation. As we have seen before in Chapter 6, exact reconstruction is possible in the case of biorthogonal FIR linear phase filters if we use different filters for the reconstruction stage. The decomposition formulas remain the same, but the reconstruction formulas employ operators different from the adjoints of the decomposition operators. We leave it as an exercise for the reader to reformulate the two-dimensional wavelet algorithm in the biorthogonal case.

7.3 Multiscale Edge Representation of Images

Edges in images are characterized by sharp variations in intensity values. However, these variations can occur at several scales, ranging from edges of large objects (at low resolutions) and contours of smaller objects (at higher resolutions) to texture (at even higher resolutions). The distinction between edge information and texture information in an image is largely contextual in the sense that what might appear as edges at a particular resolution may appear as texture at a lower resolution. In any given image, this distinction is largely based on the psychophysics of human vision. Nevertheless, edges and texture are the two most important characteristics of images, from the point of human visual perception. Edges are more important than texture for image understanding and object recognition, and distortion or degradation of edge information markedly affects the quality of the image and the recognizability of various features in it. Texture carries contextual information about lighting, surface features, depth and other perceptual cues of objects in an image, and while being less structured and harder to characterize than edge information, it affects the perception and quality of an image to a significant degree. However, distortions introduced in texture are not as perceptible to the human eye as edge distortions, and this aspect is taken advantage of to obtain high compressions by encoding texture information less accurately, while at the same time introducing distortions that are not very apparent to the human eye. Contours of objects, while being crucial to characterizing the objects, can be encoded more economically than the objects themselves as they are spatially much sparser than the image, and since the range of intensities is considerably reduced, far fewer bits are required to encode the contours. The question, then, is, can one faithfully represent objects and images by means of edge and texture information alone? In other words, how close is the image recovered from its edge-texture representation to the original image? Another issue is that of distinguishing "important" edges corresponding to contours of objects from those corresponding to noise or texture.

As mentioned before, an edge in an image is characterized by a change in image intensity. To examine the nature of this change, we may for the sake of simplicity (and without loss of generality) consider the case of one-dimensional signals. A transition in the signal corresponding to an (ideal) edge may be modeled by an appropriately scaled Heaviside function (Fig 7.4(a)), whose discontinuity is located at the place where the transition in the signal occurs. However, in reality, changes in intensity do not occur abruptly at a single point, as modeled by a Heaviside function, but rather over a transition band or an interval (Fig 7.4(b)). The width of this band may be very narrow or very wide, depending on the scale at which the intensity change takes place.

Typically, image intensity changes occur over a wide range of scales, and a gradual change may go undetected at a fine scale, while sharp transitions may be hard to localize, or appear as noise or texture, at a coarse scale. Thus, it does not make much sense to talk of intensity changes without reference to the scale at which these changes are taking place. In order to analyze an image at different scales, it is necessary to smooth the image with filters of appropriate time-scale characteristics. The effect of smoothing an image with a low-pass filter is that of taking local averages of the image intensities. This results in a low-pass filtered image in which the range of scales over which intensity changes take place is decreased (e.g., halved.) The smoothed (and downsampled) image may now be examined for intensity changes occurring in its range of scales. This process is repeated over and over again to obtain a multiscale analysis of intensity variations of the image. Since the purpose of filtering the image is to reduce the range of scales over which its intensity variations occur, the filter function must be smooth and well localized in the Fourier (scale/frequency) domain. While the averaging is done to reduce the range of scales over which the variations are observed, it is also important to keep these averages local, since transitions in intensity values at each scale correspond to phenomena that are spatially localized at that scale. This requirement imposes that the corresponding filter function be well localized and smooth in the spatial domain as well. Thus, the filter function is so chosen that both it and its Fourier transform are smooth and well-localized functions.

At any given resolution, the edges in the smoothed image may be locally modeled by smoothed and appropriately scaled versions of the Heaviside function, as shown in Figs 7.4(c) and 7.4(f). Edge points thus correspond to inflection points of the smoothed image. These inflection points in turn correspond to local maxima of the absolute value of the first derivative (see Fig. 7.4(d)) or, equivalently, to the zero crossings of the second derivative (see Fig 7.4(e)) of the smoothed image (intensity function.) In the two-dimensional (actual) case the first and second derivatives are replaced by the first and second directional derivatives. When the smoothing function is a Gaussian, the method of detecting edge points by means of the local maxima of the modulus of the first directional derivative corresponds to a Canny edge detector , while the employment of zero crossings of the second directional derivative corresponds to a Marr-Hildreth edge detector .

The second derivative, however, also vanishes at inflection points of the image that correspond to gradual variations, *i.e.,* points at which the rate of change of intensity is very slow (see Figs 7.4(g), 7.4(h) and 7.4(i).) However, these points do not correspond to edges, and therefore, it is hard to distinguish these inflection points from the ones that correspond to rapid intensity changes using zero crossings of the second derivative alone. On the other hand, the "slow" inflection points correspond to local minima of the modulus of the first directional derivative. This means that the modulus of the first derivative of the intensity map is capable of distinguishing between the fast and the slow inflection points (corresponding to maxima and minima, respectively, of the modulus of the first derivative). Thus, one can select the inflection points corresponding to edges using the modulus of the first derivative more easily than from the zero crossings of the second derivative. Also, the zero crossings of the second derivative indicate only the position of the intensity changes and not their magnitudes. It is, therefore, not possible to distinguish small variations from prominent transitions using the zero crossings. The values of the maxima of the modulus of the first derivative, however, are a measure of the sizes of the intensity changes. Thus, the modulus maxima method of edge detection (Canny edge detection) is more advantageous than the zero crossing approach (Marr-Hildreth edge detection) for purposes of edge detection.

Edge information of an image can be obtained at various scales by extracting edges from smoothed versions of the image at various resolutions. This process is called multiscale edge detection and is very useful for object/pattern recognition and image compression.

In this section we describe the work of Mallat and Zhong [28] on characterizing images by the multiscale edge information obtained from the modulus maxima of their wavelet transforms.

Definition : An n-variable function γ is called an n-D smoothing function if $\gamma \in C_{\downarrow}^2(\mathbf{R}^n)$ and $\int_{\mathbf{R}^n} \gamma = 1$.

If $I(x,y)$ is the function corresponding to an image, and $\gamma(x,y)$ is a two-dimensional smoothing function, then for any $s > 0$, the smoothed image at scale s is given by

$$(I * \gamma_s)(x,y) = \int_{\mathbf{R}^2} I(u,v)\gamma_s(x-u,y-v)dudv,$$

where

$$\gamma_s(x,y) = \frac{1}{s^2}\gamma_s(x/s,y/s)$$

If $\tau = (\tau_x, \tau_y)$ is any unit vector in the plane, then the directional derivative of $I * \gamma$ at the point (x,y) along the direction τ is given by

$$\lim_{\delta \to 0} \frac{I * \gamma_s(x + \delta\tau_x, y + \delta\tau_y) - I * \gamma_s(x,y)}{\delta} = \tau.\nabla(I * \gamma_s)(x,y),$$

where

$$\nabla f = (\frac{\partial f}{\partial x}, \frac{\partial f}{\partial y})$$

is the gradient of f. Thus, the modulus of the directional derivative at (x, y) is maximum along the direction of the gradient vector at that point.

In a Canny edge detection scheme, the edge points of an image at scale s are taken to be those points (x_e, y_e) at which the modulus of the gradient vector $\nabla(I * \gamma_s)$ is a local maximum in the direction of the gradient vector $\nabla(I * \gamma_s)(x_e, y_e)$.

The Canny edge detection is equivalent to the detection of the local maxima of the modulus of a two-dimensional wavelet transform of an image. To illustrate this, we define a pair of two-dimensional wavelet functions from the smoothing function

$$\psi^{(1)}(x, y) = \frac{\partial \gamma(x, y)}{\partial x} \text{ and } \psi^{(2)}(x, y) = \frac{\partial \gamma(x, y)}{\partial y}$$

Since γ is smooth and rapidly decreasing, the functions $\psi^{(1)}$ and $\psi^{(2)}$ are *admissible* and their double integrals vanish and, hence, qualify for basic wavelets (see Section 7.2).

Letting $\psi_s^{(i)}(x, y) = \frac{1}{s^2}\psi_s^{(i)}(x/s, y/s)$, $i = 1, 2$, and noting that $\nabla(I * \gamma_s) = I * \nabla \gamma_s$, we have

$$\nabla(I * \gamma_s) = \frac{1}{s}(I * \psi_s^{(1)}, I * \psi_s^{(2)})$$

$(I * \psi_s^{(1)}, I * \psi_s^{(2)})$ is defined to be the two-dimensional wavelet transform of the image I, and is also denoted by $(W_s^{(1)}I, W_s^{(2)}I)$, where

$$W_s^{(i)}I(x, y) = I * \psi_s^{(i)}(x, y) = \int \int_{\mathbf{R}^2} I(u, v)\psi_s^{(i)}(x - u, y - v)dudv$$

$$= \langle I(u, v), \psi_s^{(i)}(x - u, y - v) \rangle, \quad i = 1, 2$$

This definition is at slight variance with that given in Section 7.2, but this difference is not significant and amounts to a mere change in convention (i.e., replacing $\psi(x)$ by $\psi(-x)$ in the one-dimensional definition).

Thus, the modulus of the gradient of the smoothed image at scale s is proportional to the modulus of the image's wavelet transform at scale s, as defined above. This shows that the Canny edge detection scheme is equivalent to the detection of the modulus maxima of the wavelet transform of an image.

In order to understand an image's representation and reconstruction via its multiscale edge information, we discuss and review some of the properties of the wavelet transform. As usual, in order to simplify the discussion and for ease of understanding, we will consider the one-dimensional case first, and later generalize it to the two dimensional case.

7.3.1 The one-dimensional dyadic wavelet transform

The one-dimensional equivalent of the above described edge detection scheme is obtained by replacing the image I by a one-dimensional signal/function $f \in L^2(\mathbf{R})$. The smoothing function γ is taken to be a one-dimensional smoothing function, and the wavelet function is given by $\psi(x) = \frac{d\gamma(x)}{dx}$. The gradient (derivative) of the smoothed signal $f * \gamma_s$ at scale s is given by

$$\frac{d(f * \gamma_s)(x)}{dx} = \frac{1}{s}(f * \psi_s)(x) = \frac{1}{s}\langle f(y), \psi_s(x - y)\rangle$$

(where $\varphi_s(x) = \frac{1}{s}\varphi(x/s)$). By choosing the smoothing function γ appropriately, we can ensure that ψ satisfies the stability condition

$$A \le \sum_{j \in \mathbf{Z}} |\hat{\psi}(2^j \xi)|^2 \le B \quad \forall \xi \in \mathbf{R}, \ A, B > 0$$

Such a ψ is called a *dyadic* wavelet. This in turn implies that the discrete dyadic wavelets

$$\{2^{-j}\psi(2^{-j}x - k)\}_{j \in \mathbf{Z}}$$

form a frame for $L^2(\mathbf{R})$ (see Theorem 7.16). The continuous wavelet transform of any function $f \in L^2(\mathbf{R})$ at the dyadic scale 2^j is given by

$$\mathbf{W}_{2^j} f(x) = (f * \psi_{2^j})(x) \quad \forall x \in \mathbf{R}$$

and the sequence

$$\mathbf{W}f = (\mathbf{W}_{2^j} f)_{j \in \mathbf{Z}}$$

is called the dyadic wavelet transform of f.

If we denote by $l^2(L^2(\mathbf{R}))$ the space of all sequences of functions in $L^2(\mathbf{R})$

$$\{(h_j)_{j \in \mathbf{Z}} : h_j \in L^2(\mathbf{R}) \ \forall j \in \mathbf{Z}, \text{ and } \sum_{j \in \mathbf{Z}} \| h_j \|_2^2 < \infty\},$$

then it is a linear space under componentwise addition and scalar multiplication. The linear operator $\mathbf{W} : L^2(\mathbf{R}) \to l^2(L^2(\mathbf{R}))$ given by

$$\mathbf{W}_f = (\mathbf{W}_{2^j} f)_{j \in \mathbf{Z}}$$

is called the dyadic wavelet transform operator.

The stability condition is equivalent to

$$A \parallel f \parallel_2^2 \leq \sum_{j \in \mathbf{Z}} \parallel \mathbf{W}_{2^j} f \parallel_2^2 \leq B \parallel f \parallel_2^2, \quad \forall f \in L^2(\mathbf{R})$$

Indeed,

$$(\mathbf{W}_{2^j} f)^\wedge(\xi) = \hat{f}(\xi) \hat{\psi}(2^j \xi)$$

and

$$\parallel (\mathbf{W}_{2^j} f)^\wedge \parallel_2^2 = \parallel \mathbf{W}_{2^j} f \parallel_2^2$$

along with the boundedness of

$$\sum_{j \in \mathbf{Z}} \parallel \mathbf{W}_{2^j} f \parallel_2^2$$

give

$$A \parallel \hat{f} \parallel^2 \leq \sum_{j \in \mathbf{Z}} \int_{-\infty}^{\infty} \mid \hat{f}(\xi) / \mid^2 \mid \hat{\psi}(2^j \xi) \mid^2 d\xi \leq B \parallel \hat{f} \parallel_2^2$$

i.e.,

$$A \leq \sum_{j \in \mathbf{Z}} \int_{-\infty}^{\infty} \mid \hat{f}(\xi) / \mid^2 / \parallel \hat{f} \parallel^2 \mid \hat{\psi}(2^j \xi) \mid^2 d\xi \leq B$$

or

$$A \leq \sum_{j \in \mathbf{Z}} \int_{-\infty}^{\infty} \mid \frac{1}{\parallel f \parallel_2} f(x) \psi(2^j \xi) \mid^2 dx \leq B$$

Choosing

$$f(x) = \frac{1}{\sqrt{\varepsilon}} \chi_{[-1/2,1/2]}((x - \xi)/\varepsilon),$$

we have $\parallel f \parallel_2 = 1$, and

$$\int_{-\infty}^{\infty} \mid \frac{1}{\parallel f \parallel_2} f(x) \psi(2^j x) \mid^2 dx = \int_{-1/2}^{1/2} \mid \psi(2^j (\varepsilon u + \xi)) \mid^2 du \xrightarrow{\varepsilon \to 0} \mid \psi(2^j(\xi)) \mid^2$$

The stability condition also implies that there exist dyadic duals ψ^* of ψ such that

$$\sum_{j \in \mathbf{Z}} \hat{\psi}(2^j \xi) \hat{\psi}^*(2^j \xi) = 1 \quad \forall \xi \in \mathbf{R}$$

One such dual is given by

$$\hat{\psi}^*(\xi) = \frac{\overline{\hat{\psi}(\xi)}}{\sum_{j \in \mathbf{Z}} \mid \hat{\psi}(2^j \xi) \mid^2}$$

Since $(\mathbf{W}_{2^j} f)^{\wedge}(\xi) = \hat{f}(\xi)\hat{\psi}(2^j\xi)$, we have

$$(\mathbf{W}_{2^j} f)^{\wedge}(\xi)\hat{\psi}^*(2^j\xi) = \hat{f}(\xi)\hat{\psi}(2^j\xi)\hat{\psi}^*(2^j\xi)$$

Summing both sides over all scale exponent, j, we have

$$\hat{f}(\xi) = \sum_{j\in\mathbf{Z}}(\mathbf{W}_{2^j} f)^{\wedge}(\xi)\hat{\psi}^*(2^j\xi) \quad \forall\xi \in \mathbf{R}$$

i.e.,

$$f(x) = \sum_{j\in\mathbf{Z}}(\mathbf{W}_{2^j} f)(x)^*\psi_{2^j}^*(x).$$

This reconstruction formula demonstrates the completeness of the dyadic wavelet transform, while

$$A\,\|\,f\,\|_2^2 \leq \sum_{j\in\mathbf{Z}} \|\,\mathbf{W}_{2^j} f\,\|_2^2 \leq B\,\|\,f\,\|_2^2$$

asserts that it is stable.

Another feature of the dyadic wavelet transform is its redundancy of representation.

Not every element of $l^2(L^2(\mathbf{R}))$ is the image of some function in $L^2(\mathbf{R})$ under the dyadic wavelet transform operator $\mathbf{W}$. In fact, let $\mathbf{W}^{-1}$ be the operator defined on $l^2(L^2(\mathbf{R}))$ by

$$\mathbf{W}^{-1}((h_j(x))_{j\in\mathbf{Z}}) = \sum_{j\in\mathbf{Z}}(h_j * \psi_{2^j}^*)(x)$$

By the reconstruction formula, if

$$(h_j(x))_{j\in\mathbf{Z}}$$

is the image of some function in $L^2(\mathbf{R})$ under $\mathbf{W}$, then we must have

$$\mathbf{W}(\mathbf{W}^{-1}((h_j(x))_{j\in\mathbf{Z}})) = (h_k(x))_{k\in\mathbf{Z}}$$

i.e.,

$$h_k(x) = \sum_{j\in\mathbf{Z}} h_j * (\psi_{2^j}^* * \psi_{2^k})(x) = \sum_{j\in\mathbf{Z}} h_j * K_{j,k}(x) \quad \forall k \in \mathbf{Z}$$

The functions $K_{j,k}(x) = (\psi_{2^j}^* * \psi_{2^k})(x)$ are called the reproducing kernels, as they take a dyadic wavelet transform to themselves. The L^2-norm (energy) of $K_{j,k}$ is a measure of the redundancy of representation between the continuous wavelet transforms at scales 2^j and 2^k.

7.3.2 Signal reconstruction from its one-dimensional dyadic wavelet transform

Let $f \in L^2(\mathbf{R})$, with dyadic wavelet transform

$$\mathbf{W}f = (\mathbf{W}_{2^j} f)_{j \in \mathbf{Z}}.$$

Let

$$M^j f = (x_n^j)_{n \in \mathbf{Z}}$$

be the set of all values at which

$$| \mathbf{W}_{2^j} f |$$

has local maxima. Mallat conjectured that the representation of a signal by means of the locations of the extrema of its wavelet transform and the values of the wavelet transform at these locations is stable and complete. In other words, one can recover a signal completely and uniquely from the locations and values of its wavelet transform and in such a way that small perturbations in the location and values of the extrema result in small perturbations in the recovered signal. Y. Meyer proved that the completeness of the representation depended on the nature of the specific smoothing function chosen, and even for such special choices, the representation was not stable at all frequencies. The reconstruction problem for the set of local modulus maxima of the wavelet transform of a function $f \in L^2(\mathbf{R})$ consists, then, in finding an approximation $\tilde{f}$ of f, such that the set of local modulus maxima of $\tilde{f}$ coincides with that of f at all scales 2^j. That is, the following conditions must be satisfied at each scale 2^j, $j \in \mathbf{Z}$:

(a) $\forall x_n^j \in M^j f$, $W_{2^j} f(x_n^j) = W_{2^j} \tilde{f}(x_n^j)$

(b) $x \in M^j f \Leftrightarrow x \in M^j \tilde{f}$.

If

$$\mathbf{M} = closure_{L^2(\mathbf{R})}(span(\{\psi_{2^j}(x_n^j - x)\}_{j,n \in \mathbf{Z}}))$$

where

$$(x_n^j)_{j,n \in \mathbf{Z}} = \bigcup_{j \in \mathbf{Z}} M^j f$$

is the set of all the modulus maxima points of f, and $\pi_M : L^2(\mathbf{R}) \to M$ is the orthogonal projection operator of $L^2(\mathbf{R})$ onto $\mathbf{M}$, then since

$$\mathbf{W}_{2^j} g(x_n^j) = (g * \psi_{2^j})(x_n^j) = \langle (g(x), \psi_{2^j}(x_n^j - x) \rangle \quad \forall g \in L^2(\mathbf{R}),$$

condition (a) implies that $\tilde{f}$ belongs to the subset

$$\{h \in L^2(\mathbf{R}) : \pi_M h = \pi_M f\}$$

of $L^2(\mathbf{R})$. If $M^\perp$ is the orthogonal complement of $\mathbf{M}$ in $L^2(\mathbf{R})$, then

$$\{h \in L^2(\mathbf{R}) : \pi_M h = \pi_M f\}$$

can also be written as

$$\{h \in L^2(\mathbf{R}) : h = f + g, \text{ for } g \in \mathbf{M}^\perp\}$$

denoted $f + \mathbf{M}^\perp$, which is an affine subspace of $L^2(\mathbf{R})$. Since the set of points

$$\bigcup_{j \in \mathbf{Z}} M^j f = (x_n^j)_{j,n \in \mathbf{Z}}$$

is typically randomly distributed over the real line, in general, $\mathbf{M}$ is a proper subset of $L^2(\mathbf{R})$, which in turn implies that $\mathbf{M}^\perp \neq \{0\}$. Hence, typically, there are several functions $\tilde{f}$ that satisfy condition (a).

Condition (b) says that at each scale 2^j, the set $M^j \tilde{f}$ of all points at which the modulus of the wavelet transform of the reconstructed function $\tilde{f}$ has a local maximum is identical to the corresponding set $\mathbf{M}^j f$ for the original function f. As this (nonconvex) condition is hard to enforce analytically, it is replaced by an easier (convex) constraint, which minimizes the energy (L^2-norm) of the wavelet transform at each scale. This, along with condition (a), typically tends to induce local maxima of the modulus of the approximating function. This function may have other modulus maxima apart from the ones desired. The number of these maxima at each scale depends on the oscillations of the function's wavelet transform at that scale. Thus, in order to minimize the number of undesirable maxima, the energy of the derivative of the wavelet transform at each scale is also minimized. All this can be expressed as the minimization of the following Sobolev norm of the approximating function $\tilde{f}$:

$$\| \tilde{f} \|_S^2 = | (W_{2^j} \tilde{f})_{j \in \mathbf{Z}} |^2 = \sum_{j \in \mathbf{Z}} (\| W_{2^j} \tilde{f} \|^2 + 2^{2j} \| \frac{dW_{2^j} \tilde{f}}{dx} \|^2)$$

where the weights 2^{2j} are a means of ensuring the progressive smoothness of the wavelet transform with increasing scale. Since $W_{2^j} \tilde{f} = \tilde{f} * \psi_{2^j}$, we have

$$\frac{d W_{2^j} \tilde{f}}{dx} = \frac{1}{2^j} \tilde{f} * \frac{d\psi_{2^j}}{dx} = \frac{1}{2^j} \tilde{f} * \psi'_{2^j}$$

For sufficiently smooth ψ there exist constants $0 < A' < B' < \infty$ such that

$$A' \leq \sum_{j \in \mathbf{Z}} | \hat{\psi}(2^j \xi) |^2 + \sum_{j \in \mathbf{Z}} | (\psi')^\wedge (2^j \xi) |^2 \leq B', \quad \forall \xi \in \mathbf{R}$$

This is equivalent to

$$A' \| h \|_2^2 \leq \sum_{j \in \mathbf{Z}} \| \mathbf{W}_{2^j} h \|_2^2 + \sum_{j \in \mathbf{Z}} 2^{2j} \| \frac{d\mathbf{W}_{2^j} h}{dx} \|_2^2 \leq B' \| h \|_2^2, \ \forall h \in L^2(\mathbf{R})$$

i.e.,

$$A' \| h \|_2^2 \leq \| h \|_S^2 \leq B' \| h \|_2^2, \ \forall h \in L^2(\mathbf{R})$$

That is, the Sobolev norm is equivalent to the L^2-norm. Hence, minimizing the Sobolev norm is equivalent to minimizing the L^2-norm. Thus, the problem of approximate reconstruction of f is reduced to finding an element of $f + \mathbf{M}^\perp$ with the smallest L^2-norm. Such an element is unique, and is given by $\pi_M f$. Indeed, if $f_1 = \pi_M f$ and $f_2 = f - f_1$, then clearly $f_2 \in \mathbf{M}^\perp$, $f_1 \in f + \mathbf{M}^\perp$ and $f_1 + \mathbf{M}^\perp = f + \mathbf{M}^\perp$. Thus, for any

$$h \in f_1 + \mathbf{M}^\perp \exists g \in \mathbf{M}^\perp \ni h = f_1 + g \text{ and } f_1 \perp g.$$

Hence,

$$\| h \|_2^2 = \| f_1 \|_2^2 + \| g \|_2^2 \geq \| f_1 \|_2^2, \text{ and } \| h \|_2^2 = \| f_1 \|_2^2 \Leftrightarrow \| g \|_2^2 = 0$$

i.e., $h = f_1 = \pi_M f$. Since

$$\mathbf{M} = closure_{L^2(\mathbf{R})}(span(\{\psi_{2^j}(x_n^j - x)\}_{j,n \in \mathbf{Z}})),$$

$\pi_M f$ can be stably recovered from the projection coefficients

$$(\langle f, 2^{j/2} \psi_{2^j}(x_n^j - \cdot) \rangle)_{n,j \in L^2(\mathbf{R})}$$

iff

$$(2^{j/2} \psi_{2^j}(x_n^j - \cdot))_{n,j \in L^2(\mathbf{R})}$$

is a frame for $\mathbf{M}$ (see Section 7.3) where the factor $2^{j/2}$ normalizes the L^2-norm of the functions $\psi_{2^j}(x_n^j - \cdot)$. However, since the set of points

$$M^j f = (x_n^j)_{n \in \mathbf{Z}}$$

is distributed randomly at each scale, in general, it is not possible to easily verify whether

$$(2^{j/2} \psi_{2^j}(x_n^j - \cdot))_{n,j \in L^2(\mathbf{R})}$$

is a frame for $\mathbf{M}$. Hence, the reconstruction algorithm for frames cannot be reliably used.

7.3.3   Method of alternate projections

Using a well-known alternating-projection iterative technique (see Youla and Webb[44]) one can, however, obtain a reconstruction of the wavelet transform of $\pi_M f$. The technique, displayed in Figure 7.5, can be briefly described as follows:

If an element ν in an abstract Hilbert space $\mathbf{H}$ is characterized by properties p_i, such that each property p_i constrains ν to lie in a closed convex subset $\mathbf{S}_i$ of $\mathbf{H}$, then ν lies in the intersection

$$\mathbf{S} = \bigcap_{i=1}^{n} \mathbf{S}_i$$

which is also convex and closed. Given the characterizing properties of ν, the restoration problem for ν consists of its reconstruction from its characterizing properties. The solution to the restoration problem requires that at least one element of $\mathbf{S}$ can be effectively found. This is easily done if the projection $\pi : \mathbf{H} \to \mathbf{S}$ is known, since for any $u \in \mathbf{H}, \pi u \in \mathbf{S}$. In general, however, the set $\mathbf{S}$ may be structurally more complex than each of the individual sets $\mathbf{S}_i$, and the projection π may not be easily realizable (available for direct computation). Nevertheless, if each of the projections $\pi_i : \mathbf{H} \to \mathbf{S}_i$ is realizable, then the problem of finding an element of $\mathbf{S}$ can be solved iteratively. Every element u of $\mathbf{S}$ is a fixed point of each of the operators π_i (*i.e.*, $\pi_i u = u$) since the restriction of π_i to $\mathbf{S}_i$ is the identity operator on $\mathbf{S}_i$, and

$$\mathbf{S} = \bigcap_{i=1}^{n} \mathbf{S}_i$$

Hence, each element of $\mathbf{S}$ is a fixed point of the composite operator $\tau = \pi_n \circ \pi_{n-1} \circ ... \circ \pi_1$ (in general, $\tau \neq \pi$), and if $\mathbf{S}$ is nonempty, then every fixed point of τ is also in $\mathbf{S}$. For any $u \in \mathbf{H}$, the iterations $\tau^k u, k \to \infty$ yield a fixed point of τ. If each of the $\mathbf{S}_i$ is an affine subspace of $\mathbf{H}$ (*i.e.*, $\mathbf{S}_i = w_i + \mathbf{M}_i$ for some subspace $\mathbf{M}_i$ of $\mathbf{H}$ and some element w_i of $\mathbf{H}$), then the convergence of the sequence

$$\{\tau^k u\}_{k=0}^{\infty}$$

is strong, and $\lim_{k \to \infty} \tau^k u = \pi u$.

Let

$$\mathbf{K} = \{(h_j)_{j \in \mathbf{Z}} : h_j \in L^2(\mathbf{R}) \ \ \forall j \in \mathbf{Z}$$

$$\text{and } \| (h_j)_{j \in \mathbf{Z}} \|_{\mathbf{K}}^2 = \sum_{j \in \mathbf{Z}} (\| h_j \|_2^2 + 2^{2j} \| \frac{dh_j}{dx} \|_2^2) \leq \infty \}$$

then $\mathbf{K}$ is a Hilbert space with Hilbert norm $\| \ \ \|_{\mathbf{K}}$. Here, the derivative

$$\frac{dh_j}{dx}$$

is meant in the distributional sense; *i.e.*, if $\varphi \in C_c^\infty(\mathbf{R})$ and h is any locally integrable function, then

$$\frac{d^k h}{dx^k}$$

is a linear functional

$$\frac{d^k h}{dx^k} : C_c^\infty(\mathbf{R}) \to \mathbf{C}$$

given by

$$\int_{-\infty}^\infty \frac{d^k h}{dx^k} \varphi \, dx = (-1)^k \int_{-\infty}^\infty h \frac{d^k \varphi}{dx^k} \, dx$$

If $\mathbf{V} = \mathit{Image}\,\mathbf{W}$ where $\mathbf{W} : L^2(\mathbf{R}) \to l^2(L^2(\mathbf{R}))$ is given by

$$\mathbf{W}\mathbf{f} = (\mathbf{W}_{2^j}\mathbf{f})_{j \in \mathbf{Z}},$$

then clearly, $\mathbf{V} \subset \mathbf{K}$, since the Sobolev norm and the L^2-norm are equivalent.

If

$$\Gamma = \{(h_j)_{j \in \mathbf{Z}} \in \mathbf{K} : \forall x_n^j \in \bigcup_{j \in \mathbf{Z}} M^j f = (x_n^j)_{j,n \in \mathbf{Z}}, h_j(x_n^j) = W_{2^j} f(x_n^j)\},$$

then Γ is a closed affine space contained in $\mathbf{K}$. Condition (a) implies that the reconstructed function must belong to $\Lambda = \Gamma \cap \mathbf{V}$, while (the modified) condition (b) implies that the $\mathbf{K}$-norm $\| \ \ \|_k$ of this function is minimum.

Since

$$\mathbf{W}(\mathbf{W}^{-1}((h_j(x))_{j \in \mathbf{Z}})) = (\sum_{j \in \mathbf{Z}} h_j * \mathbf{K}_{j,k}(x))_{k \in \mathbf{Z}}$$

$$= (h_k(x))_{k \in \mathbf{Z}} \quad \forall (h_j(x))_{j \in \mathbf{Z}} \in \mathbf{V},$$

the operator $\mathbf{P_V} = \mathbf{W}o\mathbf{W}^{-1}$ when restricted to $\mathbf{V}$ acts as the identity operator, and for any $(h_j)_{j \in \mathbf{Z}} \in \mathbf{K}$,

$$\mathbf{P_V}((h_j)_{j \in \mathbf{Z}}) \in \mathbf{V}.$$

Thus, $\mathbf{P_V}$ is the projection operator onto $\mathbf{V}$. If the wavelet ψ is symmetric or antisymmetric, then the kernels $\mathbf{K}_{j,k}$ are symmetric functions, which is equivalent to the orthogonality of the projection $\mathbf{P_V}$.

Let $\mathbf{P}_\Gamma$ be the orthogonal projection onto Γ. Then, the set of fixed points of the composite operator $\mathbf{P} = \mathbf{P}_\mathbf{V} o \mathbf{P}_\Gamma$ is the set $\Lambda = \Gamma \cap \mathbf{V}$. Since $\mathbf{V}$ is a subspace of $\mathbf{K}$ and Γ is an affine subspace of $\mathbf{K}$, by the above-mentioned result on alternate projections, we have

$$\lim_{n\to\infty} \mathbf{P}^n((h_j)_{j\in\mathbf{Z}}) = \mathbf{P}_\Lambda((h_j)_{j\in\mathbf{Z}})$$

Hence, iterations of the operator $\mathbf{P}$ on any element of $\mathbf{K}$ converge strongly onto the orthogonal projection of the element onto Λ, thus yielding a function satisfying condition (a). To ensure that the $\mathbf{K}$-norm $\| \quad \|_\mathbf{K}$ of this function is minimum (condition (b)), the iterations are carried out on the zero element of $\mathbf{K}$. This yields a function in Λ which is closest to the zero element of $\mathbf{K}$, and, hence, with minimum $\mathbf{K}$-norm.

If

$$(2^{j/2}\psi_{2^j}(x_n^j - \cdot))_{n,j\in L^2(\mathbf{R})}$$

is a frame for $\mathbf{M}$, and at all scales 2^j

$$\mid x_n^j - x_{n+1}^j \mid \geq C2^j \quad \forall n \in \mathbf{Z}$$

for some constant $0 \leq C \leq 1$, then the rate of convergence of the iterative alternate projection technique is exponential in the number of iterations. If

$$(2^{j/2}\psi_{2^j}(x_n^j - \cdot))_{n,j\in L^2(\mathbf{R})}$$

is not a frame, then the convergence is very slow.

We now consider the extraction of multiscale edge information from images using the two-dimensional wavelet transform and the approximate reconstruction of the image from its multiscale edge information through the iterative method of alternate projections.

7.3.4 The dyadic wavelet transform of images

We have seen that the Canny edge detection scheme for an image $\cap \in L^2(\mathbf{R}^2)$ is equivalent to the detection of the local maxima of a two-dimensional wavelet transform of the image. The smoothing function γ is a two-dimensional smoothing function, and the wavelet functions are given by

$$\psi^{(1)}(x,y) = \frac{\partial \gamma(x,y)}{\partial x} \text{ and } \psi^{(2)}(x,y) = \frac{\partial \gamma(x,y)}{\partial y}$$

The gradient of the smoothed signal $I * \gamma_s$ at scale s is given by

$$\nabla(I * \gamma_s) = \frac{1}{s}(I * \psi_s^{(1)}, I * \psi_s^{(2)}) = \frac{1}{s}(\mathbf{W}_s^1 I, \mathbf{W}_s^2 I)$$

(where $\varphi_s(x,y) = \frac{1}{s^2}\varphi(x/s, y/s)$).

By choosing the smoothing function γ appropriately, we can ensure that $\psi^{(1)}$ and $\psi^{(2)}$ satisfy the stability condition

$$A \le \sum_{j \in \mathbf{Z}} (|\, \hat{\psi}^{(1)}(2^j\xi, 2^j\eta)\,|^2 + |\, \hat{\psi}^{(2)}(2^j\xi, 2^j\eta)\,|^2) \le B \quad \forall \xi, \eta \in \mathbf{R}$$

and for some constants $A, B > 0$, where

$$\hat{f}(\xi, \eta) = \int_{-\infty}^{\infty}\int_{-\infty}^{\infty} f(x,y)e^{-i(\xi x + \eta y)}\, dx\, dy$$

is the Fourier transform of $f \in L^2(\mathbf{R}^2)$. Such a pair $\psi^{(1)}$ and $\psi^{(2)}$ are called *dyadic* wavelets, and the two-dimensional dyadic wavelet transform

$$\mathbf{WI} = (\mathbf{W}_{2^j}^1 \mathbf{I}, \mathbf{W}_{2^j}^2 \mathbf{I})_{j \in \mathbf{Z}}$$

where

$$W_{2^j}^i I(x,y) = (I * \psi_{2^j}^{(i)})(x,y) \quad \forall x, y \in \mathbf{R}, \quad i = 1, 2$$

is a stable and complete representation of the image I.

If we denote by $l^2(L^2(\mathbf{R}^2) \times L^2(\mathbf{R}^2))$ the space of all sequences of functions in $L^2(\mathbf{R}^2)$

$$\{((h_j^1, h_j^2))_{j \in \mathbf{Z}} : h_j^1, h_j^2 \in L^2(\mathbf{R}^2)\ \forall j \in \mathbf{Z},\ \text{and}\ \sum_{j \in \mathbf{Z}}(\|\,h_j^1\,\|_2^2 + \|\,h_j^2\,\|_2^2) < \infty\},$$

then it is a linear space under componentwise addition and scalar multiplication. The linear operator $\mathbf{W} : L^2(\mathbf{R}^2) \to l^2(L^2(\mathbf{R}^2) \times L^2(\mathbf{R}^2))$ given by

$$\mathbf{W}f = (W_{2^j}^1 f, W_{2^j}^2 f)_{j \in \mathbf{Z}}$$

is called the two-dimensional dyadic wavelet transform operator. As in the one-dimensional case, the stability condition is seen to be equivalent to

$$A\,\|\,f\,\|_2^2 \le \sum_{j \in \mathbf{Z}}(\|\,W_{2^j}^1 f\,\|_2^2 + \|\,W_{2^j}^2 f\,\|_2^2) \le B\,\|\,f\,\|_2^2 \quad \forall f \in L^2(\mathbf{R}^2)$$

The stability condition also implies that there exist a pair of dyadic duals $\psi^{(1)*}$ and $\psi^{(2)*}$ of $\psi^{(1)}$ and $\psi^{(2)}$, respectively, such that

$$\sum_{j \in \mathbf{Z}}(\hat{\psi}^{(1)}(2^j\xi, 2^j\eta)\hat{\psi}^{(1)*}(2^j\xi, 2^j\eta)+$$

$$\hat{\psi}^{(2)}(2^j\xi, 2^j\eta)\hat{\psi}^{(2)*}(2^j\xi, 2^j\eta)) = 1 \quad \forall \xi, \eta \in \mathbf{R}$$

One such pair is given by

$$\hat{\psi}^{(1)*}(\xi,\eta) = \frac{\overline{\hat{\psi}^{(1)}(\xi,\eta)}}{\sum_{j\in\mathbf{Z}}(|\hat{\psi}^{(1)}(\xi,\eta)|^2 + |\hat{\psi}^{(2)}(\xi,\eta)|^2)}$$

and

$$\hat{\psi}^{(2)*}(\xi,\eta) = \frac{\overline{\hat{\psi}^{(2)}(\xi,\eta)}}{\sum_{j\in\mathbf{Z}}(|\hat{\psi}^{(1)}(\xi,\eta)|^2 + |\hat{\psi}^{(2)}(\xi,\eta)|^2)}$$

Since

$$(W^i_{2^j}f)^{\wedge}(\xi,\eta) = \hat{f}(\xi,\eta)\hat{\psi}^{(i)*}(2^j\xi,2^j\eta) \quad i = 1,2$$

we have

$$(W^1_{2^j}I)^{\wedge}(\xi,\eta)\hat{\psi}^{(1)*}(2^j\xi,2^j\eta) + (W^2_{2^j}I)^{\wedge}(\xi,\eta)\hat{\psi}^{(2)*}(2^j\xi,2^j\eta)$$

$$= I^{\wedge}(\xi,\eta)(\hat{\psi}^{(1)}(2^j\xi,2^j\eta)\hat{\psi}^{(1)*}(2^j\xi,2^j\eta) + (\hat{\psi}^{(2)}(2^j\xi,2^j\eta)\hat{\psi}^{(2)*}(2^j\xi,2^j\eta)$$

Summing both sides over all scale exponents j, we have

$$\hat{I}(\xi,\eta) = \sum_{j\in\mathbf{Z}}(W^1_{2^j}I)^{\wedge}(\xi,\eta)\hat{\psi}^{(1)*}(2^j\xi,2^j\eta) + (W^2_{2^j}I)^{\wedge}(\xi,\eta)\hat{\psi}^{(2)*}(2^j\xi,2^j\eta)$$

$\forall xi,\eta \in \mathbf{R}$, *i.e.*,

$$I(x,y) = \sum_{j\in\mathbf{Z}}(W^1_2,I)(x,y) * \psi^{(1)*}_{2^j}(x,y) + (W^2_2,I)(x,y) * \psi^{(2)*}_{2^j}(x,y)$$

This reconstruction formula demonstrates the completeness of the two-dimensional dyadic wavelet transform, while

$$A\|I\|_2^2 \le \sum_{j\in\mathbf{Z}}(\|W^1_{2^j}I\|_2^2 + \|W^2_{2^j}I\|_2^2) \le B\|I\|_2^2$$

asserts that it is stable.

The two-dimensional dyadic wavelet transform representation is also redundant, that is, not every element of $l^2(L^2(\mathbf{R}^2) \times L^2(\mathbf{R}^2))$ is the image of some function in $L^2(\mathbf{R}^2)$, under the dyadic wavelet transform operator $\mathbf{W}$. If $\mathbf{W}^{-1}$ is the operator defined on $l^2(L^2(\mathbf{R}^2) \times L^2(\mathbf{R}^2))$ by

$$\mathbf{W}^{-1}((h^1_j(x,y), h^2_j(x,y))_{j\in\mathbf{Z}}) = \sum_{j\in\mathbf{Z}}((h^1_j * \psi^{(1)*}_{2^j})(x,y) + (h^2_j * \psi^{(2)*}_{2^j})(x,y)),$$

then by the reconstruction formula, if

$$(h^1_j(x,y), h^2_j(x,y))_{j\in\mathbf{Z}}$$

is the image of some function in $L^2(\mathbf{R}^2)$ under $\mathbf{W}$, we must have

$$\mathbf{W}(\mathbf{W}^{-1}((h^1_j(x,y), h^2_j(x,y))_{j\in\mathbf{Z}})) = (h^1_k(x,y), h^2_k(x,y))_{j\in\mathbf{Z}}$$

Let $I \in L^2(\mathbf{R}^2)$, with dyadic wavelet transform

$$\mathbf{W}I = (W^1_{2^j}I, W^2_{2^j}I)_{j\in\mathbf{Z}}$$

The modulus of the gradient vector of $I * \gamma_{2^j}$ at scale 2^j varies directly as the modulus of the wavelet transform at that scale, given by

$$M_{2^j}I(x,y) = (\mid W^1_{2^j}I(x,y) \mid^2 + \mid W^2_{2^j}I(x,y) \mid^2)^{1/2},$$

and the angle subtended by the gradient vector and the positive x-direction is given by

$$\Theta_{2^j}I(x,y) = \arctan(W^2_{2^j}I(x,y)/W^1_{2^j}I(x,y)).$$

Conversely, we can recover the wavelet transform

$$\mathbf{W}I = (W^1_{2^j}I, W^2_{2^j}I)_{j\in\mathbf{Z}}$$

from the modulus $M_{2^j}I$ and the angle $\Theta_{2^j}I$ as follows

$$W^1_{2^j}I(x,y) = M_{2^j}I(x,y)\cos(\Theta_{2^j}I(x,y))$$

and

$$W^2_{2^j}I(x,y) = M_{2^j}I(x,y)\sin(\Theta_{2^j}I(x,y))$$

In the case of two-dimensional signals (images), the edge information at each scale 2^j is captured by not only storing the points of local maxima

$$\mathbf{M}^j I = ((x^j_n, y^j_n))_{n\in\mathbf{Z}}$$

of the modulus $\mathbf{M}_{2^j}I$ and the values

$$(M_{2^j}I(x^j_n, y^j_n))_{n\in\mathbf{Z}}$$

of $M_{2^j}I$ at these points, but also the angles

$$(\Theta_{2^j}I(x^j_n, y^j_n))_{n\in\mathbf{Z}}$$

of the gradient vector with the positive x-direction at the maxima points. This additional information is crucial in linking together maxima points that correspond to the same edge feature. The following observations are key to understanding how multiscale edge information is obtained from the locations and values of the modulus maxima and the associated angles at each scale 2^j.

(a) The sharp intensity variations in images occur along "maxima curves," which correspond to edges and contours of objects.

(b) The local intensity variation along (the tangent to) these edge curves is very gradual, while the same is rapid in the perpendicular direction to these curves at the edge points.

(c) The gradient vector at any edge point is normal to (the tangent to) the maxima curve to which the edge point belongs.

Thus, two adjacent maxima points are linked together if their maxima values differ by less than a threshold value, and the angles of the gradients at these points also differ by less than a threshold value. That is, if (x_1^j, y_1^j) and (x_2^j, y_2^j) are neighboring pixels that are both maxima locations, such that

$$| M_{2^j} I(x_1^j, y_1^j) - M_{2^j} I(x_2^j, y_2^j) | \leq m^{(j)}$$

and

$$| \Theta_{2^j} I(x_1^j, y_1^j) - \Theta_{2^j} I(x_2^j, y_2^j) | \leq \theta^{(j)}$$

where $m^{(j)}$ and $\theta^{(j)}$ are the modulus and angle threshold constants, then the two pixels are linked together. This method of linking neighboring maxima positions results in *maxima chains* , which are curves along which the modulus varies gradually, and across which it varies rapidly. Figure 7.6 illustrates the images $W_{2^j}^i \mathcal{F}$, i =1,2: The modulus images, the angle images and the local modulus maxima images of the Lena image, for $0 \leq j \leq 4$: The top row has the original image(left) and a low frequency image (right) at level j = 4. The column (starting from the second row) from left to right are: image (1) $W_{2^j}^1 \mathcal{F}$ 2) $W_{2^j}^2 \mathcal{F}$, where the negative, zero, and positive values are indicated by black, grey, and white pixels, respectively, (3) the modulus image $M_{2^j} \mathcal{F}$, where the black pixels correspond to zero amplitudees, (4) the angle images $\theta_{2^j} \mathcal{F}$, where the angle ranges from 0 (black pixels) to 2π, and (5) the local modulus maxima images, where the white pixels correspond to points of local maxima of $M_{2^j} \mathcal{F}$ in the direction of the angles $\theta_{2^j} \mathcal{F}$. The scale increases down the columns.

Figure 7.7 illustrates the Lena image (left) of size 256×256 pixels.

7.3.5 Image reconstruction from its two-dimensional dyadic wavelet transform

The reconstruction problem for the multiscale edge information of an image $I \in L^2(\mathbf{R}^2)$ consists in finding an approximation $\tilde{I}$ of I, such that the set of local modulus maxima of I coincides with that of I at all scales 2^j. That is, the following conditions must be satisfied at each scale 2^j, $j \in \mathbf{Z}$:

(a) $\forall (x_n^j, y_n^j) \in M^j I$, $W_{2^j}^i I(x_n^j, y_n^j) = W_{2^j} \tilde{I}(x_n^j, y_n^j)$, $i = 1, 2$

(b) $(x, y) \in \mathbf{M}^j I \Leftrightarrow (x, y) \in \mathbf{M}^j \tilde{I}$.

If
$$M = closure_{L^2(\mathbf{R}^2)} (span\langle \{ 2^j \psi_{2^j}^{(i)} (x_n^j - x, y_n^j - y) \}_{\substack{i=1,2 \\ j,n \in \mathbf{Z}}} \rangle)$$
where
$$((x_n^j, y_n^j))_{j,n \in \mathbf{Z}} = \bigcup_{j \in \mathbf{Z}} \mathbf{M}^j I$$

is the set of all the modulus maxima points of I , and $\pi_{\mathbf{M}} : L^2(\mathbf{R}^2) \to \mathbf{M}$ is the orthogonal projection operator of $L^2(\mathbf{R}^2)$ onto $\mathbf{M}$, then, since

$$W_{2^j}^i g(x_n^j, y_n^j) = (g * \psi_{2^j}^{(i)})(x_n^j, y_n^j)$$

$$= \langle (g(x, y), \psi_{2^j}^{(i)}(x_n^j - x, y_n^j - y) \rangle, \quad i = 1, 2, \quad \forall g \in L^2(\mathbf{R}^2),$$

condition (a) implies that $\tilde{I}$ belongs to the subset

$$\{ h \in L^2(\mathbf{R}^2) : \pi_{\mathbf{M}} h = \pi_{\mathbf{M}} I \}$$

of $L^2(\mathbf{R}^2)$.

If $\mathbf{M}^\perp$ is the orthogonal complement of $\mathbf{M}$ in $L^2(\mathbf{R}^2)$, then

$$\{ h \in L^2(\mathbf{R}^2) : \pi_{\mathbf{M}} h = \pi_M I \}$$

can also be written as

$$\{ h \in L^2(\mathbf{R}^2) : h = I + g, \text{ for } g \in \mathbf{M}^\perp \},$$

denoted $I + \mathbf{M}^\perp$, which is an affine subspace of $L^2(\mathbf{R}^2)$.

Since the set of points

$$\bigcup_{j \in \mathbf{Z}} M^j I = ((x_n^j, y_n^j))_{j,n \in \mathbf{Z}}$$

is typically randomly distributed and not a lattice in $\mathbf{R}^2$, in general, $\mathbf{M}$ is a proper subset of $L^2(\mathbf{R}^2)$, which in turn implies that $\mathbf{M}^\perp \neq \{0\}$. Hence, typically, there are several functions $\tilde{I}$ that satisfy condition (a).

Condition (b) requires that at each scale 2^j, the set $\mathbf{M}^j \tilde{I}$ of all points at which the modulus of the wavelet transform of the reconstructed function $\tilde{I}$ has a local maximum is identical to the corresponding set $\mathbf{M}^j I$ for the original function I. As in the one-dimensional case, this nonconvex constraint is replaced by a (convex) norm minimizing constraint, which tends to induce local maxima in the modulus of the approximating function at the points

$$(x_n^j, y_n^j) \in \bigcup_{j \in \mathbf{Z}} \mathbf{M}^j I$$

by minimizing the energy ($L^2(\mathbf{R}^2)$-norm) of the wavelet transform at each scale and minimizes the number of extraneous oscillations of the wavelet transform by minimizing the energy of its derivatives. The wavelet transform components $W_{2^j}^1 I$ and $W_{2^j}^2 I$ tend to have most of their oscillations along the x and y directions, respectively, since they are obtained by convolving the image with the scaled smoothing function's partial derivatives w.r.t. x and y, respectively. It, therefore, makes sense to minimize the energies of the corresponding partial derivatives of the wavelet components only. This can be expressed as the minimization of the following Sobolev norm of the approximating function $\tilde{I}$:

$$\| \tilde{I} \|_{\mathbf{S}}^2 = | (W_{2^j}^1 \tilde{I}, W_{2^j}^2 \tilde{I})_{j \in \mathbf{Z}} |^2$$

$$\sum_{j \in \mathbf{Z}} (\| W_{2^j}^1 \tilde{I} \|^2 + \| W_{2^j}^2 \tilde{I} \|^2 + 2^{2j}(\| \frac{\partial W_{2^j}^1 \tilde{I}}{\partial x} \|^2 + \| \frac{\partial W_{2^j}^2 \tilde{I}}{\partial y} \|^2))$$

where the weights 2^{2j} are, again, a means of ensuring the progressive smoothness of the wavelet transform with increasing scale. Since $W_{2^j} \tilde{I} = \tilde{I} * \psi_{2^j}^{(i)}$, $i = 1, 2$ we have

$$\frac{\partial W_{2^j}^1 \tilde{I}}{dx} = \frac{1}{2^j} \tilde{I} * \frac{\partial \psi_{2^j}^{(1)}}{\partial x} = \frac{1}{2^j} \tilde{I} * \dot{\psi}_{2^j}^{(1)}$$

and

$$\frac{\partial W_{2^j}^2 \tilde{I}}{dy} = \frac{1}{2^j} \tilde{I} * \frac{\partial \psi_{2^j}^{(2)}}{\partial y} = \frac{1}{2^j} \tilde{I} * \dot{\psi}_{2^j}^{(2)}$$

For sufficiently smooth γ there exist constants $0 < A' < B' < \infty$ such that

$$A' \leq \sum_{j \in \mathbf{Z}} | \hat{\psi}^{(1)}(2^j \xi, 2^j \eta) |^2 + | \hat{\dot{\psi}}^{(1)}(2^j \xi, 2^j \eta) |^2 +$$

$$\sum_{j \in \mathbf{Z}} | \hat{\psi}^{(2)}(2^j \xi, 2^j \eta) |^2 + | \hat{\dot{\psi}}^{(2)}(2^j \xi, 2^j \eta) |^2 \leq B', \quad \forall \xi, \eta \in \mathbf{R}$$

This is equivalent to

$$A' \parallel h \parallel_2^2$$

$$\leq \sum_{j \in \mathbf{Z}} (\parallel W_{2^j}^1 h \parallel^2 + \parallel W_{2^j}^2 h \parallel^2 + 2^{2j}(\parallel \frac{\partial W_{2^j}^1 h}{\partial x} \parallel^2 + \parallel \frac{\partial W_{2^j}^2 h}{\partial y} \parallel^2))$$

$$\leq B' \parallel h \parallel_2^2, \quad \forall h \in L^2(\mathbf{R}^2)$$

i.e.,

$$A' \parallel h \parallel_2^2 \leq \parallel h \parallel_{\mathbf{S}}^2 \leq B' \parallel h \parallel_2^2, \quad \forall h \in L^2(\mathbf{R}^2)$$

That is, the Sobolev norm is equivalent to the $L^2(\mathbf{R}^2)$-norm. Hence, minimizing the Sobolev norm is equivalent to minimizing the $L^2(\mathbf{R}^2)$-norm. Thus, the problem of approximate reconstruction of I is reduced to finding an element of $I + \mathbf{M}^{\perp}$ with the smallest $L^2(\mathbf{R}^2)$-norm. Such an element is unique and is given by $\pi_M I$.

Since

$$\{2^j \psi_{2^j}^{(i)}(x_n^j - x, y_n^j - y)\}_{\substack{i=1,2 \\ j,n \in \mathbf{Z}}}$$

is in general not a frame, $\pi_M I$ may not be stably recoverable from the projection coefficients

$$\langle I, 2^j \psi_{2^j}^{(i)}(x_n^j - \cdot, y_n^j - \cdot) \rangle_{\substack{i=1,2 \\ n,j \in L^2(\mathbf{R}^2)}}$$

7.3.6   Method of alternate projections in two-dimensional

As in the one-dimensional case, instead of recovering $\pi_M I$, its wavelet transform is recovered by resorting to the two-dimensional version of the alternating-projection iterative technique discussed in the one dimensional case. Here,

$$\mathbf{K} = \{((h_j^1, h_j^2))_{j \in \mathbf{Z}} : h_j^1, h_j^2 \in L^2(\mathbf{R}^2) \; \forall j \in \mathbf{Z}, \text{ and } \parallel ((h_j^1, h_j^2))_{j \in \mathbf{Z}} \parallel_{\mathbf{K}}^2 \leq \infty\}$$

where

$$\parallel ((h_j^1, h_j^2))_{j \in \mathbf{Z}} \parallel_{\mathbf{K}}^2 = \sum_{j \in \mathbf{Z}} (\parallel h_j^1 \parallel_2^2 + \parallel h_j^2 \parallel_2^2 + 2^{2j} \left(\left\| \frac{\partial h_j^1}{\partial x} \right\|_2^2 + \left\| \frac{\partial h_j^2}{\partial y} \right\|_2^2 \right).$$

Then, $\mathbf{K}$ is a Hilbert space with Hilbert norm $\parallel \quad \parallel_{\mathbf{K}}$. As before, the derivatives

$$\frac{\partial h_j^1}{\partial x} \text{ and } \frac{\partial h_j^2}{\partial y}$$

are meant in the distributional sense, *i.e.*, if $\varphi \in C_c^\infty(\mathbf{R}^2)$ and h is any locally integrable function, then

$$\frac{\partial^k h}{\partial x^l \partial y^{k-1}}$$

is a linear functional

$$\frac{\partial^k h}{\partial x^l \partial y^{k-1}} : C_c^\infty(\mathbf{R}^2) \to \mathbf{C}$$

given by

$$\int_{-\infty}^\infty \int_{-\infty}^\infty \frac{\partial^k h}{\partial x^l \partial y^{k-1}} \varphi \, dx dy = (-1)^k \int_{-\infty}^\infty \int_{-\infty}^\infty h \frac{\partial^k \varphi}{\partial x^l \partial y^{k-1}} \, dx dy$$

If $\mathbf{V} = Image \mathbf{W}$, where $\mathbf{W} : L^2(\mathbf{R}^2) \to l^2(L^2(\mathbf{R}^2) \times L^2(\mathbf{R}^2))$ is given by

$$\mathbf{W} h = (W_{2^j}^1 h, W_{2^j}^2 h)_{j \in \mathbf{Z}'}$$

then clearly, $\mathbf{V} \subset \mathbf{K}$, since the Sobolev norm and the $L^2(\mathbf{R}^2)$-norm are equivalent.

If

$$\Gamma = \{((h_j^1, h_j^2))_{j \in \mathbf{Z}} \in \mathbf{K} : \forall (x_n^j, y_n^j) \in \bigcup_{j \in \mathbf{Z}} M^j I, h_j(x_n^j, y_n^j) = \mathbf{W}_{2^j} I(x_n^j, y_n^j)\},$$

then Γ is a closed affine space contained in $\mathbf{K}$. Condition (a) implies that the reconstructed function must belong to $\Lambda = \Gamma \cap \mathbf{V}$, while (the modified) condition (b) implies that the $\mathbf{K}$-norm $\| \ \|_\mathbf{K}$ of this function is minimum.

Since

$$\mathbf{W}(\mathbf{W}^{-1}(((h_j^1, h_j^2))_{j \in \mathbf{Z}})) = ((h_k^1, h_k^2))_{k \in \mathbf{Z}} \quad \forall ((h_j^1, h_j^2))_{j \in \mathbf{Z}} \in \mathbf{V},$$

the operator $\mathbf{P_V} = \mathbf{W} o \mathbf{W}^{-1}$ when restricted to $\mathbf{V}$ acts as the identity operator and for any

$$((h_j^1, h_j^2))_{j \in \mathbf{Z}} \in \mathbf{K}, \quad \mathbf{P_V}(((h_j^1, h_j^2))_{j \in \mathbf{Z}}) \in \mathbf{V}$$

Thus, $\mathbf{P_V}$ is the projection operator onto $\mathbf{V}$, which is self-adjoint (and, hence, orthogonal) if the wavelets are symmetric/antisymmetric. If $\mathcal{P}_-$ is the orthogonal projection onto Γ, then the set of fixed points of the composite operator $\mathbf{P} = \mathbf{P_V} \circ \mathbf{P_\Gamma}$ is the set $\Lambda = \Gamma \cap \mathbf{V}$. Since $\mathbf{V}$ is a subspace of $\mathbf{K}$, and Γ is an affine subspace of $\mathbf{K}$, we have

$$\lim_{n \to \infty} \mathbf{P}^n(((h_j^1, h_j^2))_{j \in \mathbf{Z}}) = \mathbf{P}_\Lambda(((h_j^1, h_j^2))_{j \in \mathbf{Z}})$$

Hence, iterations of the operator $\mathbf{P}$ on any element of K converge strongly onto the orthogonal projection of the element onto Λ, thus yielding a function satisfying condition (a). To ensure that the $\mathbf{K}$-norm $\|\ \ \|_{\mathbf{K}}$ of this function is minimum (condition (b)), the iterations are carried out on the zero element of $\mathbf{K}$. This yields a function in Λ which is closest to the zero element of $\mathbf{K}$ and, hence, with minimum $\mathbf{K}$-norm. For a proof of the convergence of the above method of alternate projections, see Youla and Webb [44]. For the numerical implementation details of the operator $\mathbf{P}_{\Gamma}$ and the convergence estimates of the method of alternate projections, the reader is referred to Mallat and Zhong [28].

7.3.7   The discrete finite dyadic wavelet transform

In practice, images are obtained by discrete sampling and, therefore, have a finite resolution. Thus, if we normalize the scale of the sampled image to 1, then the wavelet transform of the image can be computed only at scales $2^j, j \geq 0$. This constraint on the available resolution is incorporated into the analysis by defining a real-valued function $\phi \in L^2(\mathbf{R}^2)$, such that its Fourier transform satisfies

$$|\hat{\phi}(\xi,\eta)|^2 = \sum_{j\geq 1}(\hat{\psi}^{(1)}(2^j\xi,2^j\eta)\hat{\psi}^{(1)*}(2^j\xi,2^j\eta)+(\hat{\psi}^{(2)}(2^j\xi,2^j\eta)\hat{\psi}^{(2)*}(2^j\xi,2^j\eta))$$

Since

$$\sum_{j\in\mathbf{Z}}(\hat{\psi}^{(1)}(2^j\xi,2^j\eta)\hat{\psi}^{(1)*}(2^j\xi,2^j\eta)+(\hat{\psi}^{(2)}(2^j\xi,2^j\eta)\hat{\psi}^{(2)*}(2^j\xi,2^j\eta)) = 1 \ \forall \xi,\eta \in \mathbf{R},$$

it is easy to see that

$$\int_{-\infty}^{\infty}\int_{-\infty}^{\infty}\phi(x,y)dxdy = \hat{\phi}(0,0) =|\hat{\phi}(0,0)|= \lim_{j\leftarrow\infty}|\hat{\phi}(2^{-j}\xi,2^{-j}\eta)|=1,$$

i.e., is a smoothing function. One can then define a smoothing operator

$$\mathbf{S}_{2^j}I(x,y) = I * \phi_{2^j}(x,y)$$

where $\phi_{2^j}(x,y) = 2^{-j}\phi(2^{-j}x,2^{-j}y)$. Also, for any $J > 0$,

$$|\hat{\phi}(\xi,\eta)|^2 - |\hat{\phi}(2^J\xi,2^J\eta)|^2 = \sum_{\substack{i=1,2\\1\leq j\leq J}}\hat{\psi}^{(i)}(2^j\xi,2^j\eta)\psi^{(i)*}(2^j\xi,2^j\eta)$$

This implies that the difference in the detail between the smoothed images $\mathbf{S}_1 I$ and $\mathbf{S}_{2^J} I$ are supplied by the wavelet transform

$$(\mathbf{W}_{2^j}^1 I, \mathbf{W}_{2^j}^2 I)_{1\leq j\leq J}$$

between the scales 1 and 2^J. Since the sampled image is of finite size, the scale 2^J cannot exceed the dimensions of the image, and, hence, J is finite and bounded by the size of the image.

The finite sequence of functions

$$((\mathbf{W}^1_{2^j}I, \mathbf{W}^2_{2^j}I)_{1\leq j\leq J}, \mathbf{S}_{2^J}I)$$

is called the finite dyadic wavelet transform of the image I. The Sobolev norm with respect to this finite scale transform is modified to

$$
\begin{aligned}
\| I \|^2_{\mathbf{S}} &= \| ((\mathbf{W}^1_{2^j}I, \mathbf{W}^2_{2^j}I)_{1\leq j\leq J}, \mathbf{S}_{2^J}I) \|^2 \\
&= \sum_{1\leq j\leq J} (\| \mathbf{W}^i_{2^j}I \|^2 + \| \mathbf{W}^i_{2^j}I \|^2) \\
&\quad + 2^{2j} \sum_{1\leq j\leq J} (\| \frac{\partial \mathbf{W}^1_{2^j}I}{\partial x} \|^2 \\
&\quad + \| \frac{\partial \mathbf{W}^2_{2^j}I}{\partial x} \|^2) + \| \mathbf{S}_{2^J}I \|^2,
\end{aligned}
$$

and the Hilbert space $\mathbf{K}$ is redefined as

$$\mathbf{K} = \{((h^1_j, h^2_j)_{1\leq j\leq J}, g_J) : h^1_j, h^2_j, g_J \in L^2(\mathbf{R}^2) \ \forall j \in \mathbf{Z}$$

$$\text{and } \| ((h^1_j, h^2_j)_{1\leq j\leq J}, g_J) \|^2_{\mathbf{K}} \leq \infty\}$$

where

$$
\begin{aligned}
\| ((h^1_j, h^2_j)_{1\leq j\leq J}, g_J) \|^2_{\mathbf{K}} &= \sum_{1\leq j\leq J} (\| h^1_j \|^2 + \| h^2_j \|^2 \\
&\quad + 2^{2j}(\| \frac{\partial h^1_j}{\partial x} \|^2 + \| \frac{\partial h^2_j}{\partial x} \|^2)) + \| g_J \|^2_2
\end{aligned}
$$

The subspace $\mathbf{V}$ of $\mathbf{K}$ then consists of all finite dyadic wavelet transforms of functions in $L^2(\mathbf{R}^2)$.

If

$$(c_{n,m})_{1\leq n,m\leq N}$$

is a discrete sample representing an image, then one can find a (nonunique) function $I \in L^2(\mathbf{R}^2)$ such that

$$\mathbf{S}_1 I(n, m) = c_{n,m} \ \forall 1 \leq n, m \leq N$$

For an appropriate choice of the wavelet functions, the discrete finite dyadic wavelet transform

$$((\mathbf{W}^{1,d}_{2^j}I, \mathbf{W}^{2,d}_{2^j}I)_{1\leq j\leq J}, \mathbf{S}^d_{2^J}I)$$

can be computed from $(S_1 I(n,m))_{1\leq n,m\leq N}$, where, for each scale 2^j

$$\mathbf{W}_{2^j}^{i,d} I = (\mathbf{W}_{2^j}^{i,d} I(n+s, m+s))_{1\leq n,m\leq N} \quad i=1,2$$

and

$$\mathbf{S}_{2^j}^d I = (\mathbf{S}_{2^j}^d I(n+s, m+s))_{1\leq n,m\leq N}$$

Here s is a sampling shift that depends on the wavelet functions chosen. If $J = \log N + 1$, then $S_{2^J}^d I$ is constant and equal to the uniform average of the discrete image

$$(c_{n,m})_{1\leq n,m\leq N}$$

Since the image is finite, the one-sidedness of information at the borders of the image is avoided by symmetrizing the image about each of its border edges and continuing it indefinitely in the plane as a biperiodic function of period $(2N, 2N)$. That is, the sequence

$$(c_{i,j})_{1\leq i,j\leq N}$$

is mapped to the biperiodic sequence

$$(\tilde{c}_{i,j})_{i,j\in \mathbf{Z}'}$$

given by

$$\tilde{c}_{i,j} = \tilde{c}_{2N-i,j} = \tilde{c}_{i,2N-j'}$$

and

$$\tilde{c}_{i+2N,j} = \tilde{c}_{i,j+2N} = c_{i,j}$$

The computation of the discrete finite dyadic wavelet transform from the sequence

$$(\tilde{c}_{i,j})_{i,j\in \mathbf{Z}'}$$

and the reconstruction of the discrete image from the discrete wavelet transform are both achieved by fast wavelet algorithms with complexity $O(N^2 \log N)$. For details regarding the discrete and numerical implementation of the fast wavelet algorithms, the reader is referred to Mallat and Zhong [28].

7.4 Double-Layered Image encoding

As pointed out earlier, from the point of image transmission and storage, it is highly desirable to be able to encode the image with as few bits per pixel as possible, as this economizes on the time needed to transmit/retrieve the image as well as the amount of space (memory) needed to store the image.

This means that some amount of information/detail has to be discarded in such a way that it does not degrade the image quality significantly, while achieving a cost effective compression of the image.

Mallat and Zhong [28] implemented a compact coding algorithm based on the close approximate characterization of an image by its multiscale edge information, described above. While compression factors of as high as 30 are attained, the algorithm ignores texture information completely, which is a disadvantage in some applications, since texture carries important cues and supplementary image information.

Froment and Mallat [15] developed a double-layered image encoding technique wherein the image is first encoded using multiscale edge information only, and the difference between the original image and the edge encoded image, which contains the texture information, is encoded using an orthogonal wavelet pyramidal algorithm. The edge-based encoding is more accurate than the texture-based encoding, which employs coarse quantization. A compression factor of 100 is achieved by them for a 512×512 pixel image.

In this section, we will describe the multiscale edge-based compact image encoding algorithm, followed by the texture encoding algorithm.

7.4.1 Multiscale edge-based image encoding

In this encoding technique, the most significant edge (maxima) curves are first selected and then efficiently encoded. The significance of an edge curve is, in general, a contextual valuation and depends to a considerable extent on the importance of the semantic information provided by the curve from the point of pattern or object recognition. In general, it is hard to evaluate a curve's significance, as it is image dependent. In this technique, however, the significance of an edge curve is based on whether its length is above a certain threshold value, the rationale behind this criterion being that the contours of the salient objects in an image give rise to long edge curves. The efficient coding capitalizes on the positional and structural similarity of the important edge curves across scales. Since most of the frequency (detail) information is concentrated at the finer scales, the edge encoding is restricted to these scales. Thus, only the scales 2^j, $1 \leq j \leq J_0$ for some appropriately chosen J_0 (say $J_0 = 3$) are considered for edge extraction, while the larger scale information is stored in the low frequency image $\mathbf{S}_{2^{J_0}}^d I$. The edge information at very fine scales contains a large number of "insignificant" edges corresponding to texture and noise and is computationally expensive at the edge selection stage. Hence, an appropriate intermediate scale 2^i, $1 \leq i \leq J_0$ (say $i = 2$) is chosen from which to select the important edge curves. First, the edge curves whose lengths are below a threshold value are eliminated. Next, from the rest of the curves, only those that correspond to rapid variations in image intensity are retained. This operation is performed by computing the average value of the modulus $\mathbf{M}_{2^j} I$ and thresholding it below. The positional information (maxima locations) of these selected edge curves is recorded at scale 2^i and taken to be the same at all other scales 2^j, $1 \leq j \leq J_0$. These edge curves are encoded predictively by encoding the first pixel of each curve and encoding only the difference in position of each successive pixel from that of the previous pixel. This method of encoding is shown (Carlsson [2]) on an average to take 10 bits for the first pixel and 1.3 bits for each successive pixel. The cost of this encoding is further halved by performing it after downsampling the edge image by 2 and linearly interpolating the edge point locations after upsampling the encoded edge image. Since at any scale the gradient at the edge points is orthogonal to the tangent to the curve to which the edge point belongs, the angles $\Theta_{2^j} I$ at the edge points can be directly approximated from the tangents to the encoded curves at all scales 2^j, $1 \leq j \leq J_0$, and, hence, are not recorded. Thus, the position and angle information of the edge points at the scales 2^j, $1 \leq j \leq J_0$ are the same. The values of the modulus $\mathbf{M}_{2^j} I$ are encoded at each scale 2^j, $1 \leq j \leq J_0$ along the edge curves in a predictive fashion employing a coarse quantization of the predicted values. The energy of the low frequency image $\mathbf{S}^d_{2^{J_0}} I$ is largely concentrated in the frequency domain over a region whose size is 2^{-2J_0} times smaller than the bandwidth of the image I. Hence, it is downsampled by a factor of 2^{J_0} in each direction to obtain a low frequency image that is 2^{-2J_0} times smaller than the original image. This method of encoding results in high compression ratios that vary with the number of edge points remaining after edge selection. Figure 7.8 illustrates the Lena image (left) reconstructed by the method of alternate projections from the edge image (right), obtained by selecting only those maxima chains whose lengths and average modulus values are above certain threshold values.

7.4.2 Texture-based image encoding

The difference between the original image and the image reconstructed from the multiscale edge-based encoding yields an error image. Since the edge-based encoding disregarded texture information, the error image contains the texture information of the image along with certain object features whose edges were omitted. Due to the high compression achieved in the edge-based encoding, the error image tends to contain more information than the edge-encoded image. The texture information in the form of the error image is encoded using an orthonormal wavelet pyramidal technique. As most of the edge features are encoded, one can achieve high compression ratios in encoding the error image without creating rippling effects (Gibbs phenomenon) at points of rapid intensity variations. To economize on the number of wavelet coefficients to be encoded, each of the three high-pass images at each scale is partitioned into smaller square blocks, and in each block the total energy of the wavelet coefficients in that block is computed (by evaluating the sum of the squares of these coefficients); if the energy is below a threshold value, all the coefficients in that block are reset to zero. The justification for this operation is that there is no texture in this block and any nonzero coefficient is either due to noise or due to some insignificant fluctuation in image intensity. The rest of the coefficients' values are quantized based on the frequency of occurrence. This encoding preserves most of the image's texture.

The reconstructed edge-based encoding and the reconstructed texture-based encoding are added to obtain the approximate image with edge and texture information. The compression factors obtained by this double-layered encoding technique are impressive, ranging from 40 for a 256×256 pixel image to 100 for a 512×512 image.

This technique may be thought of as a broad strategy, as there seems to be scope for improvement and refinement of the various strategies employed along with possible variations on some of the ideas involved. Although edge and texture alogorithms have been proposed in the past (Carlsson [2], Kunt et al. [23]), the methods of Mallat, Zhong and Froment described here are unique in that they efficiently use the multiscale nature of the wavelet transform and are easily adapted to multichannel image filtering and other image processing applications.

7.5 Additional Wavelet Applications

In the final section, we present a brief sampling of the varied application domains in image processing and computer graphics to which wavelets have been applied. For instance, in Simhadril et al. [39], a variation of the discrete wavelet transform is used to detect weak edges in satellite oceanographic imagery by noise suppression. This problem is particularly tricky since the presence of noise makes the determination of weak edges difficult, and most noise suppression algorithms tend to suppress weak edges simultaneously. In Lee [24], neurophysiological constraints are combined with wavelets to derive a 2D Gabor wavelet family used to completely represent any image. This derivation is useful in understanding how the brain interprets imagery. In the realm of computer graphics, wavelets as a computational tool have seen much success. Haar wavelet bases have been applied to speedup radiosity computations in global illumination calculations [17]. Semi-orthogonal and biorthogonal cubic B-spline wavelets have been applied to interactive parametric curve design [14, 16]. For a detailed survey of the role wavelets play in computer graphics applications, the reader is referred to Schröder [37].

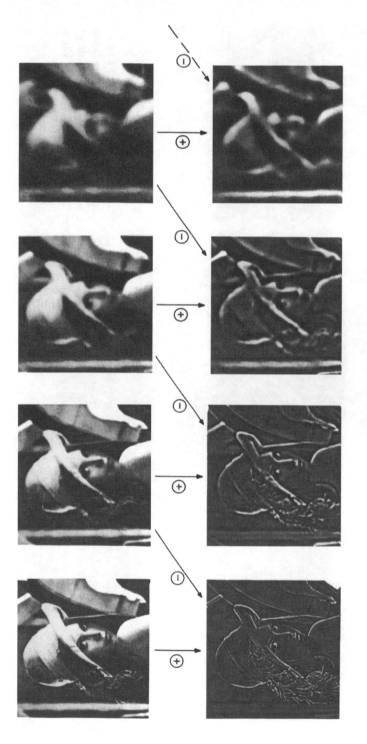

Figure 7.2: The first four levels of the Gaussian and Laplacian pyramidal decompositions for the Lena image. (c) IEEE 1983. (Adapted from Burt, P. J. and Adelson, E. H., The Laplacian pyramid as a compact image code, *IEEE Trans. Commun.* 31(4): 535-536, April, 1983.)

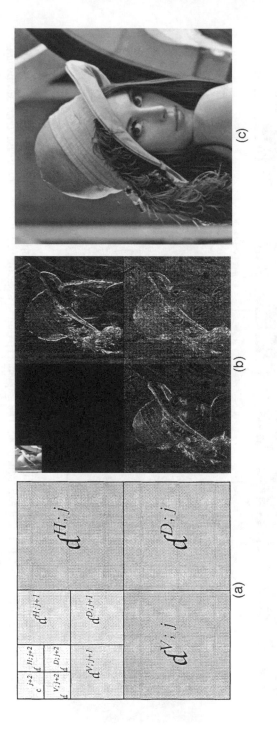

Figure 7.3: Example of the two-dimensional fast wavelet algorithm as a pyramidal algorithm. (a) The pyramidal structure of the two-dimensional fast wavelet algorithm. (b) The pyramidal decomposition of the two-dimensional fast wavelet algorithm. (c) The reconstruction of the image from (b). (Adapted from Froment, J. and Mallat, S., Second generation compact image coding with wavelets: wavelet analysis and its applications, vol. 2, in *Wavelets: A Tutorial in Theory and Applications*, Chui, C. K., ed., Academic Press, Boston, 1992.)

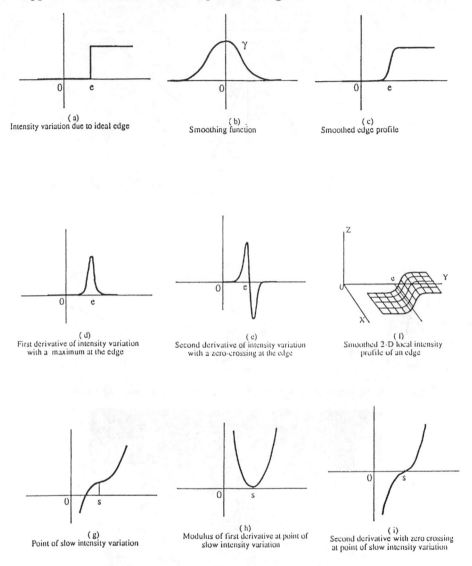

Figure 7.4: Multiscale edge representation signals in the one-dimensional case.

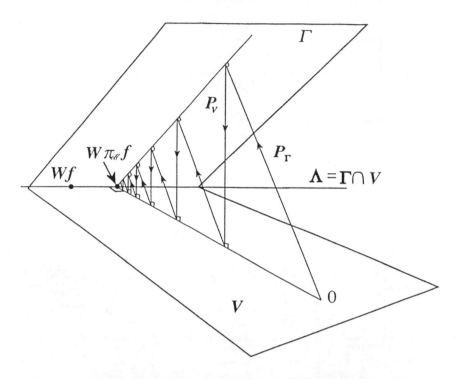

Figure 7.5: Schematic diagram of the method of alternate projections

Figure 7.6: Example of maxima chains. (Adapted from Froment, J. and Mallat, S.,Second generation compact image coding with wavelets: wavelet analysis and its applications, vol. 2, in *Wavelets: A Tutorial in Theory and Applications*, Chui, C. K., ed., Academic Press, Boston, 1992.)

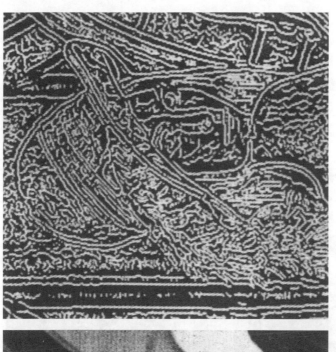

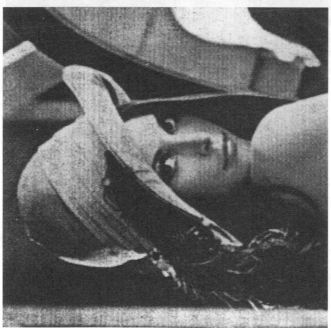

Figure 7.7: The Lena image (left) of size 256 X 256 pixels and the corresponding edge image obtained from the local modulus maxima. (Adapted from Froment, J. and Mallat, S., Second generation compact image coding with wavelets: wavelet analysis and its applications, vol. 2, in *Wavelets: A Tutorial in Theory and Applications*, Chui, C. K., ed., Academic Press, Boston, 1992.)

Figure 7.8: Reconstruction of Lena image (left) from edge image (right) by the method of alternate projections. (Adapted from Froment, J. and Mallat, S., Second generation compact image coding with wavelets: wavelet analysis and its applications, vol 2, in *Wavelets: A Tutorial in Theory and Applications*, Chui, C. K., ed., Academic Press, Boston, 1992.)

Appendix A

Bezout's Theorem

Bezout's Theorem Let $P_1(x)$ and $P_2(x)$ be any two polynomials in one variable of degrees p_1 and p_2, respectively, such that they have no common root. Then, there exists a unique pair of polynomials $Q_1(x)$ and $Q_2(x)$ in one variable of degrees $\leq p_2 - 1$ and $p_1 - 1$, respectively, satisfying the equation

$$P_1(x)Q_1(x) + P_2(x)Q_2(x) \equiv 1.$$

Proof: The proof is a straightforward application of Euclid's division algorithm and mathematical induction. Without a loss of generality, assume $p_2 \leq p_1$. Then, by Euclid's division algorithm $\exists$ polynomials $A_1(x)$ and $P_3(x)$ such that

$$P_1(x) = A_1(x)P_2(x) + P_3(x),$$

where $p_3 = \deg(P_3(x)) < p_2$ and $\deg(A_1(x)) = p_1 - p_2$. Repeating this procedure recursively, we get

$$P_m(x) = A_m(x)P_{m+1}(x) + P_{m+2}(x),$$

where $p_{m+2} = \deg(P_{m+2}(x)) < p_{m+1}$ and $\deg(A_m(x)) = p_m - p_{m+1}$. Since $p_2 > p_3 > \ldots > p_m > p_{m+1} > \ldots$, $\exists N$ such that $P_{N+1}(x) \equiv 0$. Thus, the last equation in the recursive application of Euclid's division algorithm is given by

$$P_{N-1}(x) = A_{N-1}(x)P_N(x).$$

Now,

$$P_N(x) \mid P_{N-1}(x), \quad (P_N(x) \text{divides} P_{N-1}(x)),$$

and since $P_{N-2}(x) = A_{N-2}(x)P_{N-1}(x) + P_N(x)$, it follows that

$$P_N(x) \mid P_{N-2}(x).$$

Proceeding thus, we see that $P_N(x) \mid P_2(x)$ and $P_N(x) \mid P_1(x)$ also. However, since $P_1(x)$ $P_2(x)$ have no common roots, $P_N(x) = P_N \equiv$ constant $\neq 0$. Since $P_{N-2}(x) = A_{N-2}(x)P_{N-1}(x) + P_N$, we can write

$$P_N = -A_{N-2}(x)P_{N-1}(x) + P_{N-2}(x),$$

and since $P_{N-3}(x) = A_{N-3}(x)P_{N-2}(x) + P_{N-1}(x)$, we can write

$$P_N = A_{N-2}(x)\left(A_{N-3}(x)P_{N-2}(x) - P_{N-3}(x)\right) + P_{N-2}(x),$$

or

$$
\begin{aligned}
P_N &= \left(A_{N-2}(x)A_{N-3}(x) + 1\right)P_{N-2}(x) - A_{N-2}(x)P_{N-3}(x) \\
&= R_2(x)P_{N-2}(x) + S_3(x)P_{N-3}(x),
\end{aligned}
$$

where $R_2(x) = (A_{N-2}(x)A_{N-3}(x) + 1)$, and $S_3(x) = -A_{N-2}(x)$.

Let $P_N = R_k(x)P_{N-k}(x) + S_{k+1}(x)P_{N-(k+1)}(x)$. Since

$$P_{N-(k+2)}(x) = A_{N-(k+2)}(x)P_{N-(k+1)}(x) + P_{N-k}(x),$$

we have

$$
\begin{aligned}
P_N &= R_k(x)\left(P_{N-(k+2)}(x) - A_{N-(k+2)}(x)P_{N-(k+1)(x)}\right) \\
&\quad + S_{k+1}(x)P_{N-(k+1)}(x) \\
&= \left(S_{k+1}(x) - A_{N-(k+2)}(x)R_k(x)\right)P_{N-(k+1)}(x) \\
&\quad + R_k(x)P_{N-(k+2)}(x) \\
&= R_{k+1}(x)P_{N-(k+1)}(x) + S_{k+2}(x)P_{N-(k+2)}(x).
\end{aligned}
$$

$\therefore$

$$R_{k+1}(x) = S_{k+1}(x) - A_{N-(k+2)}(x)R_k(x)$$

and

$$S_{k+2}(x) = R_k(x).$$

$\therefore$

$$R_{k+1}(x) = -A_{N-(k+2)}(x)R_k(x) + R_{k-1}(x)$$

and

$$P_N = R_k(x)P_{N-k}(x) + R_{k-1}(x)P_{N-(k+1)}(x)$$

with $R_0(x) = 1$ and $R_1(x) = -A_{N-2}(x)$.

Now,

$$
\begin{aligned}
\deg(R_{k+1}) &= \deg(A_{N-(k+2)}) + \deg(R_k) \\
&= p_{N-(k+2)} - p_{N-(k+1)} + \deg(R_k)
\end{aligned}
$$

i.e., $\deg(R_k) - \deg(R_{k-1}) = p_{N-k-1} - p_{N-k}$, $1 \leq k \leq N-1$. $\therefore$ $\deg(R_{N-2}) = p_1 - p_{N-1} < p_1$ since $\deg(P_{N-1}(x) \geq 1$, and $\deg(R_{N-3}) = p_2 - p_{N-1} < p_2$. Thus,

$$
\begin{aligned}
P_N &= R_{N-2}(x) P_2(x) + R_{N-3}(x) P_1(x), \ P_N \neq 0 \\
&\Rightarrow \left(\frac{R_{N-3}}{P_N}\right) P_1(x) + \left(\frac{R_{N-2}}{P_N}\right) P_2(x) \equiv 1
\end{aligned}
$$

or, letting $Q_1(x) = \frac{R_{N-3}(x)}{P_N}$ and $Q_2(x) = \frac{R_{N-2}(x)}{P_N}$, we have

$$
P_1(x) Q_1(x) + P_2(x) Q_2(x) \equiv 1,
$$

where $\deg(Q_1) \leq p_2 - 1$ and $\deg(Q_2) \leq p_1 - 1$. This proves the existence part of the theorem.

To prove uniqueness, assume $\exists$ polynomials $\tilde{Q}_1$ and $\tilde{Q}_2$ different from Q_1 and Q_2, respectively, such that

$$
P_1(x) \tilde{Q}_1(x) + P_2(x) \tilde{Q}_2(x) \equiv 1
$$

with $\deg(\tilde{Q}_1) \leq p_2 - 1$ and $\deg(\tilde{Q}_2) \leq p_1 - 1$. Then,

$$
P_1(x)(Q1(x) - \tilde{Q}_1(x)) + P_2(x)(Q_2(x) - \tilde{Q}_2(x)) \equiv 0.
$$

Since $P_1(x)$ and $P_2(x)$ have no common roots, this implies

$$
P_1(x) \mid \left(Q_2(x) - \tilde{Q}_2(x)\right)
$$

and

$$
P_2(x) \mid \left(Q_1(x) - \tilde{Q}_1(x)\right).
$$

However,

$$
\deg\left(Q_2(x) - \tilde{Q}_2(x)\right) < p_1 = \deg(P_1(x))
$$

and

$$
\deg\left(Q_1(x) - \tilde{Q}_1(x)\right) < p_2 = \deg(P_2(x)).
$$

$\therefore$

$$
Q_2(x) \equiv \tilde{Q}_2(x) \text{ and } Q_1(x) \equiv \tilde{Q}_1(x).
$$

This proves the uniqueness part of the theorem.

Bibliography

[1] P. J. Burt and E. H. Adelson. The laplacian pyramid as a compact image code. *IEEE Trans. Commun.*, com-31(4), April 1983.

[2] S. Carlsson. Sketch based coding of greylevel images. *Signal Processing*, 15(1), 1988.

[3] C. K. Chui. *An Introduction to Wavelets*. Academic Press, Boston, 1992.

[4] C. K. Chui, editor. *Wavelets: A Tutorial in Theory and Applications*. Academic Press, Boston, 1992.

[5] C. K. Chui and G. Chen. *Signal Processing and Systems Theory-Selected Topics*. Springer-Verlag, 1992.

[6] A. Cohen and I. Daubechies. A stability criterion for biorthogonal wavelet bases and their related subband coding schemes. Technical report, AT&T Bell Laboratories, 1991.

[7] A. Cohen, I. Daubechies, and J. C. Feauveau. Biorthogonal bases of compactly supported wavelets. *Comm. Pure and Appl. Math.*, 45, 1992.

[8] I. Daubechies. Orthonormal bases of compactly supported wavelets. *Comm. Pure and Appl. Math.*, 41, 1988.

[9] I. Daubechies. *Ten Lectures on Wavelets*. SIAM, 1992.

[10] I. Daubechies, editor. *Different Perspectives on Wavelets*. AMS, 1993.

[11] I. Daubechies and J. Lagarias. Two-scale difference equations i: Existence and global regularity of solutions. *SIAM J. Math. Anal.*, 22, 1991.

[12] G. David. *Wavelets and Singular Integrals on Curves and Surfaces*. Springer-Verlag, 1992.

[13] H. Dym and McKean. *Fourier Series and Integrals.*

[14] A. Finklestein and D. H. Salesin. Multiresolution curves. In *Computer Graphics Proc., Conf. Series*, 1994.

[15] J. Froment and S. Mallat. Second generation compact image coding with wavelets. *Wavelet Analysis and its Applications*, 2, 1992. Academic Press.

[16] S. J. Gortler and M. F. Cohen. Hierarchical and variational geometric modeling with wavelets. Technical Report MSR-TR-95-25, Microsoft Research Advanced Technology Division, 1995.

[17] S. J. Gortler, P. Schröoder, M. F. Cohen, and P. Hanrahan. Wavelet radiosity. In *Computer Graphics Proc., Conf. Series*, 1993.

[18] P. R. Halmos. *Measure Theory.* D. Van Nostrand Company, 1950.

[19] K. Hoffman and R. Kunze. *Linear Algebra.* Prentice-Hall, 1971.

[20] M. Holschneider. *Wavelets, An Analysis Tool.* Oxford Science, 1995.

[21] Y. Katznelson. *An Introduction to Harmonic Analysis.* Dover, 1976.

[22] A. N. Kolmogorov and S. V. Fomin. *Introductory Real Analysis.* Dover, 1975.

[23] M. A. Kunt, Ikonomopoulos, and M. Kochev. Second-generation image coding techniques. *Proc. IEEE*, 73(1), 1985.

[24] T. S. Lee. Image representation using 2d gabor wavelets. *IEEE Trans. Pattern Anal. Machine Intell.*, 18(10), 1996.

[25] P.G. Lemarié and G. Malouyres. Support des fonctions de base dans une analyse multirésolution. *Comptes Rendes*, 1992.

[26] M. Lounsbery. *Multiresolution Analysis for Surfaces of Arbitray Topological Type.* PhD thesis, University of Washington, 1994.

[27] S. Mallat. Multifrequency channel decompositions of images and wavelet models. *IEEE Trans. Acous., Speech, Signal Process.*, 37(12), 1989.

[28] S. Mallat and S. Zhong. Characteristics of signals from multiscale images. Technical report, NYU Computer Science, 1991.

[29] Y. Meyer. *Wavelets and Operators.* Cambridge University Press, 1992.

[30] A. N. Netravali and B. G. Haskell. *Digital Pictures: Representation and Compression.* Plenum Press, 1988.

[31] A. V. Oppenheim and R. W. Schafer. *Discrete-Time Signal Processing.* Prentice-Hall, 1989.

[32] A. Rosenfeld, editor. *Multiresolution Image Processing and Analysis.* Springer-Verlag, 1984.

[33] H. L. Royden. *Real Analysis.* Macmillan, 1963.

[34] W. Rudin. *Principles of Mathematical Analysis.* McGraw-Hill, 1964.

[35] W. Rudin. *Real and Complex Analysis.* McGraw-Hill, 1966.

[36] W. Rudin. *Functional Analysis.* McGraw-Hill, 1973.

[37] P. Schröder. Wavelets in computer graphics. *Proc. of the IEEE*, 84(4), 1996.

[38] G. E. Shilov and B. L. Gurevich. *Integral, Measure and Derivative: A Unified Approach.* Dover, 1977.

[39] K. Simhadril, S. S. Iyengar, R. Hoyler, M. Lybanon, and J. M. Zachary. Wavelet based feature extraction from oceanographic images. In submission to IEEE Transaction on Geoscience and Remote Sensing, 1996.

[40] G. F. Simmons. *Topology and Modern Analysis.* McGraw-Hill, 1963.

[41] J. L. Starck, F. Murtagh, and A. Bijaoui. Image restoration with noise suppresion using a wavelet transform and a multiresolution support constraint. *Image Reconstruction and Restoration, SPIE Proceedings*, 2302, 1994.

[42] G. Strang and T. Nguyen. *Wavelets and Filter Banks.* Wellesley-Cambridge Press, 1996.

[43] B. Z. Vulikh. *A Brief Course in the Theory of Functions of a Real Variable.* Mir Publishers, Moscow, 1976.

[44] D. Youla and A. H. Webb. Image restoration by the method of convex projections. *IEEE Trans. Med. Imaging*, 1, 1982.

Index

276

Dr. Lakshman Prasad obtained his master's degree in Mathematics from the Indian Institute of Technology in Kanpur and later did research in Mathematics at the Tata Institute of Fundamental Research. He obtained his Ph.D. in Computer Science at Louisiana State University in 1995.

Dr. Prasad is currently with the Theoretical Division of the Los Alamos National Laboratory. His areas of interest include signal processing, applied harmonic analysis, wavelet transform theory, mathematical morphology with applications to pattern recognition and machine vision, and computational geometry. He has published one book and many research papers in these and other areas.

Dr. S.S. Iyengar is Chairman and Professor of the Department of Computer Science at Louisiana State University and an IEEE Fellow. Since receiving his Ph.D. from Mississippi State University in 1974, he has served as a principal investigator on research projects funded by the Office of Naval Research, National Science Foundation, U.S. Army Research Office, Department of Energy, Naval Research Laboratory, NASA Jet Propulsion Laboratory, and various agencies of the State of Louisiana.

Dr. Iyengar has made significant contributions to the areas of high performance image processing algorithms, sensor fusion, parallel models of computation, robot motion planning, and computer vision. He has published several books and monographs in addition to over 200 research articles.

Dr. Iyengar has served as guest editor for *IEEE Computer, IEEE Transactions of Data and Knowledge Engineering, IEEE Transactions on Systems, Man, and Cybernetics, IEEE Transactions on Software Engineering, the Journal of Theoretical Computer Science, the Journal of Computers and Electrical Engineering, and the Journal of the Franklin Institute*. In addition to being an IEEE Fellow, he is also a distinguished visitor of the IEEE and a member of the New York Academy of Sciences. In 1996, he was awarded the LSU Distinguished Faculty Award and the Tiger Athletic Foundation Teaching Award. He has supervised over 26 Ph.D. dissertations and over 40 M.S. projects and theses while at LSU.